T0074706

Birkhäuser

Applied and Numerical Harmonic Analysis

Jeffrey A. Hogan • Joseph D. Lakey

Duration and Bandwidth Limiting

Prolate Functions, Sampling, and Applications

Jeffrey A. Hogan
School of Mathematical
and Physical Sciences
University of Newcastle
Callaghan, NSW 2308
Australia
Jeff.Hogan@newcastle.edu.au

Joseph D. Lakey
Department of Mathematical Sciences
New Mexico State University
Las Cruces, NM 88003-8001
USA
jlakey@nmsu.edu

ISBN 978-0-8176-8306-1 e-ISBN 978-0-8176-8307-8
DOI 10.1007/978-0-8176-8307-8
Springer New York Dordrecht Heidelberg London

Library of Congress Control Number: 2011942155

Mathematics Subject Classification (2010): 42A10, 42C05, 65T50, 94A12, 94A20

Printed on acid-free paper

Birkhäuser Boston is part of Springer Science+Business Media (www.birkhauser.com)

In memory of our fathers, Ron Hogan and Frank Lakey

ANHA Series Preface

The *Applied and Numerical Harmonic Analysis (ANHA)* book series aims to provide the engineering, mathematical, and scientific communities with significant developments in harmonic analysis, ranging from abstract harmonic analysis to basic applications. The title of the series reflects the importance of applications and numerical implementation, but richness and relevance of applications and implementation depend fundamentally on the structure and depth of theoretical underpinnings. Thus, from our point of view, the interleaving of theory and applications and their creative symbiotic evolution is axiomatic.

Harmonic analysis is a wellspring of ideas and applicability that has flourished, developed, and deepened over time within many disciplines and by means of creative cross-fertilization with diverse areas. The intricate and fundamental relationship between harmonic analysis and fields such as signal processing, partial differential equations (PDEs), and image processing is reflected in our state-of-the-art *ANHA* series.

Our vision of modern harmonic analysis includes mathematical areas such as wavelet theory, Banach algebras, classical Fourier analysis, time–frequency analysis, and fractal geometry, as well as the diverse topics that impinge on them.

For example, wavelet theory can be considered an appropriate tool to deal with some basic problems in digital signal processing, speech and image processing, geophysics, pattern recognition, biomedical engineering, and turbulence. These areas implement the latest technology from sampling methods on surfaces to fast algorithms and computer vision methods. The underlying mathematics of wavelet theory depends not only on classical Fourier analysis, but also on ideas from abstract harmonic analysis, including von Neumann algebras and the affine group. This leads to a study of the Heisenberg group and its relationship to Gabor systems, and of the metaplectic group for a meaningful interaction of signal decomposition methods. The unifying influence of wavelet theory in the aforementioned topics illustrates the justification for providing a means for centralizing and disseminating information from the broader, but still focused, area of harmonic analysis. This will be a key role of *ANHA*. We intend to publish the scope and interaction that such a host of issues demands.

Along with our commitment to publish mathematically significant works at the frontiers of harmonic analysis, we have a comparably strong commitment to publish major advances in the following applicable topics in which harmonic analysis plays a substantial role:

Antenna theory	*Prediction theory*
Biomedical signal processing	*Radar applications*
Digital signal processing	*Sampling theory*
Fast algorithms	*Spectral estimation*
Gabor theory and applications	*Speech processing*
Image processing	*Time–frequency and*
Numerical partial differential equations	*time-scale analysis*
	Wavelet theory

The above point of view for the *ANHA* book series is inspired by the history of Fourier analysis itself, whose tentacles reach into so many fields.

In the last two centuries, Fourier analysis has had a major impact on the development of mathematics, on the understanding of many engineering and scientific phenomena, and on the solution of some of the most important problems in mathematics and the sciences. Historically, Fourier series were developed in the analysis of some of the classical PDEs of mathematical physics; these series were used to solve such equations. In order to understand Fourier series and the kinds of solutions they could represent, some of the most basic notions of analysis were defined, e.g., the concept of "function". Since the coefficients of Fourier series are integrals, it is no surprise that Riemann integrals were conceived to deal with uniqueness properties of trigonometric series. Cantor's set theory was also developed because of such uniqueness questions.

A basic problem in Fourier analysis is to show how complicated phenomena, such as sound waves, can be described in terms of elementary harmonics. There are two aspects of this problem: first, to find, or even define properly, the harmonics or spectrum of a given phenomenon, e.g., the spectroscopy problem in optics; second, to determine which phenomena can be constructed from given classes of harmonics, as done, e.g., by the mechanical synthesizers in tidal analysis.

Fourier analysis is also the natural setting for many other problems in engineering, mathematics, and the sciences. For example, Wiener's Tauberian theorem in Fourier analysis not only characterizes the behavior of the prime numbers, but also provides the proper notion of spectrum for phenomena such as white light; this latter process leads to the Fourier analysis associated with correlation functions in filtering and prediction problems, and these problems, in turn, deal naturally with Hardy spaces in the theory of complex variables.

Nowadays, some of the theory of PDEs has given way to the study of Fourier integral operators. Problems in antenna theory are studied in terms of unimodular trigonometric polynomials. Applications of Fourier analysis abound in signal processing, whether with the fast Fourier transform (FFT), or filter design, or the adaptive modeling inherent in time–frequency-scale methods such as wavelet theory. The coherent states of mathematical physics are translated and modulated

Fourier transforms, and these are used, in conjunction with the uncertainty principle, for dealing with signal reconstruction in communications theory. We are back to the raison d'être of the *ANHA* series!

University of Maryland
College Park

John J. Benedetto
Series Editor

Preface

John von Neumann developed much of the rigorous formalism for quantum mechanics during the late 1920s [339], including a theory of unbounded operators in a Hilbert space that led to a mathematical justification of Heisenberg's uncertainty principle. In his autobiography [356, p. 97 ff.], Norbert Wiener gave a poignant account of his own role in advancing mathematical ideas parallel to von Neumann's during Wiener's visit to Göttingen in 1925. In contrast to Heisenberg, Wiener was motivated by limitations on simultaneous localization in time and frequency, or *duration and bandwidth limiting* pertaining to macroscopic physical systems. Meanwhile, across the Atlantic, Nyquist [246] and Hartley [137], working at Bell Labs, were laying the foundations for Shannon's communication theory in their accounts of transmission of information over physical communications channels. Hartley also quantified limitations on joint localization in time and frequency, perhaps in a more positive light than suggested by Wiener, by defining a bound on the rate at which information might be transmitted over a band-limited channel. Eventually this bound, expressed in terms of the *time–bandwidth product*, was formalized in the *Bell Labs theory*—a series of papers by Landau, Slepian, and Pollak [195, 196, 303, 309] appearing in the *Bell System Technical Journal* in the early 1960s, with additional follow-on work extending through 1980.

The Bell Labs theory identified those band-limited functions—the *prolate spheroidal wave functions* (PSWFs)—that have the largest proportions of their energies localized in a given time interval. It also quantified the eigenvalue behavior of the time- and frequency-localization operator "$P_\Omega Q_T$" that gives rise to these eigenfunctions. There are approximately $2\Omega T$ orthogonal functions that are band limited to the frequency band $[0,\Omega]$, while having most of their energies localized in the time interval $[-T,T]$. Perhaps the most satisfying account of this $2\Omega T$ principle, at least from a purely mathematical perspective, is due to Landau and Widom (Theorem 1.3.1 in this book). Their result quantified asymptotically, but precisely, the number of eigenvalues of $P_\Omega Q_T$ larger than any given $\alpha \in (0,1)$. The masterful use of the technical machinery of compact self-adjoint operators that went into this result arguably places the Bell Labs theory, as a foundational affirmation of what is observed in the physical universe, alongside von Neumann's work. However, the theory did not lead to immediate widespread use of prolates in signal processing.

This was partly due to the fact that the appropriate parallel development in the *discrete domain* had not yet been made. During the period from the late 1970s to the early 1980s, a number of contributions provided this discrete parallel theory of time and band limiting, e.g., [307], the basic tool being the *discrete prolate spheroidal sequences* (DPSSs). Applications then emerged. Thomson [323] used this discrete theory in his *multitaper* method for spectrum estimation, and Jain and Ranganath [162] formalized the use of DPSSs in signal extrapolation. Ideas of Grünbaum [126] and others continued to provide a firm foundation for further potential applications.

While Thomson's methods caught on quickly, particularly in the geosciences, applications of the theory of time and band limiting to the modeling of communications channels, numerical analysis, and other promising areas, such as tomography, had to wait. There were several reasons for this delay: wavelets attracted much of the attention in the 1980s, and mobile communications were still in their lag phase.

Applications of time and band limiting appear now to be entering a log phase. The IEEE literature on this topic has grown quickly, largely due to potential uses in communications, where multiband signals are also playing a wider role, and in compressed sensing. At the same time, potential uses in numerical analysis are also being identified, stemming from the observation that the PSWFs have zeros that are more evenly spaced than their (Chebyshev and Legendre) polynomial counterparts, making them attractive for spectral element methods. Extension and refinement of the mathematical ideas underlying the Bell Labs theory, particularly in the context of multiband signals, but also involving higher order approximations, will continue in the coming decades.

The need to have a common source for the Bell Labs theory and its extensions, together with a resource for related methods that have been developed in the mathematics and engineering literature since 2000, motivated us to put together this monograph. Here is an outline of the contents.

Chapter 1 lays out the fundamental results of the Bell Labs theory, including a discussion of properties of the PSWFs and properties of the eigenvalues of $P_\Omega Q_T$. A discussion of the parallel theory of DPSSs is also included. Further discussion in the *chapter notes* focuses primarily on the application of the DPSSs to signal extrapolation.

The PSWFs and their values were important enough in classical physics to devote large portions of lengthy monographs to tables of their values, e.g., [54, 231, 318]. Modern applications require not only accuracy, but also computational efficiency. Since the largest eigenvalues of $P_\Omega Q_T$ are very close to one, they are also close to one another. This means that standard numerical tools for estimating finite-dimensional analogues of projections onto individual eigenspaces, such as the singular value decomposition, are numerically unstable. This is one among a raft of issues that led several researchers to reconsider numerical methods for estimating and applying PSWFs. Chapter 2 addresses such numerical analytical aspects of PSWFs, including quadrature formulas associated with prolates and approximations of functions by prolate series. Along with classical references, sources for more recent advances here include [35, 36, 41–44, 123, 178, 280, 296, 331, 364].

Chapter 3 contains a detailed review of Thomson's multitaper method. Through multitapering, PSWFs can be used to model channel characteristics; see, e.g., [124, 179, 211, 294]. However, if such a channel model is intended to be used by a single mobile device under severe power usage constraints, it might be preferable to employ a fixed prior basis in order to encode channel information in a parsimonious fashion. Zemen and Mecklenbräuker [370, 372] proposed variants of the DPS sequences for various approaches to this problem. These variants will also be outlined in Chap. 3.

Chapter 4 contains a development of parallel extensions of the Bell Labs theory to the case of multiband signals. Multiband signals play an increasing role in radio frequency (RF) communications as RF spectrum usage increases. We outline Landau and Widom's [197] extension of the $2\Omega T$ theorem to the case in which the time- and frequency-localization sets S and Σ are finite unions of intervals. Other *non-asymptotic* estimates are given for the number of eigenvalues of $P_\Sigma Q_S$ that are larger than one-half, quantified in terms of the linear distributions of S and Σ.

The second part of Chap. 4 addresses a related problem in the theory of time and multiband limiting of finite signals (i.e., signals defined on a finite set of integers). Candès, Romberg, and Tao [57–60] proved *probabilistic* estimates on the norm of the corresponding time- and band-limiting operator. The main estimate states that if the time- and frequency-localization sets are *sparse* in a suitable sense, then, with overwhelming probability, the localization operator has norm smaller than one-half. The estimates were developed for applications to compressed sensing, in particular to probabilistic recovery of signals with sparse Fourier spectra from random measurements. The techniques involve intricate combinatorial extensions of the same approach that was employed by Landau and Widom in their $2\Omega T$ theorem.

Ever since Shannon's seminal work [293], sampling theory has played a central role in the theory of band-limited functions. Aspects of sampling theory related to numerical quadrature are developed in Sect. 2.3. Chapter 5 reviews other aspects of sampling of band-limited signals that are, conceptually, closely tied to time and band limiting. This includes Landau's necessary conditions for sampling and interpolation [188, 189], phrased in terms of the Beurling densities, whose proofs use the same techniques as those employed in Chap. 4 to estimate eigenvalues for time and band limiting. Results quantifying the structure of sets of sampling and interpolation for Paley–Wiener spaces of band-limited functions are surveyed. The latter part of Chap. 5 addresses methods for sampling multiband signals. Several such methods have been published since the 1990s. A few of these approaches will be addressed in detail, with other approaches outlined in the chapter notes.

Chapter 6 contains several results that represent steps toward a general theory that envelops sampling and time and band limiting, both for band-limited and for multiband signals. The core of the chapter builds on work of Walter and Shen [347] and Khare and George [177], who observed that the integer samples of suitably normalized PSWFs form eigenvectors of the discrete matrix whose entries are correlations of time localizations of shifts of the sinc kernel—the reproducing kernel for the space of band-limited signals. This fact permits a method for constructing approximate time- and band-limiting projections from samples. Other discrete methods for building time- and multiband-limiting projection operators from constituent components are also presented in Chap. 6.

This text is primarily a mathematical monograph. While applications to signal processing play a vital motivational role, those applications also involve technical details whose descriptions would detract from the emphasis on mathematical methods. However, we also want this book to be useful to practitioners who do not prove theorems for a living—so not all of the theorems that are discussed are proved in full. In choosing which results to present in finer detail, our goal has been to maintain focus on central concepts of time and band limiting. We also emphasize real-variable methods over those of complex function theory, in several cases favoring the more intuitive result or argument over the most complete result or rigorous proof. We hope that this approach serves to make the book more accessible to a broader audience, even if it makes the reading less satisfying to a stalwart mathematician.

The core material in each chapter is supplemented to some extent with *notes and auxiliary results* that outline further major mathematical contributions and primary applications. In some cases, proofs of results auxiliary to the core theorems are also included with this supplementary material. Even so, we have made choices regarding which contributions to mention. We easily could have cited over 1000 contributions without having mentioned every important theorem and application pertinent to time and band limiting.

A number of colleagues contributed to this work in substantial ways. We are particularly indebted to Chris Brislawn, Chuck Creusere, Scott Izu, Hans Feichtinger, Charly Gröchenig, Ned Haughton, Chris Heil, Tomasz Hrycak, Richard Laugesen, Kabe Moen, Xiaoping Shen, Adam Sikora and Mark Tygert for the generosity with which they gave advice and technical assistance.

This book would not have been possible without the hospitality and support of the Department of Mathematics at Washington University in St. Louis, where Lakey was on sabbatical in 2009-2010 and Hogan enjoyed an extended visit in early 2010. Both authors are especially grateful to Guido Weiss and Ed Wilson for arranging these visits and for their guidance on a number of matters. Hogan is appreciative of support provided by the University of Newcastle's Centre for Computer-Assisted Research Mathematics and its Applications as well as the Centre for Complex Dynamic Systems and Control.

The staff at Birkhäuser did a wonderful job of keeping us on track through all stages of this project. Special thanks to Ben Cronin and Tom Grasso.

Thanks to Roy and HG for determined drive from the backline. Thanks Max, for always being there. Finally, thanks to Mysie and to Ellen for encouragement and understanding. We owe you.

Newcastle, NSW, Australia *Jeff Hogan*
Las Cruces, New Mexico *Joe Lakey*

Contents

Chapter 1
The Bell Labs Theory

1.1 Information Theory of Time and Band Limiting

Duration limiting, or *time limiting*, refers to restricting a signal by setting its values equal to zero outside of a finite time interval or, more generally, outside of a compact set. Bandwidth limiting, or *band limiting*, refers to restricting a signal by setting its amplitudes equal to zero outside of a finite frequency interval or again, more generally, outside of a compact set. This book addresses primarily the theory of time and band limiting whose core was developed by Landau, Pollak, and Slepian in a series of papers [195, 196, 303, 309] appearing in the *Bell System Technical Journal* in the early to middle part of the 1960s, and a broader body of work that grew slowly but steadily out of that core up until around 1980, with a resurgence since 2000, due in large part to the importance of time and band limiting in wireless communications. The 1960s *Bell Labs theory* of time and band limiting is but one aspect of the Bell Labs information theory. The foundations of this encompassing theory were laid, in large part, in Nyquist's fundamental papers "Certain Topics in Telegraph Transmission Theory" [247], which appeared in the *Transactions of the American Institute of Electrical Engineers* in 1928, and "Certain Factors Affecting Telegraph Speed," published in April 1924 in the *Bell System Technical Journal*, along with Hartley's paper "Transmission of Information," which also appeared in the *Bell System Technical Journal* in 1928 [137]. These papers quantified general ideas that were in the air, though certain specific versions were attributed to Kelvin and Wiener among others. Of course, Claude Shannon's seminal work, "A Mathematical Theory of Communication," which appeared in the *Bell System Technical Journal* in July and October 1948 [293], is often cited as providing the basis for much of modern communications theory. His sampling theory plays a central role in Chap. 5 of this monograph. The works of Nyquist and Hartley however remain, in some ways, more germane to the study at hand.

Although we will comment briefly on other aspects of this early work, the emerging principle that is most fundamental here is summarized in the following quote from Hartley [137]:

> *the total amount of information which may be transmitted over such a* [band-limited, but otherwise distortionless] *system is proportional to the product of the frequency-range which it transmits by the time during which it is available for the transmission.*

Hartley also introduced the fundamental concept of *intersymbol interference* and considered its role in limiting the possible rate at which information could be intelligibly transmitted over a cable.

In his 1928 paper, Nyquist first laid the information-theoretic groundwork for what eventually became known as the Shannon sampling theorem. Therein, Nyquist showed that, in order to reconstruct the original signal, the sampling rate must be at least twice the highest frequency present in the sample. However, the bounds on the rate by which information could be transmitted across a *bandwidth constrained* channel were established in his 1924 paper. Specifically, Nyquist considered the problem of what waveform to use in order to transmit information over such a channel and what method might be optimal for encoding information to be transmitted via such a waveform. When phrased in these terms, the work of Landau, Slepian, and Pollak provided a precise quantitative bound on the capacity of a clear, band-limited channel defined in terms of an ideal lowpass filter to transmit a given amount of information per unit time by providing, in turn, a precise quantification of the dimension of the space of essentially time- and band-limited signals and an identification of what those signals are. As Slepian [308] made abundantly clear, this is an interpretation that applies only to an idealization of physical reality. Nevertheless, the mathematical Bell Labs theory of time and band limiting that will be laid out in this chapter is both elegant and satisfying. The information-theoretic details and consequences of this theory extend well beyond the brief comments just made. Some of the additional sources for these details and further general discussion include [80, 84, 108, 135, 161, 191, 194, 229, 302, 304, 305, 308, 310, 311, 359–361].

1.2 Time- and Frequency-localization Operators

1.2.1 Paley–Wiener Spaces on $\mathbb{R}$ and Localization Projections

We normalize the Fourier transform $\mathscr{F}f$ of $f \in L^2(\mathbb{R})$ as the limit, in $L^2(\mathbb{R})$,

$$(\mathscr{F}f)(\xi) = \widehat{f}(\xi) = \lim_{N\to\infty} \int_{-N}^{N} f(t)\,\mathrm{e}^{-2\pi i t\xi}\,\mathrm{d}t. \tag{1.1}$$

The Paley–Wiener (PW) spaces PW_I are subspaces of $L^2(\mathbb{R})$ consisting of functions whose Fourier transforms vanish off the interval I. If $f \in \mathrm{PW}_I$ then f has an extension to an entire function that can be expressed in terms of its absolutely convergent inverse Fourier integral,

$$f(z) = \int_{-\infty}^{\infty} \widehat{f}(\xi)\, \mathrm{e}^{2\pi i z\xi}\, \mathrm{d}\xi, \quad (z \in \mathbb{C}).$$

Although function theory underlies much of what follows, its role will not be emphasized. One can associate a Paley–Wiener space PW_Σ to any compact frequency support set Σ by defining it as the range of the *band-limiting* orthogonal projection operator P_Σ defined by

$$(P_\Sigma f)(t) = (\widehat{f}\, \mathbb{1}_\Sigma)^\vee(t), \tag{1.2}$$

where $\mathbb{1}_\Sigma$ denotes the indicator function of the set Σ and $g^\vee$ denotes the inverse Fourier transform of g. In the special case in which $\Sigma = [-\Omega/2, \Omega/2]$ is an interval centered at the origin, we will abbreviate $P_{[-\Omega/2,\Omega/2]}$ as P_Ω and, when $\Omega = 1$, we will simply write $P_1 = P$.

The *time-limiting* operator Q_S is defined by the orthogonal projection

$$(Q_S f)(t) = \mathbb{1}_S(t)\, f(t). \tag{1.3}$$

When S is an interval centered about the origin of length $2T$, we will abbreviate $Q_{[-T,T]}$ to Q_T. This notation is slightly asymmetrical from the band-limiting case. In particular, the time–bandwidth product corresponding to the operator $P_\Omega Q_T$ is $2T\Omega$. There are two important differences between our normalizations and those used in the Bell Labs papers [195, 196, 303, 309]. The first is that Slepian et al. used the Fourier kernel $\mathrm{e}^{-it\xi}/\sqrt{2\pi}$ as opposed to the kernel $\mathrm{e}^{-2\pi it\xi}$ used here. The second is that their band-limiting operator was defined in terms of multiplication by $\mathbb{1}_{[-\Omega,\Omega]}$ while their time-limiting operator was given by multiplication by $\mathbb{1}_{[-T/2,T/2]}$.

Time and band limiting then refers to a composition of the form $P_\Sigma Q_S$. Since the separate operators P_Σ and Q_S do not commute, except in trivial cases, $P_\Sigma Q_S$ itself is not a projection. However, it is still self-adjoint as an operator on PW_Σ. Sometimes we will consider the operator $P_\Sigma Q_S P_\Sigma$, which is self-adjoint on all of $L^2(\mathbb{R})$. When Σ is compact, PW_Σ is a reproducing kernel Hilbert space (RKHS) with kernel $K_\Sigma(t,s) = K_\Sigma(t-s)$ where $K_\Sigma(t) = (\mathbb{1}_\Sigma)^\vee(t)$. The kernel of $P_\Sigma Q_S$, as an operator on PW_Σ, is $K_{S,\Sigma}(t,s) = \mathbb{1}_S(s)(\mathbb{1}_\Sigma)^\vee(t-s)$. Since $P_\Sigma Q_S$ is positive definite, it follows from general theory (e.g., [277]) that if S and Σ are both compact then $P_\Sigma Q_S$ is a trace-class operator with

$$\mathrm{tr}\,(P_\Sigma Q_S) = \int_{\mathbb{R}} K_{S,\Sigma}(s,s)\, \mathrm{d}s = \int_S |\Sigma|\, \mathrm{d}s = |S||\Sigma|,$$

since $K_{S,\Sigma}(s,s) = \mathbb{1}_S(s)(\mathbb{1}_\Sigma)^\vee(0) = |\Sigma|\mathbb{1}_S(s)$. Let $\{\psi_n\}$ be an orthonormal basis of PW_Σ, that is, $\{\widehat{\psi_n}\}$ is an orthonormal basis of $L^2(\Sigma)$. The Hilbert–Schmidt norm $\|P_\Sigma Q_S\|_{\mathrm{HS}}$ of $P_\Sigma Q_S$ satisfies

$$\|P_\Sigma Q_S\|_{\mathrm{HS}}^2 = \sum_n \|(P_\Sigma Q_S)\psi_n\|^2 = \sum_n \|\widehat{Q_S\psi_n}\|_{L^2(\Sigma)}^2$$
$$= \sum_n \int_\Sigma \Big|\int_\Sigma \widehat{\mathbb{1}_S}(\xi-\eta)\,\widehat{\psi_n}(\eta)\,\mathrm{d}\eta\Big|^2 \mathrm{d}\xi = \sum_n \int_\Sigma |\langle \widehat{\mathbb{1}_S}(\xi-\cdot),\,\widehat{\psi_n}\rangle|^2\,\mathrm{d}\xi$$
$$= \int_\Sigma \sum_n |\langle \widehat{\mathbb{1}_S}(\xi-\cdot),\,\widehat{\psi_n}\rangle|^2\,\mathrm{d}\xi = \int_\Sigma\int_\Sigma |\widehat{\mathbb{1}_S}(\xi-\eta)|^2\,\mathrm{d}\eta\,\mathrm{d}\xi\,.$$

When S is symmetric, $(P_S Q_\Sigma)^* = Q_\Sigma P_S$ is unitarily equivalent to the operator $\mathscr{F}^{-1}Q_\Sigma\mathscr{F}\mathscr{F}^{-1}P_S\mathscr{F} = P_\Sigma Q_S$, and

$$\|P_\Sigma Q_S\|_{\mathrm{HS}}^2 = \|K_{S,\Sigma}\|_{L^2(S\times S)}^2 = \int_S\int_S |(\mathbb{1}_\Sigma)^\vee(t-s)|^2\,\mathrm{d}s\,\mathrm{d}t\,.$$

The Bell Labs papers studied the spectral decomposition of $P_\Omega Q_T$ and several mathematical and physical consequences of it. This theory and some extensions will be presented in the rest of this chapter. We begin with a brief review of the eigenfunctions of $P_\Omega Q_T$, the *prolate spheroidal wave functions*.

1.2.2 *Prolate Spheroidal Wave Functions*

Definition and Commentary

When used in the context of time and band limiting, the term *prolate spheroidal wave function* (PSWF) is not particularly resonant. The notion of a PSWF originated, rather, in the context of solving the wave equation in prolate spheroidal coordinates by means of separation of variables. The prolate spheroidal coordinates (ξ,η,ϕ) are related to the standard Euclidean (x,y,z)-coordinates by the equations

$$\begin{aligned} x &= a\sinh\xi\,\sin\eta\,\cos\phi\\ y &= a\sinh\xi\,\sin\eta\,\sin\phi\\ z &= a\cosh\xi\,\cos\eta\,. \end{aligned}$$

Upon substituting $u=\cosh\xi$ and $v=\cos\eta$, the Laplacian takes the form

$$\nabla^2 f = \frac{1}{a^2(u^2-v^2)}\Big\{\frac{\partial}{\partial u}(u^2-1)\frac{\partial f}{\partial u} + \frac{\partial}{\partial v}(v^2-1)\frac{\partial f}{\partial v} + \frac{u^2-v^2}{(u^2-1)(1-v^2)}\frac{\partial^2 f}{\partial\phi^2}\Big\}.$$

After multiplying by $a^2(u^2-v^2)$, the Helmholtz equation $(\nabla^2+k^2)f=0$ becomes

$$\Big\{\frac{\partial}{\partial u}(u^2-1)\frac{\partial}{\partial u} + \frac{\partial}{\partial v}(v^2-1)\frac{\partial}{\partial v} + \frac{u^2-v^2}{(u^2-1)(1-v^2)}\frac{\partial^2}{\partial\phi^2} + c^2(u^2-v^2)\Big\} f = 0 \tag{1.4}$$

where $c=ak$. Suppose that, for fixed c, one can express f as a product

$$f = R(u)S(v)\begin{cases}\cos m\phi \\ \sin m\phi\end{cases} \qquad m = 0, 1, \ldots .$$

Such an f can satisfy (1.4) if there exists an eigenvalue $\chi = \chi_{mn}$ such that

$$\frac{\mathrm{d}}{\mathrm{d}u}(u^2-1)\frac{\mathrm{d}R}{\mathrm{d}u} - \left(\chi - c^2u^2 + \frac{m^2}{u^2-1}\right)R(u) = 0 \quad \text{and} \tag{1.5a}$$

$$\frac{\mathrm{d}}{\mathrm{d}v}(v^2-1)\frac{\mathrm{d}S}{\mathrm{d}v} - \left(\chi - c^2v^2 + \frac{m^2}{v^2-1}\right)S(v) = 0. \tag{1.5b}$$

Solutions $R = R_{mn}(u;c)$ in (1.5) are called *radial* PSWFs and solutions $S = S_{mn}(v;c)$ are called *angular* PSWFs. The equations (1.5) are the same except that the variable $u = \cosh(\xi)$ is strictly defined on $[1,\infty)$ while $v = \cos\eta$ is strictly defined in $[-1,1]$. However, as solutions of (1.5), R_{mn} and S_{mn} can be defined on $\mathbb{R}$ and differ from one another by a scale factor. The parameter m is called the *order* of the PSWF. *For time and band limiting, we will be concerned only with the case of the angular functions and* $m = 0$. We will use the variable t for *time* and will be interested in solutions $S_n(t) = S_{0n}(t;c)$ that are eigenfunctions of the prolate operator $\mathscr{P}$,

$$\mathscr{P}S_n(t) = \chi_n S_n(t); \quad \mathscr{P} = \frac{\mathrm{d}}{\mathrm{d}t}(t^2-1)\frac{\mathrm{d}}{\mathrm{d}t} + c^2t^2. \tag{1.6}$$

Further basic properties of the PSWFs $S_{mn}(\cdot;c)$ and $R_{mn}(\cdot;c)$, especially those asymptotic properties useful for applications in mathematical physics, are developed in the monographs of Flammer [101], Meixner and Schäfke [231], Stratton et al. [318], and Morse and Feshbach [240], among other classical sources. To use Slepian and Pollak's phrase [309], "We will draw freely from this literature"—particularly regarding features of PSWFs that are incidental but not central to further discussion.

Various normalizations will be of fundamental importance. These normalizations include endpoint values of the PSWF solutions of (1.6) and normalizations of the eigenvalues. First, the solutions S_{0n} of (1.6) are defined, initially, on $[-1,1]$. These functions extend analytically both in t and in c. Because they are eigenfunctions of a positive definite, self-adjoint operator, basic Sturm–Liouville theory (e.g., [373]) shows that they are orthogonal over $[-1,1]$ and are also complete in $L^2[-1,1]$. It is worth just mentioning for now (see Chap. 2) that the eigenvalues $\chi_n(c)$ are nondegenerate and that S_{0n} is indexed in such a way that χ_n is strictly increasing. In the $c = 0$ limit, S_{0n} (suitably normalized) becomes the nth Legendre polynomial P_n, as will be discussed in more detail momentarily.

In what follows, we will let $\phi_n = \phi_n(t;c)$ be a multiple of $S_{0n}(t;c)$ normalized such that $\|\phi_n\|_{L^2(\mathbb{R})} = 1$. Plots of PSWFs ϕ_n normalized to $\|\phi_n\|_{L^2[-1,1]} = 1$ are shown in Fig. 1.1. We will usually employ the *variant phi*, $\varphi_n = \varphi_n(t;T,\Omega)$, to denote the nth eigenfunction of $P_\Omega Q_T$, also having L^2-norm one. When the values of Ω and T are clear from context or unspecified we will typically omit reference to Ω and T and simply write $\varphi_n(t)$. As will be seen momentarily, when $T = 1$, this notation gives $\varphi_n(t;1,c/\pi) = \phi_n(t;c)$.

Scaling and the Fourier Transform

Suppose that $f:\mathbb{R}\to\mathbb{C}$ is absolutely integrable and that the same applies to the image of f under any second-order, linear differential operator of the form $\rho_0(t)\frac{\mathrm{d}^2}{\mathrm{d}t^2}+\rho_1(t)\frac{\mathrm{d}}{\mathrm{d}t}+\rho_2(t)$ in which the ρ_i are polynomials of degree at most two. Under this hypothesis, the Fourier transform (1.1) of the derivative of f is $2\pi\mathrm{i}\xi\widehat{f}(\xi)$. This follows upon integrating by parts. Interchanging the roles of f and $\widehat{f}$ and using the fact that the kernel of the inverse Fourier transform is the conjugate of that of the forward transform,

$$\left(\frac{\mathrm{d}\widehat{f}}{\mathrm{d}\xi}\right)^{\vee}=-2\pi\mathrm{i}t\,f(t)\quad\text{or}\quad\widehat{(tf)}(\xi)=\frac{\mathrm{i}}{2\pi}\frac{\mathrm{d}\widehat{f}}{\mathrm{d}\xi}.$$

Formally, these facts readily generalize to

$$\mathscr{F}P\Big(t,\frac{\mathrm{d}}{\mathrm{d}t}\Big)=P\Big(\frac{\mathrm{i}}{2\pi}\frac{\mathrm{d}}{\mathrm{d}\xi},2\pi\mathrm{i}\xi\Big)\mathscr{F}\tag{1.7}$$

whenever P is a polynomial function of two variables. Differentiation and multiplication by t do not commute. In particular, $(tf)'-tf'=f$ where $f'=\mathrm{d}f/\mathrm{d}t$. Therefore, $P(t,\mathrm{d}/\mathrm{d}t)$ has to be interpreted strictly as a linear combination of monomials in which the operator of the second argument is applied first, and that of the first argument is then applied.

The *Lucky Accident*

One of the fundamental observations made by Slepian and Pollak in [309] is that the differential operator $\mathscr{P}$ in (1.6) commutes with the time-localization operator $Q=Q_1$ and the frequency-localization operator P_a, for a an appropriate fixed multiple of c. Slepian and Pollak simply observed that this "lucky accident," as Slepian [308] called it, followed from general properties relating differential operators to corresponding integral operators. Considerably later, Walter [346] viewed this accident as a characterization of certain differential operators of the form

$$P\Big(t,\frac{\mathrm{d}}{\mathrm{d}t}\Big)=\rho_0(t)\frac{\mathrm{d}^2}{\mathrm{d}t^2}+\rho_1(t)\frac{\mathrm{d}}{\mathrm{d}t}+\rho_2(t)\tag{1.8}$$

that commute with multiplication by the indicator function of an interval.

Lemma 1.2.1. *A second-order linear differential operator (1.8) commutes with multiplication by the indicator function of $I=[a,b]$ in the sense that, for any $f\in C^2(\mathbb{R})$,*

$$P\Big(t,\frac{\mathrm{d}}{\mathrm{d}t}\Big)(\mathbb{1}_I f)=\mathbb{1}_I P\Big(t,\frac{\mathrm{d}}{\mathrm{d}t}\Big)(f),$$

if and only if ρ_0 vanishes at the endpoints of I and the values of ρ_1 and $\mathrm{d}\rho_0/\mathrm{d}t$ coincide at the endpoints of I.

Proof. The distributional derivative of $\mathbb{1}_{[-T,T]}$ is $\delta_{-T}-\delta_T$ and its second derivative is $\delta'_{-T}-\delta'_T$ where the prime denotes differentiation with respect to t. To prove the lemma, first it suffices to assume that I is symmetric, of the form $[-T,T]$ for some $T>0$, so multiplication by $\mathbb{1}_{[-T,T]}$ is Q_T. From the product rule for derivatives, therefore,

$$P\Big(t,\frac{\mathrm{d}}{\mathrm{d}t}\Big)(Q_T f)=\rho_0 f\,(\delta'_{-T}-\delta'_T)+(2\rho_0 f'+\rho_1 f)(\delta_{-T}-\delta_T)+Q_T\Big(P\Big(t,\frac{\mathrm{d}}{\mathrm{d}t}\Big)f\Big).$$

The commutation condition then takes the form

$$\rho_0 f\,(\delta'_{-T}-\delta'_T)+(2\rho_0 f'+\rho_1 f)(\delta_{-T}-\delta_T)=0,\quad (f\in C^2(\mathbb{R})).$$

Restricting to those $f\in C^2(\mathbb{R})$ that are compactly supported in $(0,\infty)$, the commutation condition implies that

$$\rho_0 f\,\delta'_T+(2\rho_0 f'+\rho_1 f)\delta_T=0\,.$$

As distributions, $f\,\delta_T=f(T)\delta_T$ and $f\,\delta'_T=f(T)\delta'_T-f'(T)\delta_T$. Therefore,

$$\rho_0(T)\,f(T)\,\delta'_T-\rho'_0(T)\,f(T)\,\delta_T+(\rho_0(T)\,f'(T)+\rho_1(T)\,f(T))\delta_T=0\,.$$

Taking f such that $f'(T)=0$ but $f(T)\neq 0$ gives

$$\rho_0(T)\,f(T)\,\delta'_T+(\rho_1(T)-\rho'_0(T))\,f(T)\,\delta_T=0\,.$$

Treating δ'_T and δ_T as independent then leads to $\rho_0(T)=0$ and $\rho'_0(T)=\rho_1(T)$. A similar argument gives the corresponding conditions at $-T$ also. □

For quadratic P, the criteria can be reformulated as follows.

Corollary 1.2.2. *If ρ_0 and ρ_1 are quadratic polynomials then $P\big(t,\frac{\mathrm{d}}{\mathrm{d}t}\big)$ in (1.8) commutes with $\mathbb{1}_{[-T,T]}$ if and only if there exist constants a, b such that*

$$\rho_0(t)=a(t^2-T^2)\quad\text{and}\quad\rho_1(t)=2at+b(t^2-T^2). \tag{1.9}$$

Consider now the special case in which P is quadratic in both of its arguments. Temporarily, we will use the shorthand $\partial_\xi=\frac{\mathrm{i}}{2\pi}\frac{\mathrm{d}}{\mathrm{d}\xi}$. Consider the action of $P\big(\frac{\mathrm{i}}{2\pi}\frac{\mathrm{d}}{\mathrm{d}\xi},2\pi\mathrm{i}\xi\big)$ when P is as in Corollary 1.2.2:

$$P(\partial_\xi,2\pi\mathrm{i}\xi)=a(\partial_\xi^2-T^2)\,(2\pi\mathrm{i}\xi)^2+\big(2a\partial_\xi+b(\partial_\xi^2-T^2)\big)\,(2\pi\mathrm{i}\xi)+c_1\partial_\xi^2+c_2\partial_\xi+c_3.$$

As operators,

$$\partial_\xi^2(2\pi\mathrm{i}\xi)^2 = \left(\frac{\mathrm{d}^2}{\mathrm{d}\xi^2}\right)\xi^2 = \frac{\mathrm{d}}{\mathrm{d}\xi}\left(2\xi+\xi^2\frac{\mathrm{d}}{\mathrm{d}\xi}\right) = 2+4\xi\frac{\mathrm{d}}{\mathrm{d}\xi}+\xi^2\frac{\mathrm{d}^2}{\mathrm{d}\xi^2},$$
$$\partial_\xi(2\pi\mathrm{i}\xi) = -\frac{\mathrm{d}}{\mathrm{d}\xi}\xi = -1-\xi\frac{\mathrm{d}}{\mathrm{d}\xi}, \quad \text{and}$$
$$\partial_\xi^2(2\pi\mathrm{i}\xi) = -\frac{\mathrm{i}}{2\pi}\left(\frac{\mathrm{d}^2}{\mathrm{d}\xi^2}\right)\xi = -\frac{\mathrm{i}}{2\pi}\frac{\mathrm{d}}{\mathrm{d}\xi}\left(1+\xi\frac{\mathrm{d}}{\mathrm{d}\xi}\right) = -\frac{\mathrm{i}}{2\pi}\left(2\frac{\mathrm{d}}{\mathrm{d}\xi}+\xi\frac{\mathrm{d}^2}{\mathrm{d}\xi^2}\right).$$

Substitution of the corresponding terms in $P(\partial_\xi, 2\pi\mathrm{i}\xi)$ with $c_3 = 0$ gives

$$\begin{aligned}
P(\partial_\xi, 2\pi\mathrm{i}\xi) &= a(\partial_\xi^2 - T^2)(2\pi\mathrm{i}\xi)^2 + \left(2a\partial_\xi + b(\partial_\xi^2 - T^2)\right)(2\pi\mathrm{i}\xi) + c_1\partial_\xi^2 + c_2\partial_\xi \\
&= a\left(2+4\xi\frac{\mathrm{d}}{\mathrm{d}\xi}+\xi^2\frac{\mathrm{d}^2}{\mathrm{d}\xi^2}\right) - aT^2(2\pi\mathrm{i}\xi)^2 - 2a\left(1+\xi\frac{\mathrm{d}}{\mathrm{d}\xi}\right) \\
&\quad - \frac{ib}{2\pi}\left(2\frac{\mathrm{d}}{\mathrm{d}\xi}+\xi\frac{\mathrm{d}^2}{\mathrm{d}\xi^2}\right) - bT^2(2\pi\mathrm{i}\xi) - \frac{c_1}{(2\pi)^2}\frac{\mathrm{d}^2}{\mathrm{d}\xi^2} + \frac{ic_2}{2\pi}\frac{\mathrm{d}}{\mathrm{d}\xi} \\
&= \left(a\xi^2 - \frac{ib\xi}{2\pi} - \frac{c_1}{4\pi^2}\right)\frac{\mathrm{d}^2}{\mathrm{d}\xi^2} + \left(2a\xi + \mathrm{i}\frac{c_2-2b}{2\pi}\right)\frac{\mathrm{d}}{\mathrm{d}\xi} + T^2(4\pi^2 a\xi^2 - 2\pi\mathrm{i}b\xi) \\
&\equiv Q\left(\xi, \frac{\mathrm{d}}{\mathrm{d}\xi}\right)
\end{aligned}$$

where $Q(\xi, \frac{\mathrm{d}}{\mathrm{d}\xi}) = \sigma_0(\xi)\frac{\mathrm{d}^2}{\mathrm{d}\xi^2} + \sigma_1(\xi)\frac{\mathrm{d}}{\mathrm{d}\xi} + \sigma_2(\xi)$ with

$$\begin{aligned}
\sigma_0(\xi) &= a\xi^2 - \frac{ib}{2\pi}\xi - \frac{c_1}{4\pi^2}, \\
\sigma_1(\xi) &= 2a\xi + \mathrm{i}\frac{c_2-2b}{2\pi}, \quad \text{and} \\
\sigma_2(\xi) &= 4\pi^2 a\xi^2 T^2 - 2\pi\mathrm{i}b\xi T^2.
\end{aligned}$$

Applying the criterion of Corollary 1.2.2, one finds that in order for $Q(\xi, \mathrm{d}/\mathrm{d}\xi)$ to commute with the characteristic function of an interval $[-\Omega, \Omega]$, it is necessary that $\sigma_0(\xi) = \alpha(\xi^2 - \Omega^2)$ and then that $\sigma_1(\xi) = 2\alpha\xi + \beta(\xi^2 - \Omega)^2$. Since σ_1 has no quadratic term, it follows that $\beta = 0$ and, consequently, that $\alpha = a$ and $c_2 - 2b = 0$. The required form of σ_0 implies that $b = 0$ (so $c_2 = 0$ also) and then that $c_1 = 4\pi^2 a\Omega^2$. Putting these observations together and allowing for an additive constant term in σ_2, we conclude the following.

Theorem 1.2.3. *Suppose that $P(t, \mathrm{d}/\mathrm{d}t)$ is a differential operator of the form $\rho_0\frac{\mathrm{d}^2}{\mathrm{d}t^2} + \rho_1\frac{\mathrm{d}}{\mathrm{d}t} + \rho_2$ with quadratic coefficients ρ_i such that $P(t, \mathrm{d}/\mathrm{d}t)$ commutes with multiplication by the characteristic function of $[-T, T]$ and such that $\mathscr{F}(P(t, \mathrm{d}/\mathrm{d}t)) = P(\partial_\xi, 2\pi\mathrm{i}\xi)$ commutes with multiplication by the characteristic function of $[-\Omega, \Omega]$. Then there exist constants a and b such that*

$$\rho_0(t) = a(t^2 - T^2) \quad \text{and} \quad \rho_1(t) = 2at \quad \text{while} \quad \rho_2(t) = 4\pi^2 a\Omega^2 t^2 + b. \tag{1.10}$$

Returning to (1.6) and writing

$$\mathscr{P} = \frac{\mathrm{d}}{\mathrm{d}t}(t^2-1)\frac{\mathrm{d}}{\mathrm{d}t} + b^2t^2 = (t^2-1)\frac{\mathrm{d}^2}{\mathrm{d}t^2} + 2t\frac{\mathrm{d}}{\mathrm{d}t} + b^2t^2,$$

one recovers the particular case of (1.10) in which $a=1$, $T=1$, and $b=2\pi\Omega$. We conclude that *any second-order differential operator with quadratic coefficients that commutes with the time- and band-limiting operators is a multiple of a rescaling of the differential operator* $\mathscr{P}$ *for prolate spheroidal functions of order zero, plus a multiple of the identity*. In particular, PSWFs are eigenfunctions of time- and band-limiting operators.

The PSWF Parameter c and the Time–Bandwidth Product $\mathfrak{a}$

The parameter c in the operator $\mathscr{P}$ in (1.6) is closely related to the time–bandwidth product, and is often used in the literature to index quantities that depend on this product, but c is not equal to this product. The operator $\mathscr{P}$ commutes with both Q and $P_{2\Omega}$ when $c=2\pi\Omega$. Since the time–bandwidth product of $[-1,1]$ and $[-\Omega,\Omega]$ is 4Ω, this tells us that *c equals the time–bandwidth product multiplied by $\pi/2$*. The area of the product of the time-localization and frequency-localization sets plays such a vital role in what follows that we introduce here the special symbol "$\mathfrak{a}$" in order to reference this quantity. Thus, to the operator $P_\Sigma Q_S$, with P_Σ in (1.2) and Q_S in (1.3), one associates the area $\mathfrak{a}(S,\Sigma)=|S||\Sigma|$. Abusing notation as before, to the operator $P_\Omega Q_T$, where $P_\Omega = P_{[-\Omega/2,\Omega/2]}$ and $Q_T = Q_{[-T,T]}$, we associate $\mathfrak{a} = \mathfrak{a}(T,\Omega) = 2T\Omega$. When ϕ is a unitary dilate of $S_n(t;c)$ in (1.6) that also happens to be an eigenfunction of $P_\Omega Q_T$, one will have $c=\pi\mathfrak{a}/2$. In the Bell Labs papers, Landau, Slepian and Pollak defined the Fourier transform as the integral operator with kernel $\mathrm{e}^{-\mathrm{i}t\xi}$. We will denote this Fourier operator by $\sqrt{2\pi}\mathscr{F}_{2\pi}$. The conversion from $\mathscr{F}$ to $\mathscr{F}_{2\pi}$ can be regarded as a units conversion from hertz to radians per second. The *Bell Labs* inverse Fourier transform has kernel $\mathrm{e}^{\mathrm{i}t\xi}/2\pi$. Also, the Bell Labs papers denoted by $Df(t) = \mathbb{1}_{[-T/2,T/2]}(t)f(t)$ the time-limiting operator of duration T, and by $B = \mathscr{F}_{2\pi}^{-1}\mathbb{1}_{[-\Omega,\Omega]}\mathscr{F}_{2\pi}$ the band-limiting operation of bandwidth 2Ω. The basic Bell Labs time- and band-limiting operator then was BD. Thus, in the Bell Labs notation, with c as in (1.6), $2c=\Omega T$, whereas $2c=\pi\Omega T$ in our notation.

1.2.3 Time–Frequency Properties of PSWFs

In this section we discuss certain analytical properties of eigenfunctions of $P_\Omega Q_T$. Specifically, the eigenvalues are simple, the eigenfunctions are orthogonal in L^2 over $[-T,T]$ as well as over $\mathbb{R}$, the eigenfunctions are complete in $L^2[-T,T]$, and the eigenfunctions are covariant under the Fourier transform, meaning that their Fourier transforms are dilations of their cutoffs. For later reference, Table 1.1 lists several relationships involving time limiting, band limiting, and scaling.

Table 1.1 Scaling relations for time and band limiting

Operation	Formula	Relations
D_α	$(D_\alpha f)(t) = \sqrt{\alpha}\, f(\alpha t)$	$D_{\alpha_1} D_{\alpha_1} = D_{\alpha_1 \alpha_2}$
$\mathscr{F}$	$(\mathscr{F} f)(s) = \int_{-\infty}^{\infty} f(t)\, \mathrm{e}^{-2\pi i s t}\, \mathrm{d}t$	$(\mathscr{F}^2 f)(t) = f(-t)$
Q_β	$(Q_\beta f)(t) = f(t)\, \mathbb{1}_{[-\beta,\beta]}(t)$	$Q_{\beta_1} Q_{\beta_2} = Q_{\min\{\beta_1,\beta_2\}}$
P_γ	$(P_\gamma f)(t) = (\mathscr{F}^{-1} Q_{\gamma/2} \mathscr{F} f)(t)$	$P_{\gamma_1} P_{\gamma_2} = P_{\min\{\gamma_1,\gamma_2\}}$
		$D_\alpha \mathscr{F} = \mathscr{F} D_{1/\alpha}$
		$Q_\beta D_\alpha = D_\alpha Q_{\alpha\beta}$
		$P_{\alpha\gamma} D_\alpha = D_\alpha P_\gamma$
		$\mathscr{F} P_\gamma Q_\beta = Q_{\gamma/2} P_{2\beta} \mathscr{F}^{-1}$

Time–Bandwidth Product: Scaling and Covariance

With $Q_T f = f\,\mathbb{1}_{[-T,T]}$ and $P_\Omega f = (\widehat{f}\,\mathbb{1}_{[-\Omega/2,\Omega/2]})^{\vee}$, the time–bandwidth product associated with the time- and band-limiting operator $P_\Omega Q_T$ is $a = 2T\Omega$. Ultimately, we want to describe the behavior of the eigenvalues of $P_\Omega Q_T$ in terms of a. We will see that, when the Fourier transform is normalized as in (1.1), the number of eigenvalues of $P_\Omega Q_T$ close to one is, essentially, the time–bandwidth product $2\Omega T$. Define the unitary dilation operator $(D_a f)(t) = \sqrt{a} f(at)$ where $a > 0$, and set $(\mathscr{F}_a f)(\xi) = (\mathscr{F} D_a f)(\xi)$ where $\mathscr{F}$ is the Fourier transform as defined in (1.1). Many authors *define* the Fourier transform as the operator $\mathscr{F}_{2\pi}$, in our notation, whose kernel is $\mathrm{e}^{-it\xi}/\sqrt{2\pi}$. In this normalization, the operator with time cutoff $[-T,T]$ followed by frequency cutoff $[-\Omega/2,\Omega/2]$ has essentially $\Omega T/\pi$ eigenvalues close to one.

Using either Fourier kernel, a dilation by $a > 0$ in time becomes a dilation by $1/a$ in frequency, for example, $(D_\alpha f)^\wedge = D_{1/\alpha}(\widehat{f})$. The PSWFs are also covariant under dilation, so the eigenvalues and eigenfunctions can be regarded as functions of the time–bandwidth product. To make this precise, observe that $P = D_{1/\Omega} P_\Omega D_\Omega$ and, similarly, $Q_T = D_{1/T} Q D_T$. Thus,

$$PQ_T = D_{1/T} D_T P D_{1/T} Q D_T = D_{1/T}\,(P_T Q)\, D_T\ .$$

Since D_T is unitary, PQ_T and $P_T Q$ have the same eigenvalues and any eigenfunction of PQ_T is obtained by applying $D_{1/T}$ to an eigenfunction of $P_T Q$ having the same eigenvalue. Consequently, $P_{\Omega_1} Q_{T_1}$ and $P_{\Omega_2} Q_{T_2}$ share the same eigenvalues whenever $\Omega_1/\Omega_2 = T_2/T_1$ and their eigenfunctions are unitary dilations of one another.

Fourier Covariance

A PSWF eigenfunction ψ of PQ_T is real analytic and has a Fourier transform supported in $[-1/2, 1/2]$.

Proposition 1.2.4. *If $\psi = \varphi_n$ is an eigenfunction of PQ_T with eigenvalue $\lambda > 0$, then*

$$\widehat{\psi}\Big(\frac{\xi}{2T}\Big) = \lambda \mathrm{i}^n \mathbb{1}_{[-T,T]}\,\psi(\xi)\,.$$

Proof. Since $(\mathbb{1}_{[-1/2,1/2]})^{\vee}(t) = \sin \pi t/\pi t$,

$$\begin{aligned}
\lambda\sqrt{2T}\,D_{1/(2T)}\widehat{\psi}(\xi) &= (PQ_T\psi)^{\wedge}(\xi/2T)\\
&= \int_{-\infty}^{\infty} \mathrm{e}^{-2\pi \mathrm{i} x\xi/(2T)} \int_{-T}^{T} \frac{\sin\pi(x-t)}{\pi(x-t)}\,\psi(t)\,\mathrm{d}t\,\mathrm{d}x\\
&= \int_{-T}^{T} \psi(t)\,\mathrm{e}^{-2\pi \mathrm{i} t\xi/(2T)} \int_{-\infty}^{\infty} \frac{\sin\pi x}{\pi x}\,\mathrm{e}^{-2\pi \mathrm{i} x\xi/(2T)}\,\mathrm{d}x\,\mathrm{d}t\\
&= \mathbb{1}_{[-T,T]}(\xi) \int_{-T}^{T} \psi(t)\,\mathrm{e}^{-2\pi \mathrm{i} t\xi/(2T)}\mathrm{d}t\\
&= \mathbb{1}_{[-T,T]}(\xi) \int_{-\infty}^{\infty} \widehat{\psi}\Big(\frac{\eta}{2T}\Big)\frac{\sin\pi(\xi-\eta)}{\pi(\xi-\eta)}\,\mathrm{d}\eta\\
&= \sqrt{2T}\,Q_T P D_{1/(2T)}\widehat{\psi}(\xi)\,.
\end{aligned}$$

Therefore, $D_{1/(2T)}\widehat{\psi}$ is an eigenfunction of $Q_T P$ with eigenvalue λ. Since $D_{1/(2T)}\widehat{\psi}$ is supported in $[-T,T]$ this shows that, in fact, it is the restriction to $[-T,T]$ of an eigenfunction of PQ_T with eigenvalue λ. Assuming for the time being that PQ_T has a simple spectrum, $D_{1/(2T)}\widehat{\psi}$ then is a unimodular multiple of $Q_T\psi$. Since $(\mathscr{F}^2 f)(t) = f(-t)$, the unimodular multiple has a square equal to 1 if $\psi = \varphi_n$, where n is even, and equal to -1 if n is odd. □

Balancing Time and Frequency

Fix $\alpha > 0$ and $\beta > 0$ and consider $P_{2\beta}Q_\alpha = \mathscr{F}^{-1}Q_\beta \mathscr{F} Q_\alpha$. Based on the commutation rules in Table 1.1,

$$\begin{aligned}
\mathscr{F}^{-1}Q_\beta\mathscr{F}Q_\alpha D_{\sqrt{\beta/\alpha}} &= \mathscr{F}^{-1}Q_\beta\mathscr{F}D_{\sqrt{\beta/\alpha}}Q_{\sqrt{\alpha\beta}} = \mathscr{F}^{-1}Q_\beta D_{\sqrt{\alpha/\beta}}\mathscr{F}Q_{\sqrt{\alpha\beta}}\\
&= \mathscr{F}^{-1}D_{\sqrt{\alpha/\beta}}Q_{\sqrt{\alpha\beta}}\mathscr{F}Q_{\sqrt{\alpha\beta}} = D_{\sqrt{\beta/\alpha}}\mathscr{F}^{-1}Q_{\sqrt{\alpha\beta}}\mathscr{F}Q_{\sqrt{\alpha\beta}}\,.
\end{aligned}$$

That is,

$$\mathscr{F}^{-1}Q_\beta\mathscr{F}Q_\alpha D_{\sqrt{\beta/\alpha}} = D_{\sqrt{\beta/\alpha}}\mathscr{F}^{-1}Q_{\sqrt{\alpha\beta}}\mathscr{F}Q_{\sqrt{\alpha\beta}}, \tag{1.11}$$

which can also be expressed as $P_{2\beta}Q_\alpha(D_{\sqrt{\beta/\alpha}}\varphi) = D_{\sqrt{\beta/\alpha}}P_{2\sqrt{\alpha\beta}}Q_{\sqrt{\alpha\beta}}\,\varphi$. In particular, if ψ is a λ-eigenfunction of $P_{2\beta}Q_\alpha$ then $D_{\sqrt{\alpha/\beta}}\,\psi$ is a λ-eigenfunction of

$P_{2\sqrt{\alpha\beta}}Q_{\sqrt{\alpha\beta}}$, which is a time- and band-limiting operator in which the time and frequency intervals both have equal length $2\sqrt{\alpha\beta}$.

To rephrase the observation just made, for a fixed $a>0$, one has $P_{\sqrt{a}}Q_{\sqrt{a}/2} = \mathscr{F}^{-1}Q_{\sqrt{a}/2}\mathscr{F}Q_{\sqrt{a}/2}$. This suggests that $\mathscr{F}Q_{\sqrt{a}/2}$ can be regarded as a *square root* of $P_{\sqrt{a}}Q_{\sqrt{a}/2}$, provided that one takes into account the parity of functions on which it acts. When f is even, the action of $P_{\sqrt{a}}Q_{\sqrt{a}/2}$ on f is the same as that of $(\mathscr{F}Q_{\sqrt{a}/2})^2$. When f is odd, its action is the same as that of $(\mathrm{i}\mathscr{F}Q_{\sqrt{a}/2})^2$. A calculation along the lines of that in the proof of Proposition 1.2.4 and using the known parities of the PSWFs gives the following.

Proposition 1.2.5. *If ψ is an eigenfunction of $P_{\sqrt{a}}Q_{\sqrt{a}/2}$ with eigenvalue $\lambda_n = \lambda_n(a)$, then ψ is an eigenfunction of $\mathscr{F}Q_{\sqrt{a}/2}$ with eigenvalue $\mathrm{i}^n\sqrt{\lambda_n(a)}$.*

The eigenfunctions of $P_{\sqrt{a}}Q_{\sqrt{a}/2}$ are dilates by $\sqrt{a}$ of those of $PQ_{a/2}$. Since $PQ_T = \mathscr{F}^{-1}Q_{1/2}\mathscr{F}Q_T$, Proposition 1.2.4 also suggests that $\mathscr{F}Q_T$ is *half* of a time- and band-limiting operation. By (1.11), $(D_a\mathscr{F}Q)^2 = \pm P_{2a}Q$ when acting, respectively, on real-valued even or odd functions. One obtains an equivalent version of Proposition 1.2.5 that will be useful in what follows.

Corollary 1.2.6. *If ψ is an eigenfunction of $P_{a/2}Q$ with eigenvalue $\lambda = \lambda_n(a)$, then ψ is an eigenfunction of F_a with $\mu = \mu_n(a) = 2\mathrm{i}^n\sqrt{\lambda_n(a)/a}$, where* [1]

$$(F_a f)(t) = \int_{-1}^{1} \mathrm{e}^{\frac{\pi a}{2}\mathrm{i}st} f(s)\,\mathrm{d}s = \frac{2}{\sqrt{a}}(D_{a/4}\mathscr{F}^{-1}Qf)(t)\,.$$

One has $P_{a/2}Q = \frac{a}{4}F_a^*F_a$ so that $\lambda_n(a) = \frac{a}{4}|\mu_n(a)|^2$.

Double Orthogonality

The orthogonality of the PSWFs over $\mathbb{R}$, that is, $\langle\varphi_n,\varphi_m\rangle = \delta_{n,m}$, follows from the fact that they are eigenfunctions of the self-adjoint operator PQ_TP, defined on $L^2(\mathbb{R})$, having different eigenvalues:

$$\lambda_n\langle\varphi_n,\varphi_m\rangle = \langle PQ_TP\varphi_n,\varphi_m\rangle = \langle\varphi_n,PQ_TP\varphi_m\rangle = \lambda_m\langle\varphi_n,\varphi_m\rangle.$$

The proof that $\langle Q_T\varphi_n,\varphi_m\rangle = \lambda_n\,\delta_{n,m}$ uses the reproducing kernel property:

[1] Many authors denote by F_c the operator given by integration against the kernel $\mathrm{e}^{-\mathrm{i}cst}\mathbb{1}_{[-1,1]}(s)$. This is the same as F_a when $c = a\pi/2$ as we assume here.

$$
\begin{aligned}
\delta_{n,m} &= \int_{-\infty}^{\infty} \varphi_n(t)\,\varphi_m(t)\,\mathrm{d}t \\
&= \frac{1}{\lambda_n\,\lambda_m}\int_{-\infty}^{\infty}\int_{-T}^{T} \mathrm{sinc}(t-s)\,\varphi_n(s)\,\mathrm{d}s \int_{-T}^{T} \mathrm{sinc}(t-u)\,\varphi_m(u)\,\mathrm{d}u\,\mathrm{d}t \\
&= \frac{1}{\lambda_n\,\lambda_m}\int_{-T}^{T} \varphi_n(s) \int_{-T}^{T} \varphi_m(u) \int_{-\infty}^{\infty} \mathrm{sinc}(u-t)\mathrm{sinc}(t-s)\,\mathrm{d}t\,\mathrm{d}u\,\mathrm{d}s \\
&= \frac{1}{\lambda_n\,\lambda_m}\int_{-T}^{T} \varphi_n(s) \int_{-T}^{T} \varphi_m(u)\,\mathrm{sinc}(u-s)\,\mathrm{d}u\,\mathrm{d}s \\
&= \frac{1}{\lambda_n}\int_{-T}^{T} \varphi_n(s)\,\varphi_m(s)\,\mathrm{d}s\,.
\end{aligned}
$$

The proof actually shows more. If one replaces $\mathrm{sinc}(t-s)$ by a real-valued, symmetric reproducing kernel $\rho(t,s) = \rho(t-s) = \rho(s-t)$ then one has the identity

$$
\int_{-\infty}^{\infty} \rho(t-u)\,\rho(t-s)\,\mathrm{d}t = \int_{-\infty}^{\infty} \rho(u-t)\,\rho(t-s)\,\mathrm{d}t = \rho(u-s)\,.
$$

Following the proof above with sinc replaced by ρ one obtains the following.

Corollary 1.2.7. *If ρ is the inverse Fourier transform of $\mathbb{1}_\Sigma$ where Σ is compact and symmetric and if $Rf = \int \rho(t-s)\,f(s)\,\mathrm{d}s$, then the eigenfunctions of RQ_T belonging to different eigenvalues are orthogonal over $[-T,T]$ as well as over $\mathbb{R}$.*

Completeness in $L^2[-T,T]$

There are a number of ways in which to see that the PSWF eigenfunctions of PQ_T are complete in $L^2[-T,T]$. The fact follows from Sturm–Liouville theory as is proved in Chap. 2; see Theorem 2.1.16. Alternatively, the spectral theorem for compact self-adjoint operators implies that the eigenfunctions of PQ_T are complete in PW. From Parseval's theorem and Proposition 1.2.4, it follows that the PSWFs are also complete in $L^2[-T,T]$.

PQ_T has a Simple Spectrum

Theorem 1.2.8. *For each $T > 0$ the operator $PQ_T : \mathrm{PW} \to \mathrm{PW}$ has a simple spectrum. If the eigenvalues $\lambda_n(T)$ are listed in decreasing order $\lambda_0(T) > \lambda_1(T) > \cdots$ then the nth eigenvector φ_n is a multiple of a dilation of the prolate spheroidal wave function belonging to the eigenvalue $\chi_n(c)$, $c = \pi T$, in the natural ordering $\chi_0 < \chi_1 < \cdots$ of the eigenvalues of (1.6).*

Theorem 1.2.8 will be proved, in effect, in Sect. 2.1.3 by showing that the nondegeneracy of λ_n follows from that of χ_n. The nondegeneracy of χ_n will also be explained in Chap. 2. That the χ_n are nondegenerate for small values of c also follows from a continuation argument, starting with the observation that the $c = 0$ limit of the operator $\mathscr{P}$ in (1.6) generates the Legendre polynomials.

1.2.4 Further Analytical and Numerical Properties

The Case $c = 0$: The Legendre Polynomials

The nth Legendre polynomial $P_n(t)$ is a solution of (1.6) when $c = 0$:

$$\frac{\mathrm{d}}{\mathrm{d}t}(t^2-1)\frac{\mathrm{d}P_n}{\mathrm{d}t} = \chi_n(0)P_n(t) \quad \text{or} \quad (t^2-1)\frac{\mathrm{d}^2P_n}{\mathrm{d}t^2} + 2t\frac{\mathrm{d}P_n}{\mathrm{d}t} - \chi_n(0)P_n = 0. \quad (1.12)$$

The constant functions are solutions with $\chi_0(0) = 0$, and the function $P_1(t) = t$ with $\chi_1(0) = 2$. The Legendre polynomials can be defined iteratively by means of the Gram–Schmidt process in order that the polynomials be orthogonal over $[-1,1]$ with respect to $\langle f, g\rangle = \int_{-1}^{1} f(t)\,g(t)\,\mathrm{d}t$. It is standard to normalize P_n so that $P_n(1) = 1$.

Among algebraic relationships satisfied by the Legendre functions, the most important for us will be *Bonnet's recurrence formula*

$$(n+1)P_{n+1}(t) - (2n+1)tP_n(t) + nP_{n-1}(t) = 0. \quad (1.13)$$

This formula is useful in computing the eigenvalues $\chi_n(0)$ in (1.12), namely

$$\frac{\mathrm{d}}{\mathrm{d}t}(t^2-1)\frac{\mathrm{d}P_n}{\mathrm{d}t} = n(n+1)P_n(t), \quad \text{i.e.,} \quad \chi_n(0) = n(n+1). \quad (1.14)$$

Starting with $n = 1$ one obtains $P_0 = 1$ and $P_1(t) = t$, so that, by (1.13) $P_2(t) = 3t^2/2 - 1/2$ and so on.

The Legendre functions are also coefficients of the generating function

$$g(x,t) = (1-2xt+x^2)^{-1/2} = \sum_{n=0}^{\infty} P_n(t)x^n. \quad (1.15)$$

Upon differentiating (1.15) with respect to x one obtains

$$\frac{t-x}{\sqrt{1-2xt+x^2}} = (1-2xt+x^2)\sum_{n=1}^{\infty} nP_n(t)x^{n-1}.$$

Replacing the inverse square root on the left side by the series in (1.15) and equating the coefficients of corresponding powers of x leads to the recurrence formula (1.13). One also has the identity

$$(t^2-1)\frac{\mathrm{d}P_n}{\mathrm{d}t} = n(tP_n(t) - P_{n-1}(t)). \quad (1.16)$$

Computing the Prolates for Small c: Bouwkamp's Method

Bouwkamp [40] made use of the similarity between the differential equations satisfied by the prolates and the Legendre polynomials and the recurrence formulas satisfied by the Legendre polynomials to compute the coefficients of the prolates in their Legendre series. Since the prolates are in fact Legendre polynomials when $c = 0$, and their dependence on c is analytic, this approach makes good sense and one would expect, at least for small values of c, that the coefficients in these expansions would be *diagonally dominant*.

Suppose now that $\bar{\phi}_n = \bar{\phi}_n^{(c)}$ is the nth $L^2[-1,1]$-normalized solution of the eigenvalue equation $\mathscr{P}\phi = \chi\phi$ (where $\mathscr{P}$ is the differential operator of (1.6)) ordered according to the eigenvalues: $\chi_0 < \chi_1 < \cdots < \chi_n < \cdots$. Since the Legendre polynomials $\{P_n\}_{n=0}^{\infty}$ form an orthogonal basis for $L^2[-1,1]$, each $\bar{\phi}_n^{(c)}$ admits an expansion of the form

$$\bar{\phi}_n^{(c)} = \sum_{m=0}^{\infty} b_m P_m.$$

By equation (1.14),

$$\chi_n \bar{\phi}_n = \mathscr{P}\bar{\phi}_n = \sum_{m=0}^{\infty} b_m \mathscr{P} P_m = \sum_{m=0}^{\infty} b_m [m(m+1)P_m + c^2 x^2 P_m]. \tag{1.17}$$

However, two applications of the recurrence formula (1.13) will give

$$x^2 P_m = \frac{m^2+3m+2}{4m^2+8m+3} P_{m+2} + \frac{2m^2+2m-1}{4m^2+4m-3} P_m + \frac{m^2-m}{4m^2-1} P_{m-2}.$$

Substituting this into (1.17) gives

$$\begin{aligned}\chi_n \sum_{m=0}^{\infty} b_m P_m = \sum_{m=0}^{\infty} \Big[c^2 \frac{m(m-1)}{(2m-3)(2m-1)} b_{m-2} \\ + \left(m(m+1) + c^2 \frac{(2m^2+2m-1)}{(2m+3)(2m-1)} \right) b_m + \frac{(m+1)(m+2)}{(2m+5)(2m+3)} b_{m+2} \Big] P_m\end{aligned}$$

with the understanding that $b_m = 0$ if $m < 0$. The orthogonality of the Legendre polynomials enables us to equate the coefficients of both sides of this equation to obtain the following recurrence formula for $\{b_m\}_{m=0}^{\infty}$:

$$\begin{aligned}c^2 \frac{m(m-1)}{(2m-3)(2m-1)} b_{m-2} + \left(m(m+1) + c^2 \frac{(2m^2+2m-1)}{(2m+3)(2m-1)} - \chi_n \right) b_m \\ + \frac{(m+1)(m+2)}{(2m+5)(2m+3)} b_{m+2} = 0.\end{aligned}$$

The first task is to compute the eigenvalues χ_n, after which we can apply the recurrence formula to compute the coefficients. With

$$\alpha_m = m(m+1) - \frac{c^2}{2}\left(1+\frac{1}{(2m-1)(2m+3)}\right) \qquad (m \geq 0)$$

$$\beta_m = c^4 \frac{m^2(m-1)^2}{(2m-1)^2(2m-3)(2m+1)} \qquad (m \geq 2)$$

$$N_m = c^2 \frac{c^2 m(m-1)}{(2m-1)(2m+1)} \frac{b_m}{b_{m-2}} \qquad (m \geq 2),$$

we have

$$N_m = \frac{\beta_m}{\alpha_m - \chi + N_{m+2}}; \qquad N_{m+2} = \alpha_m - \chi - \frac{\beta_m}{N_m}.$$

Expressing the two formulas for N_{m+2} as continued fractions and setting them equal to one another gives the continued fractions equation

$$\alpha_m - \chi - \cfrac{\beta_m}{\alpha_{m-2} - \chi - \cfrac{\beta_{m-2}}{\alpha_{m-4} - \chi - \cfrac{\beta_{m-4}}{\ddots}}} = \cfrac{\beta_{m+2}}{\alpha_{m+2} - \chi - \cfrac{\beta_{m+4}}{\alpha_{m+4} - \chi - \cfrac{\beta_{m+6}}{\ddots}}}.$$

The equation is satisfied when χ is an eigenvalue χ_n. Such a value can be approximated (after initial estimation) by Newton iterations. Bouwkamp [40] also gave an asymptotic formula for χ_n valid for small values of c.

Computing the Prolates for Small c: Boyd's Method

A more up-to-date approach was outlined by Boyd [44]. Let $\bar{P}_m = \sqrt{m+1/2}P_m$ $(m \geq 0)$ be the L^2-normalized basis of Legendre polynomials. As in the Bouwkamp method, each prolate $\bar{\phi}_n$ is expanded in this basis,

$$\bar{\phi}_n = \sum_{m=0}^{\infty} \beta_{nm}\bar{P}_m, \tag{1.18}$$

and the prolate differential operator $\mathscr{P}$ is applied to both sides of this equation. Let $A = \{a_{mk}\}_{m,k=0}^{\infty}$ be the doubly infinite tridiagonal matrix with nonzero elements

$$a_{m,k} = \begin{cases} \dfrac{c^2 m(m-1)}{(2m-1)\sqrt{(2m-3)(2m+1)}} & \text{if } m \geq 2,\, k = m-2 \\ m(m+1) + \dfrac{c^2(2m^2+2m-1)}{(2m+3)(2m-1)} & \text{if } m = k \geq 0 \\ \dfrac{c^2(m+2)(m+1)}{(2m+3)\sqrt{(2m+5)(2m+1)}} & \text{if } m \geq 0,\, k = m+2 \\ 0 & \text{else.} \end{cases}$$

Then $\mathscr{P}\bar{P}_m = \sum_{k=0}^{\infty} a_{mk}\bar{P}_k$. Applying $\mathscr{P}$ to both sides of (1.18) gives

$$\chi_n \bar{\phi}_n = \sum_{k=0}^{\infty} \left(\sum_{m=0}^{\infty} \beta_{nm} a_{mk} \right) \bar{P}_k. \tag{1.19}$$

For fixed $n \geq 0$, let $\mathbf{b}_n = (\beta_{n0}, \beta_{n1}, \beta_{n2}, \dots)^T$. Equating coefficients in (1.19) gives the matrix equation

$$A^T \mathbf{b}_n = \chi_n \mathbf{b}_n, \tag{1.20}$$

that is, the vector $\mathbf{b}_n$ (whose entries are the L^2-normalized Legendre coefficients of $\bar{\phi}_n$) is an eigenvector of A^T with eigenvalue χ_n.

Truncating the sum in (1.18) after N_{tr} terms and following the procedure outlined above yields a finite matrix eigenvalue problem

$$(A^{\mathrm{tr}})^T \mathbf{b}_n^{\mathrm{tr}} = \chi_n^{\mathrm{tr}} \mathbf{b}_n^{\mathrm{tr}} \tag{1.21}$$

with $\mathbf{b}_n^{\mathrm{tr}} \in \mathbb{R}^{N_{\mathrm{tr}}}$ and A^{tr} the top left $N_{\mathrm{tr}} \times N_{\mathrm{tr}}$ submatrix of the matrix A above. Because of the structure of A^{tr} (whereby the only nonzero entries are of the form $a_{m,m-2}$, $a_{m,m}$, and $a_{m,m+2}$), equation (1.21) may be written as a pair of uncoupled eigenproblems for even and odd values of n respectively:

$$(A_{\mathrm{e}}^{\mathrm{tr}})^T \mathbf{b}_{\mathrm{e}}^{(2n)} = \chi_{2n}^{\mathrm{tr}} \mathbf{b}_{\mathrm{e}}^{(2n)}; \qquad (A_{\mathrm{o}}^{\mathrm{tr}})^T \mathbf{b}_{\mathrm{o}}^{(2n+1)} = \chi_{2n+1}^{\mathrm{tr}} \mathbf{b}_{\mathrm{o}}^{(2n+1)} \tag{1.22}$$

with A_{o}, A_{e} both $(N_{\mathrm{tr}}/2) \times (N_{\mathrm{tr}}/2)$ tridiagonal matrices (assuming N_{tr} even) given by $(A_{\mathrm{e}})_{mk} = a_{2m,2k}$ and $(A_{\mathrm{o}})_{mk} = a_{2m+1,2k+1}$. Once solved, the eigenproblems (1.22) yield approximations of the Legendre coefficients β_{nm} of the prolates $\bar{\phi}_0^{(c)}$, $\bar{\phi}_1^{(c)}, \dots$, $\bar{\phi}_{N-1}^{(c)}$ as in (1.18) and the associated eigenvalues $\chi_0, \chi_1, \dots, \chi_{N-1}$. Plots of $\bar{\phi}_n^{(c)}$ for $c = 5$ and $n = 0, 3, 10$ using Boyd's method can be found in Fig. 1.1. Boyd [44] reported that for all N and c, the worst approximated prolate is that of highest order $\bar{\phi}_{N-1}^{(c)}$ so that if the truncation N_{tr} is chosen large enough so that $\bar{\phi}_{N-1}^{(c)}$ is computed with sufficient accuracy, then so too will $\bar{\phi}_n^{(c)}$ with $0 \leq n \leq N-2$. Numerical evidence was given to suggest that if $N_{\mathrm{tr}} \approx 30 + 2N$, then $\beta_{N-1,N_{\mathrm{tr}}} < 10^{-20}$, and it was claimed that if $N_{\mathrm{tr}} > 30 + 2N$ then the approximations to $\{\bar{\phi}_n^{(c)}\}_{n=0}^{N-1}$ are accurate as long as $c \leq c_*(N) = \pi(N + 1/2)/2$.

Estimating the Prolates for Larger c: Hermite Functions

As before, let $\mathscr{P}$ denote the operator of (1.6). With the substitution $u = \sqrt{c}t$, one rescales $\mathscr{P}$ to

$$\mathscr{P}_s = (u^2 - c)\frac{\mathrm{d}^2}{\mathrm{d}u^2} + 2u\frac{\mathrm{d}}{\mathrm{d}u} + cu^2; \quad \mathscr{P}_s \Phi_n = \chi_n \Phi_n \,. \tag{1.23}$$

When $c \gg u^2$, upon dividing by $u^2 - c$, the equation $\mathscr{P}_s \Phi = \chi \Phi$ approximately reduces to

$$\mathscr{H}\Phi = \frac{\chi}{c}\Phi \quad \text{where} \quad \mathscr{H} = -\frac{\mathrm{d}^2}{\mathrm{d}u^2} + u^2 \,. \tag{1.24}$$

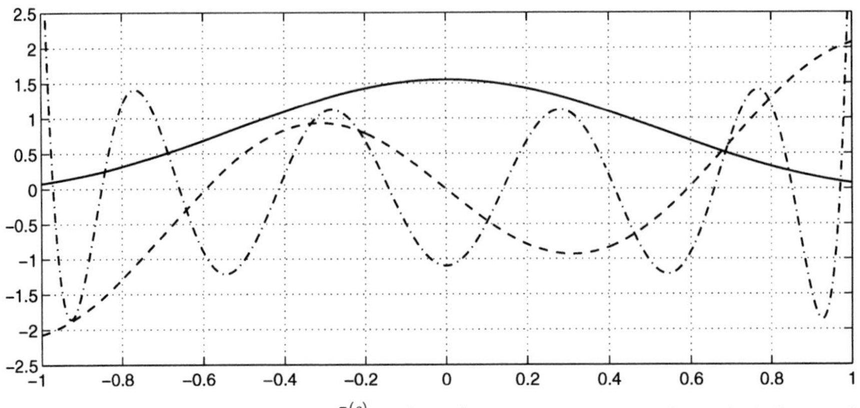

Fig. 1.1 *PSWFs for* $c = 5$. Prolates $\bar{\phi}_n^{(c)}$ on $[-1,1]$ generated using Boyd's method, for $n = 0$ (solid), $n = 3$, and $n = 10$. $\bar{\phi}_n^{(c)}$ has n zeros in $[-1,1]$

The operator $\mathscr{H}$ is called the *Hermite operator*. Its $L^2(\mathbb{R})$-normalized eigenfunctions are the Hermite functions

$$h_n(u) = \frac{1}{\sqrt{n!2^n\sqrt{\pi}}}\mathrm{e}^{-u^2/2}H_n(u);\quad H_n(u) = (-1)^n\mathrm{e}^{u^2}\frac{\mathrm{d}^n}{\mathrm{d}u^n}\mathrm{e}^{-u^2}. \tag{1.25}$$

The polynomial $H_n(u)$ is called the "physicists' Hermite polynomial" of degree n, as opposed to the "probabilists' Hermite polynomial," $H_n^{\text{prob}}(u) = 2^{-n/2}H_n(u/\sqrt{2})$. The eigenvalue property is $\mathscr{H}h_n = (2n+1)h_n$. The functions $h_n(u)$ form a complete orthonormal basis for $L^2(\mathbb{R})$.

Hermite Recurrence and Hermite Expansions of Scaled PSWFs

The Hermite functions h_n obey the recurrence relation

$$h_n'(t) = \sqrt{\frac{n}{2}}h_{n-1}(t) - \sqrt{\frac{n+1}{2}}h_{n+1}(t).$$

This allows for an expansion of the rescaled PSWFs Φ_n in (1.23) by means of the recursion

$$\Phi_m(t) = \sum_{n=0}^{\infty} s_{n,m}h_{n+m}(t)$$

where

$$\begin{aligned}- s_{n-4,m} + [4\chi_n(c) - 8c(n+m+1/2) + 2(n+m)^2 + 2(n+m) + c]s_{n,m}\\ - (n+m+1)(n+m+2)(n+m+3)(n+m+4)s_{n+4,m} = 0\,. \end{aligned} \tag{1.26}$$

The Hermite functions are eigenfunctions of the Fourier transform $\mathscr{F}_{2\pi}$ which acts by integration against $\mathrm{e}^{-\mathrm{i}x\xi}/\sqrt{2\pi}$. To see this, note that

$$\mathscr{F}_{2\pi}\left(\mathrm{e}^{-\frac{x^2}{2}+2xt-t^2}\right)(\xi)=\mathrm{e}^{-\frac{\xi^2}{2}-2\mathrm{i}\xi t+t^2}.$$

Equating powers of t in the generating function relationship

$$\mathrm{e}^{-\frac{x^2}{2}+2xt-t^2}=\sum_{n=0}^{\infty}\mathrm{e}^{-\frac{x^2}{2}}H_n(x)\frac{t^n}{n!}$$

then shows that $\mathscr{F}_{2\pi}(h_n)=(-\mathrm{i})^n h_n$.

PSWFs and Hermite Functions

The asymptotic resemblance between the $\sqrt{c}$-scaled operator $\mathscr{P}_s$ and the operator $\mathscr{H}$ suggests that $\phi_n(t;c)\approx D_{\sqrt{c}}h_n(t)$ as $c\to\infty$ as the following lemma, proved in [231, Theorem 9, p. 243] (cf. also [43, Theorem 5.2]), indicates.

Lemma 1.2.9. *If* $|t|<(\log\sqrt{c})^{1/3}$ *then*

$$D_{\sqrt{c}}\phi_n(t;\sqrt{c})(t)=h_n(t)+o(1)\quad\text{and}\quad\frac{\mathrm{d}}{\mathrm{d}t}D_{\sqrt{c}}\phi_n(t;\sqrt{c})=\frac{\mathrm{d}}{\mathrm{d}t}h_n(t)+o(1).$$

In addition, the nth eigenfunction $\phi_n(t;c)$ of $\mathscr{P}$ has n simple zeros in $[-1,1]$ for any given value of c (see Chap. 2) while the nth Hermite polynomial H_n and hence, h_n, has its n real roots in $[-n/2,n/2]$, with its largest zero approximated by $\pi\sqrt{n}/4$, e.g., [1, p. 787]. The eigenvalues χ_n of $\mathscr{P}$ can be estimated by $\chi_n\sim(2n+1)c-(n^2+n+3/2)/2+O(1/c)$ [43].

1.2.5 Eigenvalues for Time and Band Limiting: Asymptotics

In this section we review very briefly some aspects of the asymptotic behavior of $\lambda_n(c)$ when one of n or c is fixed and the other tends to infinity. Prior to 2000, most eigenvalue and eigenfunction estimates for PSWFs were asymptotic in nature. Quoting Miles [232] in his work on eigenfunction estimates,

> *The emphasis is on qualitative accuracy (such as might be useful to the physicist), rather than on efficient algorithms for very accurate numerical computation, and the error factor for most of the approximations is* $1+O(1/c)$ *as* $c\uparrow\infty$.

In numerical applications, it is important to have precise estimates of λ_n when n is on the order of the time–bandwidth product or when n and c both tend to infinity with some functional relationship between the two. Such estimates will be considered in Chap. 2. The results described here require expressing λ_n in terms of the asymptotic behavior of the functional values $\varphi_n(t;c)$ when t tends to infinity along the imaginary axis. Key estimates along these lines are found in the works of Fuchs [103], Widom [354], and Slepian [304] which also make technically sophisticated uses of Wentzel–Kramers–Brillouin (WKB) methods and Sturm–Liouville theory. While the methods in Sect. 2.6, due largely to Rokhlin et al. [280], do not always recover precisely the full range of asymptotic estimates of λ_n, and certainly do not reproduce the pointwise estimates on PSWFs exhibited in the works of Fuchs, Widom, and Slepian and subsequent works of Miles [232], Dunster [90], and others, they do nevertheless come surprisingly close to recovering the more elaborate results noted using only fairly elementary methods.

Behavior of λ_n as $c \to \infty$ (n Fixed)

As of 1961, the precise rate of decay of $1-\lambda_n(c)$ as $c\to\infty$ was not known. Quoting Landau and Pollak [195, p. 78], "$\lambda_0(c)\to 1$ quite rapidly as $c\to\infty$; the approach is exponential, but the exact rate has not been proved." The first specific quantification of the rate of convergence of $\lambda_n \to 1$ as $c\to\infty$ was made by Fuchs [103], who normalized $\lambda_n(c)$ to be the nth eigenvalue of the operator $Q_cP_{1/\pi}Q_c$ with kernel $k_c(x,y) = \mathbb{1}_{[-c,c]}(x)\sin(x-y)/(\pi(x-y))\mathbb{1}_{[-c,c]}(y)$ and, hence, $a = 2c/\pi$. Thus, up to a dilation, the corresponding eigenfunctions are $\phi_n(t;c)$. Fuchs renormalized the PSWFs by setting

$$c = a^2, \qquad x = as, \quad \text{and} \quad y = at.$$

This normalization facilitates a differentiation with respect to a that plays a role in estimating λ_n in terms of values of ϕ_n. Under this normalization, Fuchs proved the following asymptotic formula.

Theorem 1.2.10 (Fuchs, 1964). *Set $a^2 = c$. As $a\to\infty$,*

$$1-\lambda_n \sim 4\sqrt{\pi}\,\frac{8^n a^{2n+1}}{n!}\,\mathrm{e}^{-2a^2} = 4\sqrt{\pi c}\,\frac{8^n c^n}{n!}\,\mathrm{e}^{-2c}\,.$$

The result involves a constant C that depends on n but not on a.

Behavior of λ_n as $n\to\infty$ (c Fixed)

The following result is expressed as a corollary because it is a reduction of a special case of a much more general and precise expression for $\lambda_n(c)$ obtained by Widom in [354]. Again, $\lambda_n(c)$ here denotes the nth eigenvalue of the operator $Q_cP_{1/\pi}Q_c$.

Corollary 1.2.11 (Widom, 1964). *If $c>0$ is fixed then*

$$\lambda_n(c) \sim 2\pi\left(\frac{c}{4}\right)^{2n+1}\frac{1}{(n!)^2} \quad \text{as} \quad n\to\infty\,.$$

Corollary 1.2.11 has consequences regarding approximation of a function on $[-T,T]$ that is not band limited; see also Sect. 2.5. Recall that for any band limit Ω, eigenfunctions φ_n of $P_\Omega Q_T$, when restricted to $[-T,T]$, form a complete orthogonal family in $L^2[-T,T]$. As such, any $f \in L^2[-T,T]$ has an expansion

$$f(t) = \sum_{n=0}^{\infty} \frac{1}{\lambda_n} \langle f, \mathbb{1}_{[-T,T]}\,\varphi_n \rangle\, \varphi_n(t)\,.$$

In [41, Conjecture 1], Boyd conjectured that for any fixed value of the time–bandwidth product, the eigenvalues $\lambda_n(\Omega T)$ decay super-geometrically (faster than e^{-qn} for any fixed $q > 0$). Widom's result, Corollary 1.2.11, was unfortunately overlooked until Gröchenig and Bass [17] found an application for it in 2008. As a consequence of the corollary, if $f \in L^2[-T,T]$ has an analytic extension whose coefficients in the PSWF basis decay only exponentially or slower, then the PSWF expansion is incapable of representing this extension.

The Number of Eigenvalues of PQ_T that Are Between α and $1-\alpha$

Here we fix a value of the time–frequency area and consider the number of eigenvalues that are neither close to one nor close to zero. As before, we normalize by taking $\Omega = 1$. Recall that with $\mathrm{sinc}(t) = \sin \pi t/(\pi t)$ the eigenfunctions φ_n of PQ_T satisfy

$$\lambda_n \varphi(t) = \int_{-T}^{T} \mathrm{sinc}(t-s)\, \varphi_n(s)\, \mathrm{d}s\,.$$

Thus, the kernel of PQ_T is $K(t,s) = \mathbb{1}_{[-T,T]}(t)\,\mathrm{sinc}(t-s)$ and

$$\sum_{n=0}^{\infty} \lambda_n = \mathrm{tr}(PQ_T) = \int K(t,t)\,\mathrm{d}t = \int_{-T}^{T} 1\,\mathrm{d}s = 2T$$

while the Hilbert–Schmidt norm of PQ_T satisfies

$$\sum_{n=0}^{\infty} \lambda_n^2 = \|PQ_T\|_{\mathrm{HS}}^2 = \int_{-T}^{T}\int_{-T}^{T} \mathrm{sinc}^2(t-s)\,\mathrm{d}s\,\mathrm{d}t = \int_{-T}^{T}\int_{-T-s}^{T-s} \mathrm{sinc}^2 t\,\mathrm{d}t\,\mathrm{d}s\,.$$

One can integrate by parts the *outer* integral over s, letting "u" be the output of the inside integral, and apply the fundamental theorem of calculus to get

$$\begin{aligned}\sum_{n=0}^{\infty} \lambda_n^2 &= s\int_{-T-s}^{T-s} \mathrm{sinc}^2 t\,\mathrm{d}t\,\Big|_{-T}^{T} + \int_{-T}^{T} s\,(\mathrm{sinc}^2(T-s) - \mathrm{sinc}^2(T+s))\,\mathrm{d}s \\ &= 2T\int_0^{2T} \mathrm{sinc}^2 t\,\mathrm{d}t + 2\int_0^{2T} (T-t)\,\mathrm{sinc}^2 t\,\mathrm{d}t \\ &\geq 2T - M_1 \log(2T) - M_2\end{aligned}$$

where M_1 and M_2 are independent of T. We used the facts that $\int_0^\infty \mathrm{sinc}^2 t\,dt = 1/2$, that $\int_{2T}^\infty \mathrm{sinc}^2(t)\,dt < 1/(2\pi^2 T)$, and that $\int_0^{2T} t\,\mathrm{sinc}^2(t)\,dt$ is comparable to $\log(2T)$.
Subtracting the Hilbert–Schmidt estimate from the trace identity yields

$$\sum_{n=0}^{\infty} \lambda_n(1-\lambda_n) \le M_1 \log(2T) + M_2 .$$

Since for any fixed $\alpha \in (0,1/2)$ one has

$$\sum_{\alpha<\lambda_n<1-\alpha} \lambda_n(1-\lambda_n) \ge \alpha(1-\alpha)\#\{n : \alpha < \lambda_n < 1-\alpha\},$$

it follows that

$$\#\{n : \alpha < \lambda_n < 1-\alpha\} \le \frac{\sum_{\alpha<\lambda_n<1-\alpha} \lambda_n(1-\lambda_n)}{\alpha(1-\alpha)} \le \frac{M_1 \log 2T + M_2}{\alpha(1-\alpha)} . \tag{1.27}$$

Generalizing to any pair Ω and T, one sees that several eigenvalues of $P_\Omega Q_T$ are close to one, followed by a *plunge region* of width proportional to $\log 2\Omega T$ over which the eigenvalues transition from being close to one to being close to zero, after which the remaining eigenvalues decay to zero super-exponentially. It remains to identify how many eigenvalues of $P_\Omega Q_T$ are close to one.

1.3 Landau and Widom's $2T\Omega$ Theorem

Essential Time and Band Limiting

This section presents a proof of what Slepian termed the "$2WT$" theorem—a rigorous version of the statement that the ensemble of essentially time-limited and essentially band-limited signals has a dimension equal to the time–bandwidth product, when *essentially* and *dimension* are adequately defined. The version that is presented here in some detail states that, save for a logarithmic term, the operator $P_\Omega Q_T$ has $2T\Omega$ eigenvalues that are *close to one*. This result was published by Landau and Widom in 1980 [197]. However, several earlier versions having less simple statements can be traced through the development of the Bell Labs theory. In [195], Landau and Pollak first quantified the possible proportions of energy that a signal can have in a given finite time interval and, simultaneously, in a given finite frequency band. In 1962, Landau and Pollak showed in [196] that if $f \in \mathrm{PW}_\Omega$, $\|f\| = 1$, and $\|Q_T f\|^2 \ge 1 - \varepsilon_T^2$ for an appropriate value of ε_T, then $\|f - f_{[2T\Omega]}\| \le C\varepsilon_T$ where $f_{[2T\Omega]}$ denotes the projection of f onto the span of the first $[2T\Omega]$ PSWF eigenfunctions of $P_\Omega Q_T$. In 1976, Slepian [306] (cf. [308]) offered an alternative version of the $2T\Omega$ theorem having a somewhat more concrete physical interpretation regarding the concentration of signals having finite energy. Given a projection operator R, one says that a set $\mathscr{S}$ of signals has *approximate dimension N in the range of R at* *level* η if N is the minimal integer such that, for some fixed family $\{g_1,\dots,g_N\}$, there is a linear combination $g=\sum c_i g_i$ such that $\|R(f-g)\|\leq\eta$ whenever $f\in\mathscr{S}$. Slepian defined $N(T,\Omega,\varepsilon,\eta)$ to be the *approximate dimension of* $\mathscr{G}_\varepsilon$ *at level* η, where $\mathscr{G}_\varepsilon$ consists of those signals satisfying $\|f-Q_Tf\|^2\leq\varepsilon$ and $\|f-P_\Omega f\|^2\leq\varepsilon$. He then proved that for every $\eta>\varepsilon$, $\lim_{T\to\infty}N(T,W,\varepsilon,\eta)/(2T)=\Omega$, with a similar statement when the roles of T and Ω are interchanged.

In Chap. 4 we will provide an argument due to Landau [193] showing that when $a=2T\Omega$, the operator $P_\Omega Q_T$ has $[a]$ or $[a]\pm 1$ eigenvalues greater than $1/2$. Coupled with the observation (1.27) for a large or α small, it follows that the number of eigenvalues of $P_\Omega Q_T$ that are close to one behaves like $2T\Omega$ minus a logarithmic term. This observation is the essence of the $2T\Omega$ *theorem*.

Landau and Widom [197] applied advanced techniques from spectral theory in order to estimate precisely, albeit asymptotically, the logarithmic term that is subtracted from the time–bandwidth product to estimate the number of eigenvalues of $P_\Omega Q_T$ that are close to one. Theorem 1.3.1 was actually stated and proved by Landau and Widom for time and frequency support sets consisting of finite unions of intervals. This extension will be discussed further in Chap. 4.

Theorem 1.3.1 (Landau and Widom, 1980). *Let* $a=2\Omega T$ *and set* $A(\Omega,T)=P_\Omega Q_T P_\Omega$. *Then the number* $N(A_a,\alpha)$ *of eigenvalues of* $A(\Omega,T)$ *larger than* α *satisfies*

$$N(A_a,\alpha)=a+\frac{1}{\pi^2}\log\Big(\frac{1-\alpha}{\alpha}\Big)\log a+o(\log a)\quad\text{as}\quad a\to\infty. \tag{1.28}$$

Since the spectrum of $A(\Omega,T)$ depends only on a and is invariant under time and frequency shifts, we will consider the operator $A_a=P_{[0,a]}Q_{[0,1]}P_{[0,a]}$ which has the same spectrum as $A(\Omega,T)$ when we consider the special case $\Omega=a$ and $T=1/2$. The estimate (1.28) boils down to estimating the polynomial moments of the measure $\mathrm{d}_t[-N(A_a,t)]$, which is a sum of Dirac masses at the eigenvalues λ_n so that

$$N(A_a,\alpha)=\int_\alpha^1 \mathrm{d}_t[-N(A_a,t)]\,.$$

The result then follows from approximating $\mathbb{1}_{[\alpha,1]}$ by polynomials. The key step is to show that, asymptotically, the moments of $\mathrm{d}_t[-N(A_a,t)]$ can be expressed in terms of those of an operator whose eigenvalue distribution is known. The principal tool in this reduction is a variant of Szegő's eigenvalue distribution theorem [166] also due to Landau [190]. Landau's limit theorem states in a precise sense that the eigenvalues of $A_rf=\int_{|y|<r}p(x-y)f(y)\,\mathrm{d}y$ approach the Fourier transform of the kernel as $r\to\infty$. It is, perhaps, the principal reason why Theorem 1.3.1 is necessarily asymptotic.

Proof (of Theorem 1.3.1). Here is an outline of Landau and Widom's proof. Details of some of the lemmas can be found in Sect. 1.5.3. The polynomial moments of $\mathrm{d}_t[-N(A_a,t)]$ are the traces of the powers of $A_a=P_{[0,a]}Q_{[0,1]}P_{[0,a]}$. These powers are not readily expressed in terms of a kernel having a familiar Fourier transform, but it turns out that those of the operator $A_a(I-A_a)$ do have such an expression. Landau and Widom took estimates of the powers of $A_a(I-A_a)$ as a starting point. They were then able to use symmetry properties of the corresponding spectral moments to recover corresponding estimates for A_a as we now outline.

An Equivalent Expression for Traces of Powers of $A_a(I-A_a)$

One uses idempotency and exclusion (e.g., $P_{(-\infty,0)}+P_{(1,\infty)}=I-P_{[0,1]}$) to express the Fourier transform of $[A_a(I-A_a)]^n$ as a sum of nth powers of four operators, each unitarily equivalent to the operator

$$K_a = Q_{[1,a]}P_{[0,\infty)}Q_{(-\infty,0]}P_{[0,\infty)}Q_{[1,a]},$$

plus a remainder operator term that remains uniformly bounded independent of a (expressed as "$O(1)$"). For the moment estimates, one also expresses $A_a[A_a(I-A_a)]^n$ as the sum of four operators, each unitarily equivalent to

$$Q_{[0,a]}P_{[0,1]}K_a^n + O(1).$$

In this way, one obtains the following main technical lemma, whose proof can be found in Sect. 1.5.3.

Lemma 1.3.2.

$$\begin{aligned}\operatorname{tr}[A_a(I-A_a)]^n &= 4\operatorname{tr}K_a^n + O(1), \quad \text{and}\\ \operatorname{tr}A_a[A_a(I-A_a)]^n &= 4\operatorname{tr}\left(Q_{[0,a]}P_{[0,\infty)}K_a^n\right) + O(1) = 2\operatorname{tr}K_a^n + O(1).\end{aligned}$$

Trace Kernel Estimates

Lemma 1.3.3.

$$\operatorname{tr}K_a^n = \frac{\log a}{4\pi^2}\int_0^1 (t(1-t))^n \frac{\mathrm{d}t}{t(1-t)} + o(\log a).$$

Proof (of Lemma 1.3.3). Since $P_{[0,\infty)}=\frac{1}{2}(I+\mathrm{i}H)$ where H is the Hilbert transform with principal value kernel p.v. $1/\pi x$, one has $Q_{(-\infty,0]}P_{[0,\infty)}Q_{[1,a]}=\frac{\mathrm{i}}{2}Q_{(-\infty,0]}HQ_{[1,a]}$. Since $H^*=-H$, one thus has $4K_a=Q_{[1,a]}HQ_{(-\infty,0]}HQ_{[1,a]}$. A simple direct calculation then shows that the operator K_a has integral kernel

$$\begin{aligned}k_a(x,y) &= \left(\frac{1}{4\pi^2}\int_0^\infty \frac{\mathrm{d}s}{(s+x)(s+y)}\right)\mathbb{1}_{1\le x\le y\le a}\\ &= \left(\frac{\mathrm{e}^{-u-v}}{8\pi^2}\int_{-\infty}^\infty \operatorname{sech}(r)\operatorname{sech}(u-v-r)\,\mathrm{d}r\right)\mathbb{1}_{0\le u\le v\le(\log(a))/2}\end{aligned}$$

upon changing variables to $x = \mathrm{e}^{2u}$, $y = \mathrm{e}^{2v}$, $s = \mathrm{e}^{2w}$, and $r = w - u$. Expressing a candidate μ-eigenfunction in the form $f(y)/\sqrt{y}$ one has

$$\int_1^a k_a(x,y)\,\frac{f(y)}{\sqrt{y}}\,\mathrm{d}y = \int_0^{\log a/2} k_a(\mathrm{e}^{2u},\mathrm{e}^{2v})\,\mathrm{e}^{-v} f(\mathrm{e}^{2v})\,\mathrm{d}(\mathrm{e}^{2v})$$
$$= \frac{\mathrm{e}^{-u}}{4\pi^2}\int_0^{\log a/2}\Big(\int_{-\infty}^{\infty}\mathrm{sech}\,(r)\mathrm{sech}\,(v-u-r)\,\mathrm{d}r\Big)\mathbb{1}_{0\le u\le v\le \log(a)/2}\,\widetilde{f}(v)\,\mathrm{d}v$$
$$= \mu\,\mathrm{e}^{-u}\,\widetilde{f}(u) = \mu f(x)/\sqrt{x}$$

provided that $\widetilde{f}(v) = f(\mathrm{e}^{2v})$ is a μ-eigenfunction of the operator on $L^2[0,\log a/2]$ with integral kernel $\kappa_a(u-v)\mathbb{1}_{0\le u\le v\le \log a/2}$, where

$$\kappa_a(s) = \frac{1}{4\pi^2}\int_{-\infty}^{\infty}\mathrm{sech}\,(r)\,\mathrm{sech}\,(s-r)\,\mathrm{d}r$$

when $s \ge 0$ and equal to zero when $s < 0$.

Szegő's theorem [166, 190] (see Appendix A, p. 231) can be paraphrased as the statement that the number of eigenvalues of K larger than α is equal to the measure of the set on which the Fourier transform of the kernel of K is larger than α. In the present context, this translates into

$$N(K_a,\alpha) = \frac{\log a}{2}\big(|\{\xi : |\widehat{\kappa_a}(\xi)| > \alpha\}| + o(1)\big).$$

The $\log a$ term comes from the support of κ_a. Since $\mathrm{sech}^\wedge(\xi) = \pi\,\mathrm{sech}\big(\pi^2\xi\big)$,

$$N(K_a,\alpha) = \frac{\log a}{2}(|\{\xi : \mathrm{sech}^2(\pi^2\xi) > 4\alpha\}| + o(1)) = \frac{\log a}{\pi^2}(\mathrm{sech}^{-1}(\sqrt{4\alpha}) + o(1)).$$

Since $\mathrm{sech}^{-1}(\sqrt{4\alpha})$ is defined for $0 < \alpha \le 1/4$,

$$\mathrm{tr}\,K_a^n = \int \alpha^n \mathrm{d}_\alpha[-N(K_a,\alpha)] = \frac{\log a}{2\pi^2}\int_0^{1/4}\alpha^n\frac{\mathrm{d}\alpha}{\alpha\sqrt{1-4\alpha}} + o(\log a).$$

Letting $\alpha = t(1-t)$ with $0 < t \le 1/2$ and using symmetry then gives

$$\mathrm{tr}\,K_a^n - o(\log a) = \frac{\log a}{2\pi^2}\int_0^{1/2}(t(1-t))^n\frac{\mathrm{d}t}{t(1-t)} = \frac{\log a}{4\pi^2}\int_0^1 (t(1-t))^{n-1}\,\mathrm{d}t. \quad \square$$

Applying similar reasoning to the operators corresponding to $A_a[A_a(I-A_a)]^n$ yields a corresponding moment involving $t^{n+1}(1-t)^n$.

The Measure $\mathrm{d}_t(-N(A_a,t))$

To obtain estimates on $N(A_a,\alpha)$ now one uses its definition, Lemmas 1.3.2 and 1.3.3. Upon setting $s(t)=t(1-t)$ and $S_a=A_a(I-A_a)$,

$$\begin{aligned}\mathrm{tr}[S_a]^n &= \int_{0+}^1 s^n(t)\,\mathrm{d}_t(-N(A_a,t)) = \frac{\log a}{\pi^2}\int_0^1 s^{n-1}(t)\,\mathrm{d}t + o(\log a)\\ \mathrm{tr}A_a[S_a]^n &= \int_{0+}^1 t\,s^n(t)\,\mathrm{d}_t(-N(A_a,t)) = \frac{\log a}{\pi^2}\int_0^1 t\,s^{n-1}(t)\,\mathrm{d}t + o(\log a)\,.\end{aligned}$$

The second identity follows from Lemma 1.3.2 and because $\int_0^1 t\,m(t)\mathrm{d}t = \frac{1}{2}\int_0^1 m(t)\mathrm{d}t$ when $m(t)$ is symmetric with respect to $t=1/2$. Any polynomial P vanishing at 0 and 1 can be written $P=p(s)+tq(s)$ where p and q are polynomials in $s=t(1-t)$ and divisible by $s(t)$, so, for such P,

$$\int_0^1 P(t)\,\mathrm{d}_t[-N(A_a,t)] = \frac{\log a}{\pi^2}\int_0^1 P(t)\,\frac{\mathrm{d}t}{t(1-t)} + o(\log a).$$

If $P(0)=0$ then $P(t)=tP(1)+(P(t)-tP(1))$ in which $P(t)-tP(1)$ vanishes at $t=0$ and $t=1$. Since $\mathrm{tr}(A_a)=\int_0^1 t\,\mathrm{d}_t[-N(A_a,t)]$, if $P(0)=0$ then

$$\int_0^1 P(t)\mathrm{d}_t[-N(A_a,t)] = P(1)\mathrm{tr}(A_a) + \frac{\log a}{\pi^2}\left(\int_0^1 (P(t)-tP(1))\frac{\mathrm{d}t}{t(1-t)} + o(1)\right).$$

This last identity extends from polynomials P to any function F such that $(F(t)-tF(1))/(t(1-t))$ is Riemann integrable on $[0,1]$. If $F(t)=\mathbb{1}_{[\alpha,1]}$ for $\alpha>0$ then this extension yields

$$\begin{aligned}N(A_a,\alpha) &= \int_\alpha^1 \mathrm{d}_t[-N(A_a,t)] = a + \frac{\log a}{\pi^2}\left(\int_0^1 (\mathbb{1}_{[\alpha,1]}(t)-t)\frac{\mathrm{d}t}{t(1-t)} + o(1)\right)\\ &= a + \frac{\log a}{\pi^2}\left(\int_\alpha^1 \frac{\mathrm{d}t}{t} - \int_0^\alpha \frac{\mathrm{d}t}{1-t}\right) + o(\log a)\\ &= a + \frac{\log a}{\pi^2}\log\frac{1-\alpha}{\alpha} + o(\log a)\,.\end{aligned}$$

This completes the proof of Theorem 1.3.1. □

1.4 Discrete and Finite Analogues

1.4.1 The Slepian DPS Sequences

In 1978 Slepian systematically analyzed in [307] a discrete version of the prolate spheroidal wave functions—discrete prolate spheroidal sequences (DPSSs) and discrete prolate spheroidal wave functions (DPSWFs). The sequences were recognized as eigenvectors of a certain discrete Dirichlet kernel matrix and termed *prolate sequences* by Papoulis and Bertran in 1972 [258]. In addition to unpublished work of C.L. Mallows (c. 1964) in which versions of the DPSSs and DPSWFs were defined, Slepian also cited work of Tufts and Francis [327] (1970), of Eberhard [94] (1973), and of Gori and Guattari [114] (1974) addressing various signal processing applications of DPSSs and DPSWFs. Slepian's systematic study, which paralleled the earlier Bell Labs papers, led to improved computational tools and eigenvalue asymptotics.

In the discrete setting, the Fourier transform maps a sequence $\mathbf{c} = \{c(k)\} \in \ell^2(\mathbb{Z})$ to the function $C(\omega) = \sum c(k)\,\mathrm{e}^{2\pi i k\omega}$. This series converges in $L^2(\mathbb{T})$ and $\|C\|_{L^2(\mathbb{T})} = \|\mathbf{c}\|_{\ell^2(\mathbb{Z})}$ since the exponentials $\mathrm{e}^{2\pi i k\omega}$ ($k \in \mathbb{Z}$) form an orthonormal basis for $L^2(\mathbb{T})$. The inverse Fourier transform, in turn, maps a function $F(\omega) \in L^2(\mathbb{T})$ to its sequence of Fourier coefficients $\widehat{F}[k] = \int_0^1 F(\omega)\mathrm{e}^{-2\pi i k\omega}\,\mathrm{d}\omega$. While it is common to think of the forward transform, instead, as a mapping from functions to sequences, in the present context it is preferable to regard ω as a frequency variable and the integers as discrete time indices. The *index-limiting* operator $\mathscr{Q}_{N,k_0}$ truncates a sequence to N of its nonzero values centered at k_0, while the frequency-limiting operation $\mathscr{B}_W$ multiplies a function in $L^2(\mathbb{T})$ by the periodic extension of $\mathbb{1}_{[-W,W]}$ where $0 \le W \le 1/2$. When N is odd and $k_0 = 0$, one can express index limiting $\mathscr{Q}_N$ of $\mathbf{c}$ in the frequency domain as convolution of the Fourier series $C(\omega)$ with the Dirichlet kernel D_N, the Fourier series of $\mathbb{1}_{\{-(N-1)/2,\dots,(N-1)/2\}}$:

$$D_N(\omega) = \mathrm{e}^{\pi i(1-N)\omega} \sum_{k=0}^{N-1} \mathrm{e}^{2\pi i k\omega} = \mathrm{e}^{\pi i(1-N)\omega}\left(\frac{1-\mathrm{e}^{2\pi i N\omega}}{1-\mathrm{e}^{2\pi i\omega}}\right) = \frac{\sin N\pi\omega}{\sin \pi\omega}\,.$$

The frequency- and index-limiting operator $\mathscr{L} = \mathscr{Q}_N \mathscr{B}_W$ then can be expressed as

$$\Phi \mapsto (\mathscr{L}\Phi)(\omega) = \int_{-W}^{W} \frac{\sin N\pi(\omega-\xi)}{\sin \pi(\omega-\xi)}\,\Phi(\xi)\,\mathrm{d}\xi\,. \tag{1.29}$$

When N is odd, the range of $\mathscr{L}$ is an N-dimensional space of trigonometric polynomials. There is a corresponding operation on sequences that first index limits a sequence then frequency limits. In this case, one has to convolve over $\mathbb{Z}$ the discrete-time truncation of a sequence $\mathbf{x} = \{x(k)\}$ with the sequence of Fourier coefficients of the periodization of $\mathbb{1}_{[-W,W]}$. The kth Fourier coefficient of this periodization is the discrete sinc function, $\sin 2\pi W k/\pi k$, which is defined as $2W$ when $k = 0$. It will be convenient to re-index in order to operate on indices in $\{0,1,\dots,N-1\}$ and express the discrete index- and frequency-limiting operator as the matrix operation

$$S^{(N,W)}\mathbf{x};\quad S^{(N,W)}_{k\ell} = \frac{\sin 2\pi W(k-\ell)}{\pi(k-\ell)},\ k,\ell = 0,\dots,N-1 \tag{1.30}$$

when $\mathbf{x}$ is already truncated, that is, $\mathbf{x} = [x(0),\dots,x(N-1)]^T$.

In outlining the theoretical properties of the DPSSs and DPSWFs we will reference Slepian's work [307] freely. In order to develop the theory formally, fix W, $0 < W \leq 1/2$. Fix $N \in \mathbb{R}$, *not necessarily an integer* for the moment. Set

$$\mathscr{M} = \frac{1}{4\pi^2}\frac{\mathrm{d}}{\mathrm{d}\xi}(\cos 2\pi\xi - \cos 2\pi W)\frac{\mathrm{d}}{\mathrm{d}\xi} + \frac{1}{4}(N^2-1)\cos 2\pi\xi\,. \tag{1.31}$$

The following proposition is proved later in this section.

Proposition 1.4.1. *The operators $\mathscr{L}$ and $\mathscr{M}$ commute.*

The kernel of $\mathscr{L}$ in (1.29) is real, symmetric, and Hilbert–Schmidt, that is, square-integrable over the region in which $\max\{|\xi|,|\omega|\} \leq W$. The characteristic equation $\mathscr{L}\Psi = \lambda\Psi$ has as solutions real eigenfunctions $\Psi_0, \Psi_1, \ldots$ whose restrictions to $[-W,W]$ are complete in $L^2[-W,W]$. The eigenvalues λ_n of $\mathscr{L}$ are real and those that are nonzero have finite multiplicity. The eigenvalues and eigenvectors depend continuously on N. The eigenfunctions for $\lambda \neq 0$ can be extended to $\mathbb{R}$ by extending the characteristic equation to $\mathbb{R}$. They possess derivatives of all orders.

On the other hand, the characteristic equation for $\mathscr{M}$ takes the form $\mathscr{M}\sigma = \mu\sigma$. This equation has solutions in $C^2[-W,W]$ only for discrete real values $\mu_0 \geq \mu_1 \geq \ldots$. The corresponding real eigenfunctions $\sigma_0, \sigma_1, \ldots$—the DPSWFs—form a complete orthonormal family on $[-W,W]$. The following proposition is outlined in [307].

Proposition 1.4.2. *The eigenvalues μ_n of $\mathscr{M}$ are nondegenerate. Setting $\mu_0 > \mu_1 > \ldots$, the eigenfunction σ_n associated with μ_n has exactly n zeros in the open interval $(-W,W)$ and does not vanish at $\pm W$.*

Proof (of nondegeneracy of the eigenvalues). Suppose that σ_n and σ_m are linearly independent solutions of $\mathscr{M}\sigma = \theta\sigma$ for the same value of θ. Then $\sigma_n\mathscr{M}\sigma_m = \sigma_m\mathscr{M}\sigma_n$. Equivalently,

$$\begin{aligned}\sigma_n\frac{\mathrm{d}}{\mathrm{d}\xi}(\cos 2\pi\xi - \cos 2\pi W)\frac{\mathrm{d}\sigma_m}{\mathrm{d}\xi} &- \sigma_m\frac{\mathrm{d}}{\mathrm{d}\xi}(\cos 2\pi\xi - \cos 2\pi W)\frac{\mathrm{d}\sigma_n}{\mathrm{d}\xi} \\ &= \frac{\mathrm{d}}{\mathrm{d}\xi}(\cos 2\pi\xi - \cos 2\pi W)\left(\sigma_n\frac{\mathrm{d}\sigma_m}{\mathrm{d}\xi} - \sigma_m\frac{\mathrm{d}\sigma_n}{\mathrm{d}\xi}\right) = 0\,.\end{aligned}$$

Upon integrating the right-hand side from $\xi = -W$ to $\xi_0 \in (-W,W)$, one obtains the Wronskian equation $W(\sigma_m, \sigma_n) = 0$, that is,

$$\sigma_n(\xi_0)\frac{\mathrm{d}\sigma_m}{\mathrm{d}\xi}(\xi_0) - \sigma_m(\xi_0)\frac{\mathrm{d}\sigma_n}{\mathrm{d}\xi}(\xi_0) = 0, \quad \xi_0 \in (-W,W),$$

contradicting the assumption that σ_n and σ_m are linearly independent. □

A critical part of Slepian's analysis is the expansion of σ_n in a modulated series

$$\sigma_n(\xi) = \mathrm{e}^{\pi\mathrm{i}(1-N)\xi}\sum_{k=-\infty}^{\infty} s_n(k)\,\mathrm{e}^{2\pi i k\xi} \tag{1.32}$$

which, when substituted into $\mathcal{M}\sigma_n = \mu_n\sigma_n$ results in the three-term recursion

$$\frac{1}{2}k(N-k)s_n(k-1) + \left(\left(\frac{N-1}{2}-k\right)^2\cos 2\pi W - \mu_n\right)s_n(k) + \frac{1}{2}(k+1)(N-k-1)s_n(k+1) = 0. \quad (1.33)$$

Since the coefficient of $s_n(k-1)$ in this recursion vanishes if $k=0$, and that of $s_n(k+1)$ vanishes if $k=-1$, the system uncouples, yielding finitely supported solutions, *if N is a positive integer, as we shall assume henceforth.*

Let $T = T(N,W)$ denote the *tridiagonal* matrix with entries

$$T_{k\ell} = \begin{cases} \frac{1}{2}k(N-k), & \ell = k-1 \\ \left(\frac{N-1}{2}-k\right)^2\cos 2\pi W, & \ell = k \\ \frac{1}{2}(k+1)(N-1-k), & \ell = k+1 \end{cases} \quad (1.34)$$

and vanishing at all other integer pairs. These entries are the same as the coefficients of the corresponding terms in (1.33) except for the eigenvalue μ in the $s_n(k)$-term. Consequently, it is possible to have a solution satisfying $s_n(k)=0$ for all $k<0$ and $k \geq N$ provided that, for some $\mu \in \mathbb{R}$,

$$\sum_{\ell=0}^{N-1} T_{k\ell}s_n(\ell) = \mu s_n(k), \qquad (k = 0,\dots,N-1).$$

Let T_N denote the $N \times N$ block of T corresponding to $0 \leq k,\ell < N$. One finds that T_N has N real eigenvalues which, of course, are also eigenvalues of $\mathcal{M}$. In fact, these eigenvalues correspond to the N largest eigenvalues of $\mathcal{M}$, as will be shown now.

To keep ideas straight, denote by $\mu_0 > \mu_1 \cdots > \mu_{N-1}$ the N largest eigenvalues of $\mathcal{M}$. Denote the positive real eigenvalues of the reduced matrix T_N by $\nu_0 \geq \cdots \geq \nu_{N-1}$. The eigenvector corresponding to ν_m is denoted $\mathrm{s}_m = [s_m(0),\dots,s_m(N-1)]^T$, with "s" for "Slepian," although Slepian used v_n to denote these vectors and U_n to denote the eigenfunctions σ_n. Here, s_n is normalized so that $\sum s_n(k) \geq 0$ and $\|\mathrm{s}_n\| = 1$. For $n = 0,1,\dots,N-1$, the corresponding eigenfunction σ_n of $\mathcal{M}$ can be expressed as (cf. [307, p. 1380])

$$\sigma_n(\omega) = \varepsilon_n \sum_{k=0}^{N-1} s_n(k)\,\mathrm{e}^{-\pi\mathrm{i}(N-1-2k)\omega}, \quad \varepsilon_n = \begin{cases} 1, & n \text{ even} \\ \mathrm{i}, & \text{n odd}. \end{cases} \quad (1.35)$$

Symmetry properties of the sequences s_n analogous to those of the eigenfunctions of PQ_T are listed in [307, p. 1378] as follows:

$$\mathrm{s}_n^W(k) = (-1)^n\mathrm{s}_n^W(N-1-k) = (-1)^n\mathrm{s}_{N-1-n}^{1/2-W}(N-1-k), \quad n = 0,1,\dots,N-1.$$

Here, the superscript indicates the frequency-limiting parameter. The sequence s_n is symmetric with respect to the index $(N-1)/2$ if n is even and antisymmetric if n is odd. As in [307], $\mathscr{G}_N$ will denote the space of modulated polynomials of the form

$$g(\omega) = \sum_{k=0}^{N-1} g_k \mathrm{e}^{-\pi \mathrm{i}(N-1-2k)\omega}, \qquad g_k \in \mathbb{C}.$$

Since $\{\mathbf{s}_0, \cdots, \mathbf{s}_{n-1}\}$ forms a complete orthonormal family for $\mathbb{R}^N$, an element $\mathbf{g} = [g_0, \dots, g_{N-1}]^T$ of $\mathbb{R}^N$ can be expressed in the form $\mathbf{g} = \sum_{n=0}^{N-1} \gamma_n \mathbf{s}_n$. Thus, $\mathscr{G}_N$ is invariant under $\mathscr{M}$, since

$$(\mathscr{M} g)(\omega) = \sum_{n=0}^{N-1} \gamma_n (\mathscr{M} \sigma_n)(\omega) = \sum_{n=0}^{N-1} \gamma_n \mu_n \sigma_n(\omega).$$

As for $\mathscr{L}$, it also maps $\mathscr{G}_N$ into itself. As in the case of the PSWFs, the DPSWFs σ_n are doubly orthogonal. That is, they are orthonormal in $L^2(\mathbb{T})$ and, for $n \neq m$, σ_n and σ_m are orthogonal on $[-W, W]$.

The DPSWFs σ_n are indexed according to the sequences $\mathbf{s}_n$ which are indexed, in turn, according to the eigenvalues $v_0 \geq v_1 \geq \cdots$ of the matrix truncation T_N. It will be shown now that for $N \in \mathbb{N}$ fixed, if $n \in \{0, \dots, N-1\}$ then σ_n, regarded as an eigenfunction of $\mathscr{M}$, belongs to one of the N largest eigenvalues $\mu_0, \dots, \mu_{N-1}$ of $\mathscr{M}$. To do so, denote by μ_{n_v} the eigenvalue of $\mathscr{M}$ to which σ_n belongs. First, for each $n = 0, \dots, N-1$, σ_n is the product of the unimodular function $\mathrm{e}^{\pi \mathrm{i}(1-N)\omega}$ with the restriction to $\mathbb{T}$ of a polynomial of degree $N-1$ in $\mathbb{C}$. Thus, σ_n has at most $N-1$ zeros in $\mathbb{T}$, and so at most $N-1$ zeros in $(-W, W)$. Thus, $n_v \leq N-1$. By Proposition 1.4.2, $\mu_{n_v} \leq N-1$, which was to be shown.

Ordering Discrete Prolate Spheroidal Sequences

We have shown that each of $\sigma_0, \dots, \sigma_{N-1}$ must belong to one of the N largest eigenvalues of $\mathscr{M}$, but we have not established yet that $\mathscr{M} \sigma_n = \mu_n \sigma_n$, $n = 0, \dots, N-1$. We have not so far used the fact that the sinc matrix $S^{(N,W)}$ defined implicitly by (1.30) and T in (1.34) must commute as their counterparts $\mathscr{L}$ and $\mathscr{M}$ do. Consequently, the Slepian vectors $\mathbf{s}_n$, $n \in \mathbb{Z}_N$, are also the orthogonal eigenvectors of $S^{(N,W)}$. Slepian employed a continuation argument along the same lines as in Slepian and Pollak [309] in order to prove that the ordering of the nonzero eigenvalues λ_n of $\mathscr{L}$ follows that of the nonzero eigenvalues of $\mathscr{M}$. Specifically, we have already arranged the eigenvalues μ_n of $\mathscr{M}$ in decreasing order. We order λ_n in such a way that, if $\mathscr{M} \sigma_n = \mu_n \sigma_n$, then $\mathscr{L} \sigma_n = \lambda_n \sigma_n$. We claim then that, also, $\lambda_0 > \lambda_1 > \cdots > \lambda_{N-1} > 0$. We will take the nondegeneracy of λ_n for granted for the moment. Since $S^{(N,W)}$ is positive definite and shares the same nonzero eigenvalues with $\mathscr{L}$, $\lambda_{N-1} > 0$.

For any fixed positive integer N, the function $\sigma_n(\omega W)$ converges, with respect to the bandwidth parameter W, to a multiple of the Legendre polynomial $P_n(\omega)$, $|\omega| \leq 1$ as $W \to 0$. As explained in [307, p. 1404], this follows from the fact that the differential equation defining $\mathscr{M}$ becomes

$$\frac{\mathrm{d}}{\mathrm{d}\xi}(1-\xi^2)\frac{\mathrm{d}\sigma_n(\xi W)}{\mathrm{d}\xi}+\frac{1}{2}((N^2-1)-4\mu)\sigma_n(\xi W)+O(W^2)=0.$$

Consequently, $\mu=\mu_n(W)=(N^2-1)/4-n(n+1)/2+O(W)$ for $n=0,1,\ldots,N-1$, cf. (1.12). The argument in Slepian and Pollak [309], which is reproduced in Sect. 1.5 in proving Theorem 1.2.8, implies that the ordering $\lambda_0(W)>\lambda_1(W)>\cdots>\lambda_{N-1}(W)$ holds for $W>0$ sufficiently small. Since the λ_n are nondegenerate and depend continuously on W, the conclusion then applies for $0<W<1/2$.

Commutation of $\mathscr{L}$ and $\mathscr{M}$

Proposition 1.4.1 was proved by Slepian in a somewhat broader context of commutation of a differential and an integral operator satisfying certain consistency conditions. Let $\mathscr{L}$ denote the operation of integration against a kernel $K(\omega,\xi)$ over a finite interval (a,b) and let $\mathscr{M}$ be of the form

$$\mathscr{M}=\frac{\mathrm{d}}{\mathrm{d}\xi}p(\xi)\frac{\mathrm{d}}{\mathrm{d}\xi}+q(\xi).$$

Then

$$(\mathscr{M}\mathscr{L}\Phi)(\omega)=\int_a^b \mathscr{M}_\omega K(\omega,\xi)\,\Phi(\xi)\,\mathrm{d}\xi$$

where the "ω" in $\mathscr{M}_\omega$ refers to the differentiation variable. On the other hand,

$$\begin{aligned}(\mathscr{L}\mathscr{M}\Phi)(\omega)&=\int_a^b K(\omega,\xi)\mathscr{M}_\xi\,\Phi(\xi)\,\mathrm{d}\xi\\&=\Big[p(\xi)\Big\{K\frac{\mathrm{d}\Phi}{\mathrm{d}\xi}-\frac{\partial K}{\partial\xi}\Phi(\xi)\Big\}\Big]_{\xi=a}^b+\int_a^b\Phi(\xi)\mathscr{M}_\xi K(\omega,\xi)\,\mathrm{d}\xi.\end{aligned}$$

If $p(a)=p(b)=0$, then the commutator of $\mathscr{L}$ and $\mathscr{M}$ becomes

$$(\mathscr{M}\mathscr{L}\Phi-\mathscr{L}\mathscr{M}\Phi)(\omega)=\int_a^b[\mathscr{M}_\omega-\mathscr{M}_\xi]K(\omega,\xi)\,\Phi(\xi)\,\mathrm{d}\xi,$$

which vanishes if $\mathscr{M}_\omega K(\omega,\xi)=\mathscr{M}_\xi K(\omega,\xi)$. When $K(\omega,\xi)=K(|\omega-\xi|)$, as is the case when K is a symmetric convolution kernel, this vanishing condition can be expressed as

$$(p(\omega)-p(\xi))\frac{\partial^2 K}{\partial\omega^2}+(p(\omega)+p(\xi))\frac{\partial K}{\partial\omega}+(q(\omega)-q(\xi))\,K=0. \tag{1.36}$$

In the case of time and frequency limiting, $p(\omega)=\pi^2(\cos 2\pi\omega-\cos 2\pi W)/4$, $q(\omega)=(N^2-1)(\cos 2\pi\omega)/4$, and $b=-a=W$. Thus, $p(a)=p(b)=0$. Additionally,

$$p(\omega)-p(\xi) = -\frac{1}{2\pi^2}\sin\pi(\omega-\xi)\sin\pi(\omega+\xi)$$
$$p(\omega)+p(\xi) = -\frac{1}{\pi}\cos\pi(\omega-\xi)\sin\pi(\omega+\xi)$$
$$q(\omega)-q(\xi) = -\frac{1}{2}\sin\pi(\omega-\xi)\sin\pi(\omega+\xi).$$

Dividing each term by $\sin\pi(\omega+\xi)$ and making the substitution $t=\pi(\omega-\xi)$, (1.36) takes the form

$$\sin t\frac{\mathrm{d}^2}{\mathrm{d}t^2}\frac{\sin Nt}{\sin t}+2\cos t\frac{\mathrm{d}}{\mathrm{d}t}\frac{\sin Nt}{\sin t}+(N^2-1)\sin t\frac{\sin Nt}{\sin t}=0,$$

which is readily verified.

Eigenvalue Asymptotics

Slepian developed precise asymptotic estimates for the eigenvalues λ_n of $\mathscr{L}$ both for large N and for small W. The formulas are expressed in terms of energy concentration. They are technical and are not reproduced here. However, if asymptotic behavior is prescribed in a certain way then behavior of λ_n can be inferred from the continuous, PSWF case. In particular, Slepian [307, p. 1389] showed that if $W\to 0$, $N=[c/\pi W]$ and $n=[N(1+y)/2]$ where $y\in\mathbb{R}$ is fixed, then the behavior of the eigenvalue $\lambda_n(N,W)$ of $\mathscr{L}$ is the same as that of the nth eigenvalue of P_cQ.

Finite Discrete Prolate Sequences (FDPSs)

For $N\in\mathbb{N}$, the DPS sequences are properly thought of as infinite sequences having finitely many nonzero terms. Just as time and band limiting can be defined in both the continuous and discrete settings, it can be done in the setting of $\mathbb{Z}_N$ as well. Xu and Chamzas [366] referred to the corresponding eigenvectors as *periodic discrete prolate spheroidal sequences* (P-DPSSs) and regarded them as periodic sequences. We will call them *finite discrete prolate sequences* (FDPSs), thinking of time localization and frequency localization as operations on functions $\mathbf{x}:\mathbb{Z}_N\to\mathbb{C}$. These sequences were also used in the work of Jain and Ranganath [162]. Zhou, Rushforth, and Frost [375] also defined FDP sequences, using a band-limiting matrix slightly different from that used by Xu and Chamzas. Grünbaum [126, 127] addressed the analogue of Theorem 1.2.3 for finite matrices.

In the finite, discrete Fourier transform (DFT) case, it is possible to find nonzero vectors $\mathbf{x}=[x(0),\ldots,x(N-1)]^T\in\mathbb{C}^N=\ell^2(\mathbb{Z}_N)$, whose time–bandwidth product is N. Examples include the spike vectors $\mathbf{x}=\mathbf{e}_k$ satisfying $\mathbf{e}_k(j)=\delta_{jk}$ whose DFTs are the Fourier basis vectors ω_k where $\omega_k(\ell)=\mathrm{e}^{2\pi ik\ell/N}$. More generally, if M divides N then the indicator vector $\mathbb{1}_{\ell+M\mathbb{Z}_N}$ of the coset $\mathbb{Z}_N/\mathbb{Z}_M$ has a DFT that is a modulation of $\mathbb{1}_{(N/M)\mathbb{Z}_N}$; see, e.g., [152, Chap. 4]. Time and band limiting to such cosets gives rise to operators having some number of eigenvalues equal to one.

For K fixed such that $2K+1 \le N$, define the Toeplitz matrix $A = A^K$,

$$A_{k\ell} = a_{k-\ell} = \frac{\sin(2K+1)(k-\ell)\pi/N}{N\sin((k-\ell)\pi/N)}, \quad (k,\ell = 0,\dots,N-1). \tag{1.37}$$

Vectors in the image of A will be said to be *K-band limited*, since the DFT of any such vector will be zero at an index m such that $m \mod N > K$. Since the DFT is unitary, the dimension of the range of A is $2K+1$.

Denote now by $A_M = A_M^K$ the $M \times M$ principal minor of A and consider the eigenproblem

$$A_M \mathbf{s} = \lambda \mathbf{s}, \quad \mathbf{s} = [s(0),\dots,s(M-1)]^T. \tag{1.38}$$

Assuming that $M \ge 2K+1$, the main theorem of [366] is as follows.

Theorem 1.4.3. *The ordered eigenvalues λ_n of A_M satisfy*

$$\begin{aligned} 1 = \lambda_0 = \cdots = \lambda_{M-N+2K} & \\ & > \lambda_{M-N+2K+1} > \cdots > \lambda_{2K} \\ & \qquad > \lambda_{2K+1} = \cdots = \lambda_{M-1} = 0. \end{aligned}$$

If $M + 2K < N$, the statement should be revised so that $\lambda_0 < 1$. As in the discrete case, the eigenvectors can be computed by noting that A_M commutes with a symmetric tridiagonal matrix T,

$$\begin{aligned} T_{kk} &= -\cos\frac{\pi}{N}(2k+1-M)\cos\frac{\pi}{N}(2K+1), \\ T_{k,k+1} &= \sin\frac{\pi(k+1)}{N}\sin\frac{\pi}{N}(M-1-k), \quad k = 0,\dots,M-1, \end{aligned}$$

and $T_{k-1,k} = T_{k,k+1}$ and $T_{k\ell} = 0$ if $|\ell - k| > 1$. A simple algorithm for estimating the eigenvectors of T is provided in Sect. 1.4.2. Further discussion can be found in Chap. 2; cf. also [126, 375].

The FDP sequences $\mathbf{s}_n = [s_n(0),\dots,s_n(M-1)]^T$, $n = 0,\dots,\min(M-1,2K)$ are defined now as the eigenvectors corresponding to the nonzero eigenvalues of A_M. Properly normalized extensions to $\mathbb{Z}_N$, also denoted $\mathbf{s}_n$, can be obtained by applying the matrix A in (1.37) to the vector $[s_n(0),\dots,s_n(M-1),0,\dots,0]^T$. This extension to $\mathbb{C}^N$ lies in the space of K-band-limited vectors. The resulting $\mathbf{s}_n$ are doubly orthogonal in the sense that $\sum_{k=0}^{M-1} s_n(k)s_m(k) = \lambda_n\,\delta_{m,k}$ while $\sum_{k=0}^{N-1} s_n(k)s_m(k) = \delta_{m,k}$. If $M < 2K+1$ then the restrictions of the $\mathbf{s}_n$ to their first M coordinates form a basis for $\mathbb{C}^M$, but the $\mathbb{Z}_N$ periodic extensions of $\mathbf{s}_0,\dots,\mathbf{s}_{M-1}$ are not complete in the space of K-band-limited sequences. On the other hand, if $M > 2K+1$ then these periodic extensions are complete in the space of K-band limited sequences, but the restrictions of $\mathbf{s}_0,\dots,\mathbf{s}_{2K}$ to $\mathbb{C}^M$ are not complete.

1.4.2 Numerical Computation of DPS Sequences

The DPS sequences are eigenvectors of the symmetric matrix $S^{(N,W)}$ in (1.30), which commutes with the principal $N\times N$ block T_N of the tridiagonal matrix T in (1.34). In particular, $S^{(N,W)}$ and T_N share the same eigenvectors. The matrix $S^{(N,W)}$ becomes ill conditioned for large N, so standard QR methods fail to produce its eigenvectors accurately. On the other hand, T_N remains well conditioned but, of course, its eigenvalues do not encode time–frequency localization. Gruenbacher and Hummels [124] proposed a numerical method for generating DPS sequences that takes advantage of the tridiagonal form of T_N. The method first identifies a single eigenvalue with high accuracy, then finds the eigenvector through standard iterative techniques. What follows is a brief outline of the eigenvalue method.

Given any tridiagonal, symmetric $N\times N$ matrix S, let $d_k = S_{kk}$, $k=1,\dots,N$, be the diagonal elements and $c_k = S_{k,k+1} = S_{k+1,k}$, $k=1,\dots,N-1$ denote the superdiagonal elements. A simple recursion shows that if S_M denotes the $M\times M$ principal submatrix of S, then

$$P_M(\lambda) = \det(S_M - \lambda I) = (d_M - \lambda)P_{M-1}(\lambda) - c_{M-1}^2 P_{M-2}(\lambda),\ M = 2,\dots,N,$$

with $P_0 = 1$ and $P_1(\lambda) = d_1 - \lambda$.

In order to identify the kth largest eigenvalue—the kth largest root of P_N—one needs a preliminary estimate to which one can then apply the bisection method. Such an estimate can be obtained by Sturm's theorem. If S has only simple eigenvalues then the sequence P_M generated above is a *Sturm sequence*. This means that (i) P_N has only simple roots, (ii) if $P_N(t)=0$ then sign $P_{N-1}(t) =$ sign $P_N'(t)$, and (iii) if $P_M(t)=0$ for some $0<M<N$ then sign $P_{M+1}(t) = -$sign $P_{M-1}(t)$. Sturm's theorem (e.g., see [155]) states that if $\sigma(t)$ is the number of sign changes of the ordered sequence $P_0(t), P_1(t),\dots,P_N(t)$, then the number of distinct roots of P_N in the half open interval $(a,b]$ is $\sigma(a)-\sigma(b)$. In the current context, Sturm's theorem implies that the number of eigenvalues smaller than λ is equal to the number of sign changes in the sequence $P_0(\lambda), P_1(\lambda),\dots,P_N(\lambda)$.

The spectrum of the matrix T_N defined by (1.34) lies in $[0,R]$, where $R = N(N+1)/4$. This follows from the *Gershgorin circle theorem*, which states that each eigenvalue of a matrix $A = \{a\}_{k\ell}$ lies in one of the disks centered at a diagonal a_{kk} with radius $r_k = \sum_{\ell\neq k}|a_{k\ell}|$.

One can approximate the nth largest eigenvalue of T to within an error tol through the following simple bisection.

```
while |λmax − λmin| > tol
   λn = (λmax − λmin)/2
   if σ(λn) ≥ n                    % sign changes
      λmax = λn
   else
      λmin = λn
   end
end
```

Once λ_n is estimated, one can estimate the corresponding eigenvector $v = v_n$ by reverse iteration, that is, starting with an initial unit vector v, solving for w such that $(T - \lambda_n I)w = v$, renormalizing w, replacing v by w, and iterating until $\|w - v\|$ is below a given threshold. Verma, Bilbao, and Meng [336] employed a variant using the power method to estimate λ_0 and its corresponding DPS sequence. The Matlab function `dpss` can also be used to produce numerical DPS sequences.

1.5 Notes and Auxiliary Results

1.5.1 Extrapolation of Band-limited Signals

Papoulis [254] and Gerchberg [111] introduced the "PG" algorithm for extrapolation of band-limited signals. The algorithm itself is simple to describe. Suppose that f is band limited and, for simplicity, the support of $\widehat{f}$ is $[-1/2, 1/2]$. Suppose, also, that f is known only on $[-T, T]$. One then defines $f_1 = PQ_T f$ and $g_1 = f_1 + Q_T(f - f_1)$ and then iterates, setting $f_i = Pg_{i-1}$ and $g_i = f_i + Q_T(f - f_i)$. In terms of the PSWFs φ_n, Papoulis showed that

$$f(t) - f_i(t) = \sum_{n=0}^{\infty} \langle f, \varphi_n \rangle (1 - \lambda_n)^i \varphi_n(t) . \tag{1.39}$$

The orthogonality of the PSWFs immediately gives the error bound

$$\|f - f_i\|^2 = \sum_{n=0}^{\infty} |\langle f, \varphi_n \rangle|^2 (1 - \lambda_n)^{2i} . \tag{1.40}$$

This error converges to zero rapidly if f is essentially in the span of the first $[2T]$ PSWFs, but could converge very slowly otherwise. Papoulis also analyzed aliasing errors and discussed using the *maximum entropy method* as a means of approximating the PG iterates from (over)sampled values of f. Extrapolation can be carried out in practice for the case of discrete signals. Schlebusch and Splettstösser [284] proved that, by approximating band-limited signals by trigonometric polynomials with increasing degree and period length, the solution of the discrete extrapolation problem converges to that of the corresponding continuous extrapolation problem, answering in the affirmative a question raised by Sanz and Huang [283]; cf. [338].

Jain and Ranganath [162] extended Papoulis's method to the case of discrete, sampled data in a parallel study that also considered extrapolation of noisy and cluttered data and spectrum estimation. They also considered the two-variable discrete problem and applications to image processing such as inpainting. In the discrete domain, it is possible to accelerate the convergence of the PG algorithm by using the conjugate gradient method. The discrete data $\mathbf{y} = \{y(k)\}_{k\in\mathbb{Z}}$ is W-band limited if $\sum y(k)\, e^{2\pi i k\omega}$ vanishes for $|\omega| > W$. In reality, one only has finitely many samples of a presumed band-limited sequence, and the problem is to estimate the remaining integer samples. As in (1.30), consider the discrete time- and band-limiting operator $\mathscr{B}\mathscr{T}^*\mathscr{T}$ where $\mathscr{B}$ has matrix $B_{k\ell} = 2W \operatorname{sinc} 2W(k - \ell)$ and $\mathscr{T}$ is the $(2M+1) \times \infty$ matrix with entries $T_{kk} = 1$ if $|k| \le M$ and $T_{k\ell} = 0$ otherwise. We take $M = (N-1)/2$ in the definition of the Slepian sequences. If finite sampled data $\mathbf{x} = \mathscr{T}\mathbf{y}$ for some W-band-limited sequence $\mathbf{y}$, then $\mathbf{x} = \mathscr{T}\mathscr{B}\mathbf{y}$. Jain and Ranganath proposed the minimum norm estimate of $\mathbf{y}$ defined via what they termed as *pseudoinverse extrapolation*, $\mathscr{E}(\mathbf{x}) = \mathscr{B}\mathscr{T}^*(\mathscr{T}\mathscr{B}\mathscr{T}^*)_M^{-1}(\mathbf{x})$, where $(\mathscr{T}\mathscr{B}\mathscr{T}^*)_M^{-1}$ is the inverse of the $(2M+1) \times (2M+1)$ matrix $\mathscr{T}\mathscr{B}\mathscr{T}^*$. The discrete parallel of Papoulis's algorithm takes the form

$$\mathbf{y}_{i+1} = \mathscr{B}[\mathscr{T}^*\mathbf{x} + (\mathscr{I} - \mathscr{T}^*\mathscr{T})\mathbf{y}_i], \qquad \mathbf{y}_0 = 0 \tag{1.41}$$

where $\mathscr{I}$ is the identity. We refer to this extrapolation algorithm as the Jain–Ranganath (JR) method. The error estimate for $\|\mathbf{y}_i - \mathscr{E}(\mathbf{x})\|$ is parallel to (1.40). In practice one performs a $(2M+1)$-step gradient descent, called the Jain–Ranganath conjugate gradient (JR–CG) method.

Algorithm (Jain–Ranganath (JR–CG)).

$\mathbf{d}_0 = 0, \quad \mathbf{w}_0 = -\mathbf{x}, \quad \mathbf{v}_0 = 0$
`for` $i = 0:2M$
$\quad \alpha_i = -\langle \mathbf{w}_i, \mathbf{d}_i\rangle / \langle \mathbf{d}_i, \mathscr{T}\mathscr{B}\mathscr{T}^*\mathbf{d}_i\rangle$
$\quad \beta_i = \langle \mathbf{w}_{i+1}, \mathscr{T}\mathscr{B}\mathscr{T}^*\mathbf{d}_i\rangle / \langle \mathbf{d}_i, \mathscr{T}\mathscr{B}\mathscr{T}^*\mathbf{d}_i\rangle$
$\quad \mathbf{v}_{i+1} = \mathbf{v}_i + \alpha_i \mathbf{d}_i$
$\quad \mathbf{w}_{i+1} = \mathbf{w}_i + \alpha_i \mathscr{T}\mathscr{B}\mathscr{T}^*\mathbf{d}_i$
$\quad \mathbf{d}_{i+1} = -\mathbf{w}_{i+1} + \beta_i \mathbf{d}_i$
`end`
$\mathbf{y} = \mathscr{B}\mathscr{T}^*\mathbf{v}_{2M+1}$

A parallel method for the discrete periodic case was also considered in [162]. The fact that $\mathscr{T}\mathscr{B}\mathscr{T}^*$ is ill conditioned in the sense of poor eigenvalue separation poses practical issues for this JR–CG method, even for small M.

In the discrete, periodic setting of sequences on $\mathbb{Z}_P$, one can rephrase the extrapolation problem as follows. Recall that a P-periodic sequence $\mathbf{y}$ is K-band limited if $\widehat{y}(k) = 0$ for $K < |k| < [(P+1)/2]$. If $\mathbf{y}$ is K-band limited, then

$$\mathbf{y}(k) = \frac{1}{\sqrt{P}} \sum_{\nu=-K}^{K} \widehat{\mathbf{y}}(\nu)\, e^{2\pi i k\nu/P} = p\Big(\frac{k}{P}\Big),$$

where $p(t)$ is the trigonometric polynomial of degree K with coefficients $\widehat{\mathbf{y}}(q)$. One can then rephrase the extrapolation problem of reconstructing $\mathbf{y}$ from a subset of samples to that of reconstructing a trigonometric polynomial of degree K from a set of $M \geq 2K+1$ evenly spaced samples of the form k/N. Define the *bunched sampling* operator $\mathscr{S} : C(\mathbb{T}) \to \mathbb{C}^M$ by $\mathscr{S}(p) = [p(0), p(1/P), \ldots, p((M-1)/P)]^T$ and, for a vector $\mathbf{x} = [x_0, \ldots, x_{M-1}]^T \in \mathbb{C}^M$, define the *reconstitution operator*

$$\mathscr{R}(\mathbf{x})(t) = \sum_{k=0}^{M-1} x_k D_{2K+1}\Big(t - \frac{k}{P}\Big)$$

where D_{2K+1} is the Dirichlet kernel. An analogue of the PG method for approximating p starting with its bunched samples $\mathscr{S}(p) = [p(0), p(1/P), \ldots, p((M-1)/P)]^T$ is obtained by the following *PG method with relaxation*,

$$\begin{aligned} p_0 &= \lambda \mathscr{R}(\mathscr{S}(p)) \\ p_{i+1} &= p_i + p_0 - \lambda \mathscr{R}(\mathscr{S}(p_m)). \end{aligned}$$

In the standard PG algorithm, $\lambda = 1$. Strohmer et al. [99, 319] proposed the use of conjugate gradients to accelerate convergence. The resulting algorithm is abbreviated PG–CG.

Algorithm (Strohmer et al. PG–CG). Let p be a trigonometric polynomial of degree K, where $2K+1 \leq P$, and let $M \geq 2K+1$.

$d_0 = q_0 = \mathscr{R}\mathscr{S}p$
`for` $i = 1, 2, \ldots, 2K+1$
$\quad \alpha_i = \|q_i\|^2 / \langle d_i, \mathscr{R}\mathscr{S}d_i\rangle$
$\quad \beta_i = \|q_{i+1}\|^2 / \|q_i\|^2$
$\quad q_{i+1} = q_i - \alpha_i \mathscr{R}\mathscr{S}d_i$
$\quad d_{i+1} = q_{i+1} + \beta_i d_i$
$\quad p_{i+1} = p_i + \alpha_i d_i$
`end`

Proposition 1.5.1 (Strohmer et al. PG–CG). *The p_i defined in the PG–CG algorithm converge to p at a geometric rate depending on M.*

Strohmer also proposed a variation of Jain and Ranganath's method that takes advantage of the Toeplitz structure of the matrix that has to be inverted [319, Theorem 2]. In the general case of nonuniform sampling points, the method is also adaptive, and in that case it is called the *adaptive weights conjugate gradient Toeplitz method* (ACT), which we describe in the case of uniform sampling as follows.

Algorithm (Strohmer et al. ACT). Let $\gamma_\nu = \frac{1}{\sqrt{P}} e^{\pi i\nu(M-1)/P} \frac{\sin \pi\nu M/P}{\sin \pi\nu/P}$, and let T be the Toeplitz matrix with entries $T_{k\ell} = \gamma_{k-\ell}$ for $|k| \leq K$ and $|\ell| \leq K$, where $k-\ell$ is defined modulo P.

for $|\nu| \leq K$
$$y_\nu = \sum_{k=0}^{M-1} p(k/P)\, e^{-2\pi i\nu k/P}$$
end
$$\mathbf{q}_0 = \mathbf{d}_0 = \mathbf{y} \in \mathbb{C}^{2K+1}$$
$$\mathbf{a}_0 = \mathbf{0}$$
for $i = 0,\ldots,2K$
$$\mathbf{a}_{i+1} = \mathbf{a}_i + \frac{\langle \mathbf{q}_i, \mathbf{d}_i\rangle}{\langle T\mathbf{d}_i, \mathbf{d}_i\rangle}\mathbf{d}_i$$
$$\mathbf{q}_{i+1} = \mathbf{q}_i - \frac{\langle \mathbf{q}_i, \mathbf{d}_i\rangle}{\langle T\mathbf{d}_i, \mathbf{d}_i\rangle} T\mathbf{d}_i$$
$$\mathbf{d}_{i+1} = \mathbf{q}_{i+1} - \frac{\langle \mathbf{q}_{i+1}, T\mathbf{d}_i\rangle}{\langle T\mathbf{d}_i, \mathbf{d}_i\rangle}\mathbf{d}_i$$
end

Theorem 1.5.2. *The $\mathbf{a}_i$ converge in at most $2K+1$ steps to a vector $\mathbf{a} \in \mathbb{C}^{2K+1}$ satisfying $T\mathbf{a} = \mathbf{y}$ and $p(t) = \frac{1}{\sqrt{P}} \sum_{k=-K}^{K} a_k e^{2\pi ikt}$. In particular, the ACT algorithm reconstructs any trigonometric polynomial p of degree K from its samples $p(0), p(1/P), \ldots, p((M-1)/P)$.*

Other variations of the PG algorithm have been introduced either to speed up convergence or to improve convergence in the presence of noise or aliasing. For example, Chen [70] introduced a modification of the basic algorithm

$$f_i = Pg_i, \quad g_{i+1} = Q_T f + \widetilde{Q_T} f_i, \quad g_0 = Q_T f,$$

where $\widetilde{Q_T} = I - Q_T$, by replacing f_i, expressed in terms of its sample series, by f_i^{Chen} where $f_i^{\text{Chen}} = \sum_{|k|\leq T} f(k)\,\text{sinc}(t-k) + \sum_{|k|>T} f_i(k)\,\text{sinc}(t-k)$. In other words, the terms for $|k| \leq T$ of the sampling series for f_i are replaced by those of the sampling series of the band-limited function f. Then $g_{i+1}^{\text{Chen}} = Q_T f + \widetilde{Q_T} f_i^{\text{Chen}}$. The iterates of the Chen–PG algorithm, for $i = 1, 2, \ldots$, satisfy

$$\|f - f_i^{\text{Chen}}\|^2 = \|f - f_i\|^2 - \sum_{|k|\leq T} |f_i(k) - f(k)|^2.$$

Chen's version of PG yields a smaller error when expressed in terms of $\|f - g_i\|$.

A second alternative finite PG algorithm was proposed by Marques et al. [224] as a means of performing signal extrapolation when the bandwidth of the band-limited signal is unknown. In terms of the trigonometric polynomial model discussed above, in which M (possibly irregular) samples of the polynomial are given, the problem is first to find the smallest degree Q such that there is a trigonometric polynomial of degree Q having the given sample values. Such a value of Q is taken as the bandwidth for the PG algorithm. Other methods for extrapolation of discrete signals were developed in the work of Vincenti and Volpi [338].

The problem of extrapolating a signal from inexact or noisy samples has been considered in the literature. Landau [192] proved that the accuracy of an extrapolation from Nyquist samples decreases rapidly as the inaccuracy of measurements increases, for a given observation window T, while, even with small inaccuracy, the error of the extrapolation will remain small only over a window of length proportional to the observation window. On the other hand, if the Nyquist samples of f are known exactly on an interval of length T, then the extrapolation error tends to zero geometrically in T.

1.5.2 Commutants of Symmetric, Tridiagonal Matrices

That the discrete time- and band-limiting operator commutes with a tridiagonal matrix is important in computing its eigenvectors. Grünbaum established a simple criterion under which a discrete time–frequency localization operator commutes with a tridiagonal. He also noted the parallel with the differentiation operator $\frac{\mathrm{d}}{\mathrm{d}x}p(x)\frac{\mathrm{d}}{\mathrm{d}x}+q(x)$. Specifically, any *symmetric* tridiagonal matrix having $N+1$ rows and columns can be expressed as

$$T = D_{-}AD_{+} + B$$

in which $D_{\pm}(k,\ell) = \delta_{k+1,\ell} - \delta_{k,\ell}$ and $D_{-}(k,\ell) = \delta_{k,\ell} - \delta_{k,\ell-1}$ are the matrices of the forward and reverse difference operators and A and B are diagonal matrices subject to $a_0 = b_0 = a_{N+1} = 0$. Then $T_{k,k+1} = T_{k+1,k} = a_k$, $k = 1,\ldots,N$ and $T_{k,k} = b_k - a_k - a_{k-1}$ subject to the aforementioned boundary conditions. Suppose now that K is a Toeplitz matrix, that is, $K_{k,\ell} = r(k-\ell)$. Then the commutation condition $TK = KT$ can be expressed explicitly in terms of the system of equations

$$\begin{aligned} (a_k - a_\ell)\ & [r(k-\ell+1) - 2r(k-\ell) + r(k-\ell-1)] \\ +(a_k - a_{k-1})\ & [r(k-\ell) - r(k-\ell-1)] \\ +(a_\ell - a_{\ell-1})\ & [r(k-\ell+1) - r(k-\ell)] \\ +(b_k - b_\ell)\ & [r(k-\ell)] = 0 \qquad (k,\ell = 1,\ldots,N). \end{aligned}$$

Simple row reduction operations lead to the third-order difference equation

$$(a_k - a_{k-3})(r(3)r(1) - r(2)^2) + (a_{k-1} - a_{k-2})(r(1)^2 - r(2)^2) = 0 \tag{1.42}$$

which, in turn, has the characteristic equation

$$(\lambda - 1)\left[\lambda^2 + \lambda\left(1 + \frac{r(2)^2 - r(1)^2}{r(2)^2 - r(1)r(3)}\right) + 1\right] = 0. \tag{1.43}$$

Suppose, momentarily, that the coefficient of λ in (1.43) can be reduced to the value $-2\cos(2\pi/P)$. Then (1.43) can be factored into

$$(\lambda - 1)(\lambda - \mathrm{e}^{-2\pi i/P})(\lambda - \mathrm{e}^{2\pi i/P}) = 0.$$

In turn, with the given boundary values, the coefficients a_k then become (up to multiplication by a normalizing constant $C(N,P)$)

$$a_k = 2\sin\frac{\pi}{P}(N+1)\left(\cos\frac{\pi}{P}(N+1) - \cos\frac{\pi(N+1-2k)}{P}\right),$$

and the coefficient b_k becomes

$$b_k = b_1 + 4\frac{r(1)-r(2)}{r(2)}\sin\frac{\pi}{P}(N+1)\cos\frac{\pi}{P}\left(\cos\frac{\pi N}{P} - \cos\frac{\pi(N+2-2k)}{P}\right).$$

In particular, b_k is determined by $r(1)$ and $r(2)$.

When can one verify that the coefficient of λ in (1.43) is $-2\cos(2\pi/P)$? This will constrain K and T to be time–frequency localization matrices. Specifically, fix $P \in \mathbb{N}$ and consider localization to the subsets of $\mathbb{Z}_P$ defined by

$$S = \{-M,\ldots,M\}; \qquad \Sigma = \{-N/2,\ldots,N/2\}, \quad (N \text{ even}).$$

Then $K_{k\ell}$ is the kernel with entries $r(k-\ell)$ where $r(k) = D_{2M+1}(k)$. In this specific case, the coefficient of λ in (1.42) does take the value $-2\cos 2\pi/P$.

In order to make the connection with Slepian's DPS sequences as a limiting case, Grünbaum set $M = [PW]$ with W the bandwidth parameter in the DPS sequences. Choosing the boundary term b_1 and normalizing constant $C(N,P)$ as

$$b_1 = 1 + \frac{N}{2} + \cos 2\pi W \frac{N^2}{4} \quad \text{and} \quad C(N,P) = -\frac{1}{8(N+1)}\left(\frac{P}{\pi}\right)^3,$$

and letting $P \to \infty$ one obtains, for $1 \leq k \leq N+1$,

$$a_k = \frac{1}{2}k(N+1-k) \quad \text{and} \quad b_k - a_k - a_{k-1} = \cos 2\pi W\left(\frac{N+1}{2} - k - 1\right)^2,$$

which correspond to the entries of the tridiagonal matrix $T(N,W)$ in (1.34) once one makes the substitution $N \mapsto N+1$.

Commutation with Tridiagonals: General Case

Parlett and Wu [261] gave an alternative general description of orthogonal eigenvector matrices of tridiagonal matrices. Denote by UST_N the $N \times N$ unreduced, symmetric tridiagonal matrices. Unreduced means that $T(k,k+1) \neq 0$ for $k = 1,\dots,N-1$. Such a matrix T has N distinct real eigenvalues. If $T(k,k+1) > 0$ for each $k = 1,\dots,N-1$, then (i) when the eigenvalues of T are ordered such that $\lambda_1 > \lambda_2 > \cdots > \lambda_N$, the kth eigenvector has $k-1$ sign changes, and (ii) the eigenvectors of T are completely determined by the diagonal $T(k,k+1)$, $k = 1,\dots,N-1$. Making use of these observations, Parlett and Wu characterized eigenbases of UST matrices as follows.

Theorem 1.5.3. *An orthogonal matrix $B = [\mathbf{b}_1,\dots,\mathbf{b}_N]$ is the transpose of the eigenvector matrix of some matrix $T \in \mathrm{UST}$ if and only if*

(i) $b_1(\ell) \neq 0$ *for* $\ell = 1,\dots,N$,
(ii) $b_2(\ell)/b_1(\ell) = \lambda_\ell$, $\ell = 1,\dots,N$, *with* $\lambda_k \neq \lambda_\ell$ *if* $k \neq \ell$,
(iii) $B^T \Lambda B \in \mathrm{UST}$ *where* Λ *is the diagonal matrix with* $\Lambda_{kk} = \lambda_k$.

The theorem indicates that the pair $\mathbf{b}_1$ and $\mathbf{b}_2$ and the eigenvalues λ_k determine one another, so the remaining eigenvectors are determined once the first two are chosen in such a way that the λ_k are strictly decreasing. In other words, once $\mathbf{b}_1$ is chosen (all positive, say) and $\lambda_1,\dots,\lambda_N$ are chosen, one sets $b_2(k) = \lambda_k b_1(k)$ and, for $\ell = 2,\dots,N-1$, generates $\mathbf{b}_{\ell+1}$ by setting $\beta_{\ell+1} = \Lambda \mathbf{b}_\ell - \mathbf{b}_\ell(\mathbf{b}_\ell^T \Lambda \mathbf{b}_\ell) - \mathbf{b}_{\ell-1}(\mathbf{b}_{\ell-1}^T \Lambda \mathbf{b}_\ell)$ and then normalizing, $\mathbf{b}_{\ell+1} = \beta_\ell / \|\beta_\ell\|$. Conversely, in order to check whether a given orthogonal matrix is the eigenbasis of some $T \in \mathrm{UST}$, one would proceed through the steps above, checking whether the consecutive columns are consistent with the construction steps. The construction depends critically on the ordering of the columns.

Bispectrality

The bispectral problem is often stated as asking for a classification of differential operators whose eigenfunctions satisfy a differential equation also in the spectral variable. However, a richer version of the bispectral problem can be posed for a pair of operators, one integral and one differential with variables that can be continuous, discrete, or mixed. In this broader sense, Theorem 1.2.3 is one incarnation of bispectrality, as is Proposition 1.4.1. Though it will not be emphasized, much of Chap. 2 is built on this version of bisectrality, which has its explicit origins in work of Grünbaum and Duistermaat [89, 125]. A number of explicit incarnations of this far-reaching problem can be found in the edited volume [136].

Slepian Symmetry

Brander and DeFacio [46] defined a property they called "Slepian symmetry" for an operator of the form $M_q\mathscr{F}^{-1}M_p\mathscr{F}$ in which M_q denotes pointwise multiplication by a function q. The Slepian symmetry (in a single variable) is simply that $q = p$ and q is an even function. Up to a dilation, PQ_T is such an operator with q the indicator function of an interval. Brander and DeFacio asked when the operator $M_q\mathscr{F}^{-1}M_q\mathscr{F}$ commutes with a differential operator of the form (in one variable) $-\frac{\mathrm{d}}{\mathrm{d}t}\alpha(t)\frac{\mathrm{d}}{\mathrm{d}t}+V(t)$. It was shown that α must have the form $a+bt^2$ for appropriate a and b and then that $q(t) = \alpha(t)^{-1/2}\exp\Big(-\gamma\int_0^t \frac{s}{\alpha(s)}\,\mathrm{d}s\Big)$; then $V(t)$ is determined explicitly in terms of α and q.

Convergence of Prolate Series in L^p

Barceló and Córdoba [16] considered the question: When do partial sums of the series $f = \sum_{n=0}^{\infty}\langle f,\varphi_n\rangle\varphi_n$ converge to $f \in \mathrm{PW}$ in L^p-norm? That is, for which p does it hold that $\|f - \sigma_N f\|_{L^p} \to 0$ as $N \to \infty$, where $\sigma_N f(t) = \int K_N(t,s)\, f(s)\, ds$ with $K_N(t,s) = \sum_{n=0}^{N}\varphi_n(t)\,\varphi_n(s)$? The kernel K_N can be expressed as a sum of pieces involving Bessel functions, a typical piece being $\sqrt{st}(d/dt)J_{N+1/2}(t))J_{N+1/2}(s)/(t-s)$. It turns out that the K_N are uniformly bounded on L^p if $4/3 < p < 4$, but otherwise they are unbounded and then the prolate series does not converge to f in L^p in general. A parallel result holds for $L^p(-T,T)$ with respect to the partial sum operator S_N with kernel $\sum_{n=0}^{N}\frac{1}{\lambda_n}\varphi_n(t)\,\varphi_n(s)$.

Prolates on the Sphere

With S_{n-1} the unit sphere in $\mathbb{R}^n$, the classical spherical harmonics $Y_{\ell,m}$ $(0 \le \ell < \infty,\ -\ell \le m \le \ell)$ form an orthonormal basis for $L^2(S_{n-1})$. A natural analogue of the band-limiting operator $\mathscr{P}_c$ of this chapter relevant to the sphere is the operator P_L given by

$$\begin{aligned} P_L f(\omega) &= \sum_{\ell=0}^{L}\sum_{m=-\ell}^{\ell}\Big(\int_{S_{n-1}} f(\xi)\overline{Y}_{\ell,m}(\xi)\,\mathrm{d}\sigma(\xi)\Big)Y_{\ell,m}(\omega)\\ &= \sum_{\ell=0}^{L}\int_{S_{n-1}} P_\ell^n(\langle\xi,\omega\rangle)F(\xi)\,\mathrm{d}\sigma(\xi). \end{aligned}$$

Here $\mathrm{d}\sigma$ is surface measure on the sphere, the functions $\{P_\ell^n\}_{\ell 0}^{\infty}$ are the (suitably normalized) n-dimensional Legendre polynomials associated with S_{n-1}, and the last equality is an application of the Funk–Hecke theorem [122]. When $n = 3$, $P_\ell^3 = cP_\ell$, the familiar Legendre polynomials used in this chapter. As an analogue of the time-limiting operator we consider Q_b, which restricts functions on the sphere to polar caps of radius b. Grünbaum et al. [128] gave a differential operator $\mathscr{P}_S$ (defined on the sphere) that commutes with the composition $P_L Q_b$, thus mimicking the "lucky accident" of Sect. 1.2.2. From this we find that $P_L Q_b$ shares its eigenfunctions (spherical prolates) with D_S. Chebyshev and quadrature properties of such functions follow immediately. Miranian [233] provided a survey and discussion of these results.

1.5.3 Proofs of Technical Lemmas for Theorem 1.3.1

We will rephrase Lemma 1.3.2 here for the sake of presenting the arguments in a more self-contained manner.

Lemma 1.5.4. *With $A_a = P_{[0,a]}Q_{[0,1]}P_{[0,a]}$, for each $n = 1,2,\ldots$, the operator $(A_a(I-A_a))^n$, can be expressed as the sum of nth powers of four operators each unitarily equivalent to $K_a = Q_{[1,a]}P_{[0,\infty)}Q_{(-\infty,0]}P_{[0,\infty)}Q_{[1,a]}$, plus an operator that is uniformly bounded independent of a and n. In particular, $\mathrm{tr}(A_a(I-A_a))^n = 4\mathrm{tr}(K_a^n)+O(1)$. Moreover, one also has $\mathrm{tr}(A_a(A_a(I-A_a))^n) = 2\mathrm{tr}(K_a^n)+O(1)$.*

To prove Lemma 1.5.4 we need some preliminary results.

Lemma 1.5.5. *Let J,M,K,N,L be intervals, finite or semi-infinite, and let $R = R(J,M,K,N,L) = Q_JP_MQ_KP_NQ_L$. Let $\alpha > 0$ and $\beta,\gamma\in\mathbb{R}$. Then each of the following operators is unitarily equivalent to R:*
(UED) $Q_{J/\alpha}P_{\alpha M}Q_{K/\alpha}P_{\alpha N}Q_{L/\alpha}$,
(UES) $Q_{J+\beta}P_{M+\gamma}Q_{K+\beta}P_{N+\gamma}Q_{L+\beta}$,
(UER) $Q_JP_{-M}Q_KP_{-N}Q_L$, and
(UEF) $P_JQ_MP_KQ_NP_L$.

This lemma follows directly from the facts that dilation D_α is unitary, that shifts and reflections are unitary, and that the Fourier transform is unitary.

Lemma 1.5.6. *Let J,M,K,N,L be intervals, finite or semi-infinite, and let $R = R(J,M,K,N,L) = Q_JP_MQ_KP_NQ_L$. We say that R is "O(1)" in a parameter set if $\|R\|_{\mathrm{tr}}$ is uniformly bounded independent of the parameter values. Suppose that $J\cap K = L\cap K = \emptyset$. Then R is $O(1)$ under each of the following conditions.*
(UB1) $|K|$ is uniformly bounded and either J and L are on opposite sides of K or one of J, L is uniformly separated from K.
(UB2) If $K = (-\infty,0]$, $|J|$ or $|L|$ is uniformly bounded, and J or L is uniformly separated from zero.
(UB3) If $K = (-\infty,0]$, the finite endpoints of M and N are uniformly separated from one another, and either J or L is uniformly separated from zero.
(UB4) If $K = (-\infty,0]$, $|M|$ or $|N|$ is uniformly bounded, and $|J|$ or $|L|$ is uniformly bounded.

Proof (Proof of Lemma 1.5.6). To prove (UB1), by (UED) and (UES) it suffices to take $K = [0,1]$. In this case, R is defined by integration against the kernel

$$\mathbb{1}_J(x)\,\mathbb{1}_L(y)\int_0^1 \frac{\mathrm{e}^{2\pi i n_2(u-y)}-\mathrm{e}^{2\pi i n_1(u-y)}}{2\pi i(u-y)}\,\frac{\mathrm{e}^{2\pi i m_2(x-u)}-\mathrm{e}^{2\pi i m_1(x-u)}}{2\pi i(x-u)}\,\mathrm{d}u$$

when $M = [m_1,m_2]$ and $N = [n_1,n_2]$. In the semi-infinite case, the corresponding exponentials disappear. The integrand then is the sum of at most four terms of the form $\mathrm{e}^{2\pi i[n_j(u-y)+m_k(x-u)]}/((u-y)(x-u))$. One then uses the following fact: an operator with integral kernel of the form $p(x)q(y)$ has rank one and trace norm $\|p\|\|q\|$ where $\|p\|$ is the L^2-norm of p. Since $\|\cdot\|_{\mathrm{tr}}$ satisfies the triangle inequality and $|\mathrm{tr}(A)|\le\|A\|_{\mathrm{tr}}$, it follows that if A has kernel $\int p(x,u)q(u,y)\,du$ then $|\mathrm{tr}(A)|\le\int\|p(\cdot,u)\|\|q(u,\cdot)\|\,du$. Thus, if $J\subset[1,\infty)$ and $L\subset(-\infty,0]$ then

$$\|R\|_{\mathrm{tr}}\le C\int_0^1 (u(1-u))^{-1/2}\,\mathrm{d}u = O(1)$$

while, if J and L are on the same side of $[0,1]$, on the right side, say, and $L\subset[1+\delta,\infty)$ then, with a constant depending on δ,

$$\|R\|_{\mathrm{tr}}\le C\int_0^1 ((1-u)(1+\delta-u))^{-1/2}\,\mathrm{d}u = O(1).$$

For (UB2), similar reasoning from $J=[a,b]\subset[0,\infty)$ and $L\subset[\delta,\infty)$ gives

$$\|R\|_{\mathrm{tr}}\leq C\int_{-\infty}^{0}((a-u)^{-1/2}-(b-u)^{-1/2})(\delta-u)^{-1/2}\,\mathrm{d}u=O(1)\,.$$

In the case of (UB3), R is the sum of at most four operators with kernels of the form $C\mathbb{1}_J(x)\,\mathbb{1}_L(y)\,\mathrm{e}^{2\pi\mathrm{i}(\theta_1x-\theta_2y)}\int_{-\infty}^{0}\mathrm{e}^{2\pi\mathrm{i}\alpha u}[(u-x)(y-u)]^{-1}\,du$ in which α is uniformly bounded away from zero. An integration by parts converts this kernel into one of the form

$$C\mathbb{1}_J(x)\,\mathbb{1}_L(y)\,\mathrm{e}^{2\pi\mathrm{i}(\theta_1x-\theta_2y)}\left[\frac{1}{\mathrm{i}\alpha}\frac{1}{xy}-\frac{1}{\mathrm{i}\alpha}\int_{-\infty}^{0}\frac{\mathrm{e}^{2\pi\mathrm{i}\alpha u}}{(u-x)(y-u)}\left(\frac{1}{u-x}+\frac{1}{u-y}\right)\mathrm{d}u\right].$$

If $\min\{\inf j,\inf(L)\}\geq\delta$ then the first term is rank one with trace norm at most $C/(\alpha\delta)$ while the inequality $|\mathrm{tr}(A)|\leq\int\|p(\cdot,u)\|\|q(u,\cdot)\|\,du$ shows that the second term has a uniformly bounded trace norm. Suppose now that only one of J,L is separated from zero, say $J=[\delta,\infty)$ and $L=[0,\infty)$. Then by (UES) with $\gamma=0$ and $\beta=\delta$, $R=Q_JP_MQ_KP_NQ_L$ is unitarily equivalent to

$$Q_{[2\delta,\infty)}P_MQ_{(-\infty,\delta]}P_NQ_{[\delta,\infty)}=Q_{[2\delta,\infty)}P_M[Q_{(-\infty,0]}+Q_{[0,\delta]}]P_NQ_{[\delta,\infty)}\,.$$

The first operator defined by the sum on the right-hand side is $O(1)$ by (UB2) while the second operator is $O(1)$ by (UB1).

To prove (UB4) we now take $N=[0,1]$ so that the kernel of R is

$$\frac{1}{4\pi^2}\int_{-\infty}^{0}\frac{\mathrm{e}^{2\pi\mathrm{i}(u-y)}-1}{u-y}\frac{\mathrm{e}^{2\pi\mathrm{i}m_2(x-u)}-\mathrm{e}^{2\pi\mathrm{i}m_1(x-u)}}{u-x}\,\mathrm{d}u\,.$$

The integrand is bounded by $C/(1+y-u)(x-u)$. Thus, if either J or L is contained in $[0,\infty]$ and has uniformly bounded length then R is $O(1)$ again by the inequality $|\mathrm{tr}(A)|\leq\int\|p(\cdot,u)\|\|q(u,\cdot)\|\,du$. This completes the proof of Lemma 1.5.6. □

Proof (Proof of Lemma 1.5.4). Since the operators P_Σ and Q_S are orthogonal projections one has

$$\begin{aligned}A_a-A_a^2&=Q_{[0,a]}P_{[0,1]}[Q_{(-\infty,0]}+Q_{[a,\infty)}]P_{[0,1]}Q_{[0,a]}\\&=Q_{[0,a]}P_{[0,1]}Q_{(-\infty,0]}P_{[0,1]}Q_{[0,a]}+Q_{[0,a]}P_{[0,1]}Q_{[a,\infty)}P_{[0,1]}Q_{[0,a]}\,.\end{aligned}$$

The composition of the two operators on the right is

$$Q_{[0,a]}P_{[0,1]}Q_{(-\infty,0]}P_{[0,1]}Q_{[0,a]}P_{[0,1]}Q_{[a,\infty)}P_{[0,1]}Q_{[0,a]}\,.$$

The factor $Q_{(-\infty,0]}P_{[0,1]}Q_{[0,a]}P_{[0,1]}Q_{[a,\infty)}$ and $Q_{(-\infty,0]}P_{[0,a]}Q_{[0,1]}P_{[0,a]}Q_{[1,\infty)}$ are unitarily equivalent by (UED). The latter is $O(1)$ by (UB1). The outer factors of the composition are projections having norm one, so the whole product is $O(1)$ because $\|AB\|_{\mathrm{tr}}\leq\|B\|\|A\|_{\mathrm{tr}}$. Thus, for any n one has

$$\begin{aligned}(A_a(I-A_a))^n&=(Q_{[0,a]}P_{[0,1]}Q_{(-\infty,0]}P_{[0,1]}Q_{[0,a]})^n\\&\quad+(Q_{[0,a]}P_{[0,1]}Q_{[a,\infty)}P_{[0,1]}Q_{[0,a]})^n+O(1).\end{aligned}\qquad(1.44)$$

By (UB4) the base of the first term in (1.44) can be expressed as

$$Q_{[0,a]}P_{[0,1]}Q_{(-\infty,0]}P_{[0,1]}Q_{[0,a]}=Q_{[1,a]}P_{[0,1]}Q_{(-\infty,0]}P_{[0,1]}Q_{[1,a]}+O(1)$$

while

$$\begin{aligned}
&Q_{[1,a]}P_{[0,1]}Q_{(-\infty,0]}P_{[0,1]}Q_{[1,a]} \\
&\quad = Q_{[1,a]}P_{[0,\infty]}Q_{(-\infty,0]}P_{[0,1]}Q_{[1,a]} - Q_{[1,a]}P_{[1,\infty]}Q_{(-\infty,0]}P_{[0,1]}Q_{[1,a]} \\
&\quad = Q_{[1,a]}P_{[0,\infty]}Q_{(-\infty,0]}P_{[0,\infty]}Q_{[1,a]} - Q_{[1,a]}P_{[0,\infty]}Q_{(-\infty,0]}P_{[1,\infty]}Q_{[1,a]} \\
&\quad - Q_{[1,a]}P_{[1,\infty]}Q_{(-\infty,0]}P_{[0,\infty]}Q_{[1,a]} + Q_{[1,a]}P_{[1,\infty]}Q_{(-\infty,0]}P_{[1,\infty]}Q_{[1,a]} \\
&\qquad = Q_{[1,a]}P_{[0,\infty]}Q_{(-\infty,0]}P_{[0,\infty]}Q_{[1,a]} + Q_{[1,a]}P_{[1,\infty]}Q_{(-\infty,0]}P_{[1,\infty]}Q_{[1,a]} + O(1)
\end{aligned}$$

since the terms that have both factors $P_{[0,\infty]}$ and $P_{[1,\infty]}$ are $O(1)$ by (UB3). Applying similar reasoning to the base of the second term in (1.44) gives, altogether,

$$\begin{aligned}
(A_a(I-A_a))^n &= (Q_{[1,a]}P_{[0,\infty)}Q_{(-\infty,0]}P_{[0,\infty)}Q_{[1,a]})^n \\
&+ (Q_{[1,a]}P_{[1,\infty)}Q_{(-\infty,0]}P_{[1,\infty)}Q_{[1,a]})^n \\
&+ (Q_{[0,a-1]}P_{[0,\infty)}Q_{a,\infty}P_{[0,\infty)}Q_{[1,a-1]})^n \\
&+ (Q_{[0,a-1]}P_{[1,\infty)}Q_{a,\infty}P_{[1,\infty)}Q_{[1,a-1]})^n + O(1). \qquad (1.45)
\end{aligned}$$

By combining (UED), (UES) and (UER) suitably, one finds that each of the last three operators on the right is unitarily equivalent to the first one. That $\mathrm{tr}(A_a(I-A_a))^n = 4\mathrm{tr}(K_a^n) + O(1)$ follows from linearity of trace.

Applying A_a to both sides of (1.45) expresses $A_a(A_a(I-A_a))^n$ as a sum of four operators plus a uniformly bounded term. The first of these is $Q_{[0,a]}P_{[0,1]}(Q_{[1,a]}P_{[0,\infty)}Q_{(-\infty,0]}P_{[0,\infty)}Q_{[1,a]})^n$ and the second is $Q_{[0,a]}P_{[0,1]}(Q_{[1,a]}P_{[1,\infty)}Q_{(-\infty,0]}P_{[1,\infty)}Q_{[1,a]})^n$. Applying (UES) with $\beta = 0$ and $\gamma = 1$ and the fact that $Q_J P_L Q_K = -Q_J P_M Q_K$ if $J \cap K = \emptyset$ and $L = \mathbb{R} \setminus M$ to the second operator shows that it is unitarily equivalent to the first. Similar reasoning shows that the third and fourth are also equivalent to the first. Using linearity of trace it follows that

$$\mathrm{tr}(A_a(A_a(I-A_a))^n) = 4\mathrm{tr}Q_{[0,a]}P_{[0,1]}K_a^n + O(1).$$

Replacing $P_{[0,1]}$ on the right-hand side by $P_{[0,\infty)}$, the difference between the two contains the factor $P_{[1,\infty)}Q_{[1,a]}P_{[0,\infty)}Q_{(-\infty,0]}P_{[0,\infty)}$ which is equal to $-P_{[1,\infty)}Q_{[1,a]}P_{(-\infty,0]}Q_{(-\infty,0]}P_{[0,\infty)}$ since $[0,\infty) = \mathbb{R} \setminus (-\infty,0)$. The latter operator is $O(1)$ by (UB3). One concludes that

$$\mathrm{tr}(A_a(A_a(I-A_a))^n) = 4\,\mathrm{tr}\,Q_{[0,a]}P_{[0,\infty)}K_a^n + O(1).$$

Appealing twice more to complementarity one obtains

$$K_a = Q_{[1,a]}P_{[0,\infty)}Q_{(-\infty,0]}P_{[0,\infty)}Q_{[1,a]} = Q_{[1,a]}P_{(-\infty,0]}Q_{(-\infty,0]}P_{(-\infty,0]}Q_{[1,a]}$$

and so one also has

$$\mathrm{tr}\,Q_{[0,a]}P_{[0,\infty)}K_a^n = \mathrm{tr}\,Q_{[0,a]}P_{(-\infty,0]}K_a^n = \frac{1}{2}\mathrm{tr}\,Q_{[0,a]}K_a^n = \frac{1}{2}\mathrm{tr}K_a^n$$

since $P_{(-\infty,0]} + P_{[0,\infty)} = I$. Thus, $\mathrm{tr}(A_a(A_a(I-A_a))^n) = \frac{1}{2}\mathrm{tr}K_a^n + O(1)$ as was claimed. This completes the proof of Lemma 1.5.4. □

Chapter 2
Numerical Aspects of Time and Band Limiting

This chapter is concerned with the role of the prolates in numerical analysis—particularly their approximation properties and application in the numerical solution of differential equations. The utility of the prolates in these contexts is due principally to the fact that they form a *Markov system* (see Defn. 2.1.6) of functions on $[-1,1]$, a property that stems from their status as eigenfunctions of the differential operator $\mathscr{P}$ of (1.6), and allows the full force of the Sturm–Liouville theory to be applied. The Markov property immediately gives the orthogonality of the prolates on $[-1,1]$ (previously observed in Sect. 1.2 as the *double orthogonality* property) and also a remarkable collection of results regarding the zeros of the prolates as well as quadrature properties that are central to applications in numerical analysis.

In Sect. 2.1 we give an account of the classical theory on which so many modern developments hinge. In particular, the theory of *oscillating kernels* is explored, a key result being that the eigenfunctions of such kernels form Markov systems. Much of this material is adapted from the book of Gantmacher and Krein [109]. Properties of the differential operator $\mathscr{P}$ of (1.6) are investigated. It is shown that $\mathscr{P}$ is inverted by an integral operator with oscillating kernel, from which we conclude that the eigenfunctions of $\mathscr{P}$ (i.e., the prolates) form a Markov system on $[-1,1]$.

Section 2.2 is devoted to the quadrature properties of the prolates, which are inherited from the Markov property. It is shown how generalized Gaussian quadratures for the prolates may be effectively parlayed into approximate quadrature results for band-limited functions. Of particular interest is the remarkable result of Xiao, Rokhlin, and Yarvin [364] in which the Euclidean division algorithm for polynomials is generalized to band-limited functions, where the role of the degree of the polynomial is played by the function's band limit. As a corollary we give an alternative approximate quadrature result for band-limited functions in which the nodes are the zeros of the Nth prolate rather than generalized Gaussian quadrature nodes, whose construction is cumbersome and achieved only through the approximate solution of a system of nonlinear equations.

In Sect. 2.3 we give approximate local sampling expansions of band-limited functions as corollaries of the quadrature results of the previous section. When applying the classical Shannon theorem (see Theorem 5.1.3), local approximations of a band-limited function may be obtained from uniformly spaced samples. The slow decay of the sinc function, however, means that many such samples may need to be taken (far from the region in which the approximation is required) to achieve reasonable fidelity. In contradistinction to this situation, the local sampling results of this section achieve local representations with high fidelity using only a few samples taken on the interval of interest.

The use of prolates as spectral elements in numerical schemes for the solution of differential equations is the subject of Sect. 2.4. Of special importance here is the observation that the band limit parameter c of the prolates may be adjusted to achieve optimal convergence of expansions and uniformity of quadrature nodes, this latter property giving a clear benefit in using the prolates for such purposes rather than the more familiar bases of Legendre or Chebyshev polynomials.

Section 2.5 deals with the approximation of non-band-limited functions, in particular those analytic on $[-1,1]$ and those in Sobolev spaces $H^s[-1,1]$. We outline the work done by Boyd in quantifying the performance of prolate expansions (in terms of decay of coefficients) as compared to that of Chebyshev and Legendre expansions.

A remark on notation is in order. The eigenfunction solutions $\phi_n(t;c)$ of (1.6) were normalized to have unit norm in $L^2(\mathbb{R})$. The functions $\bar{\phi}_n = \bar{\phi}_n^{(c)}$ used primarily in this chapter are normalized such that $\|\bar{\phi}_n\|_{L^2[-1,1]} = 1$. That is, $\bar{\phi}_n^{(c)}(t) = \phi_n(t;c)/\sqrt{\lambda_n(c)}$. The "bar" thus refers to normalization, not conjugation. While this notation is slightly irritating, in the limit $c \to 0$ it is consistent with the somewhat standard notation $\bar{P}_n$ for the $L^2[-1,1]$-normalized Legendre polynomials.

Since this chapter addresses primarily matters of numerical analysis, we will also define band-limitedness in this chapter in terms of the numerical analyst's Fourier transform $\sqrt{2\pi}\mathscr{F}_{2\pi}$ with kernel $\mathrm{e}^{-ix\xi}$. This usage will not be explicit. Instead, we will say that a function $f \in L^2(\mathbb{R})$ is c-band limited in this chapter only if f is in the range of the operator $F_{2c/\pi}$ as introduced on p. 12, that is, $f(t) = \int_{-1}^{1} \mathrm{e}^{icst}\, h(s)\,\mathrm{d}s$ for some $h \in L^2([-1,1])$.

2.1 Oscillation Properties

We aim to show here that the prolates $\{\bar{\phi}_n^{(c)}\}_{n=0}^{\infty}$ form a *Markov system* on $[-1,1]$, from which we deduce many properties of their zeros. The Markov property follows from the fact that the prolates are eigenfunctions of a certain integral operator with symmetric *oscillating* kernel which we show is another consequence of Slepian's "lucky accident." The results in this section are classical and many have been adapted from the book of Gantmacher and Krein [109].

2.1.1 Chebyshev and Markov Systems

Definition 2.1.1. A system of continuous functions $\{\varphi_i\}_{i=1}^n$ defined on an interval I is said to be a Chebyshev system on I if any non-trivial linear combination $\varphi(x) = \sum_{i=1}^n c_i\varphi_i(x)$ vanishes on I no more than $n-1$ times.

Given functions $\{\varphi_i\}_{i=1}^n$ defined on I as in Definition 2.1.1 and $\{x_i\}_{i=1}^n \subset I$, we denote the $n\times n$ determinant of the matrix with (i,j)th entry $\varphi_i(x_j)$ as

$$\Delta\begin{pmatrix}\varphi_1 & \varphi_2 & \dots & \varphi_n\\ x_1 & x_2 & \dots & x_n\end{pmatrix} = \det_{i,j}(\varphi_i(x_j))_{i,j=1}^n.$$

With this notation, Chebyshev systems are characterized as follows.

Lemma 2.1.2. *A system of continuous functions $\{\varphi_i\}_{i=1}^n$ defined on an interval I forms a* Chebyshev system *if and only if for all $x_1,x_2,\dots,x_n\in I$ with $x_1<x_2<\dots<x_n$ we have*

$$\Delta\begin{pmatrix}\varphi_1 & \varphi_2 & \dots & \varphi_n\\ x_1 & x_2 & \dots & x_n\end{pmatrix} \neq 0.$$

Proof. The determinant $\det_{i,j}(\varphi_i(x_j))$ vanishes if and only if there exists a non-trivial solution $(c_1,c_2,\dots,c_n)$ of the system of equations

$$\sum_{i=1}^n c_i\varphi_i(x_j) = 0 \quad (j=1,2,\dots,n).$$

Equivalently, the determinant is zero if and only if there exist constants $c_1,c_2,\dots,c_n$ not all zero such that the function $\varphi(x)=\sum_{i=1}^n c_i\varphi_i(x)$ vanishes at the distinct points $x_1,x_2,\dots,x_n$, that is, $\{\varphi_i\}_{i=1}^n$ is not a Chebyshev system. □

We classify the isolated zeros of a function as either *nodes* or *anti-nodes* as follows.

Definition 2.1.3. Let f be defined on the open interval (a,b) and $c\in(a,b)$ be a zero of f. Then c is a node of f if in every neighborhood of c there exists x_1,x_2 such that $x_1<c<x_2$ and $f(x_1)f(x_2)<0$. Also, c is an anti-node of f if there is a neighborhood U of c such that whenever $x_1,x_2\in U$ and $x_1<c<x_2$, one has $f(x_1)f(x_2)>0$.

Corollary 2.1.4. *Suppose the functions $\{\varphi_i\}_{i=1}^n$ form a Chebyshev system on (a,b) and $\varphi(x)=\sum_{i=1}^n c_i\varphi_i(x)$ with the c_i not all zero. Suppose also that φ has p nodes and q anti-nodes in (a,b). Then $p+2q\le n-1$.*

Proof. Let $a<x_1<x_2<\dots<x_m<b$. We say $X_m=\{x_k\}_{i=1}^m$ is φ-oscillating if

$$(-1)^{\ell+k}\varphi(x_k)\ge 0 \tag{2.1}$$

for some integer ℓ and all $k=1,2,\dots,m$. If X_m is φ-oscillating and α is an anti-node of φ then there exists $\varepsilon>0$ such that either $X_{m+2}^-=X_m\cup\{\alpha,\alpha-\varepsilon\}$ or $X_{m+2}^+=$

$X_m \cup \{\alpha, \alpha+\varepsilon\}$ is φ-oscillating. Let the p nodes of φ be $y_1, y_2, \dots, y_p \in (a,b)$. Then there exists $\beta_1, \beta_2, \dots, \beta_{p+1} \in (a,b)$ such that

$$\beta_1 < y_1 < \beta_2 < \cdots < \beta_p < y_p < \beta_{p+1}$$

and $\varphi(\beta_k)\varphi(\beta_{k+1}) < 0$ for $k = 1,2,\dots,p$. The collection $\{\beta_j\}_{j=1}^{p+1}$ is φ-oscillating. Since φ has q anti-nodes, we can find a φ-oscillating collection $\{x_k\}_{k=1}^{p+2q+1} \subset (a,b)$. Suppose $p+2q \geq n$. Then there is an integer ℓ such that $(-1)^{\ell+k}\varphi(x_k) \geq 0$ for $k = 1,2,\dots,n+1$. However,

$$\Delta\begin{pmatrix}\varphi_1 & \varphi_2 & \cdots & \varphi_n & \varphi \\ x_1 & x_2 & \cdots & x_n & x_{n+1}\end{pmatrix} = \sum_{i=1}^{n} c_i \Delta\begin{pmatrix}\varphi_1 & \varphi_2 & \cdots & \varphi_n & \varphi_i \\ x_1 & x_2 & \cdots & x_n & x_i\end{pmatrix} = 0 \tag{2.2}$$

since each of the determinants in the sum is zero, the last row being identical to the ith row. On the other hand, expanding the determinant on the left-hand side of (2.2) along the last row and applying (2.2) gives

$$\sum_{i=1}^{n+1}(-1)^{n+1+i}\varphi(x_i)\Delta\begin{pmatrix}\varphi_1 & \cdots & \varphi_{i-1} & \varphi_i & \cdots & \varphi_n \\ x_1 & \cdots & x_{i-1} & x_{i+1} & \cdots & x_{n+1}\end{pmatrix} = 0. \tag{2.3}$$

However, since $\{\varphi_i\}_{i=1}^{n}$ is a Chebyshev system on (a,b), each of the determinants on the left-hand side of (2.3) has the same sign. At the same time, the collection $\{x_1, x_2, \dots, x_n\}$ is φ-oscillating, so that each term in the sum on the left-hand side of (2.3) has the same sign, from which we conclude that each term is zero. Again, since $\{\varphi_i\}_{i=1}^{n}$ is a Chebyshev system, the determinants in (2.3) are not zero, hence $\varphi(x_i) = 0$ for $i = 1,2,\dots,n+1$, thus contradicting the Chebyshev property. We conclude that $p+2q \leq n-1$. □

Lemma 2.1.5. *Let f, g be continuous on (a,b) and suppose*

(i) f, g have respectively p and $p+1$ nodes in (a,b) and no other zeros inside (a,b);
(ii) For all real t, the number of zeros of $f_t(x) = f(x) - tg(x)$ in (a,b) is either p or $p+1$ and all those zeros are nodes.

Then the zeros of f and g alternate.

Proof. Let $\alpha_1 < \alpha_2 < \cdots < \alpha_{p+1}$ be the nodes of g. They partition (a,b) into $p+2$ subintervals $(a,\alpha_1), (\alpha_1,\alpha_2), \dots, (\alpha_{p+1},b)$. Inside each of these subintervals, the function $\psi(x) = f(x)/g(x)$ is continuous since f, g are continuous and g does not vanish. Furthermore, ψ is monotone on each subinterval. To see this, suppose the contrary, that is, there is a subinterval (α_i, α_{i+1}) and $x_1, x_2, x_3 \in (\alpha_i, \alpha_{i+1})$ with $x_1 < x_2 < x_3$ and $[\psi(x_2) - \psi(x_1)][\psi(x_3) - \psi(x_2)] < 0$. Without loss of generality we may assume $\psi(x_1) < \psi(x_2)$ so that $\psi(x_3) < \psi(x_2)$. Since ψ is continuous on the closed interval $[x_1, x_3]$, it attains its maximum value on $[x_1, x_3]$ at some interior point $x_0 \in (x_2, x_3)$. Let $t_0 = \psi(x_0)$. Then $\psi(x) - \psi(x_0) \leq 0$ on $[x_1, x_3]$ and

$$f_{t_0}(x) = f(x) - t_0 g(x) = f(x) - \frac{f(x_0)}{g(x_0)} g(x) = g(x)(\psi(x) - \psi(x_0))$$

has a zero but not a node at x_0 since the sign of $g(x)$ is constant on (α_i, α_{i+1}), thus contradicting the assumption that each function f_t has no anti-nodes. We conclude that ψ is monotone on each subinterval (α_i, α_{i+1}).

The monotonicity of ψ implies the existence of (possibly infinite) limits ℓ_i^+ and ℓ_i^- defined by

$$\begin{aligned} \ell_i^+ &= \lim_{x \to \alpha_i^+} \psi(x) \quad (i = 0, 1, \ldots, p+1,\ \alpha_0 = a) \\ \ell_i^- &= \lim_{x \to \alpha_i^-} \psi(x) \quad (i = 1, 2, \ldots, p+2,\ \alpha_{p+2} = b). \end{aligned}$$

We claim that none of the internal limits $\ell_1^{\pm}, \ldots, \ell_{p+1}^{\pm}$ is finite. To see this, suppose ℓ_i^- is finite for some $1 \le i \le p+1$. Then $f(\alpha_i) = 0$ so that each α_i is a common zero of f and g. Suppose ψ does not change sign as x passes through α_i, that is, the limits ℓ_i^+ and ℓ_i^- have the same sign. There are four possible cases:
Case 1: $\ell_i^+ = \pm\infty$.
Case 2: ℓ_i^+ is finite and $\ell_i^- \neq \ell_i^+$.
Case 3: $\ell_i^+ = \ell_i^-$ and ψ remains monotone as x passes through α_i.
Case 4: $\ell_i^+ = \ell_i^-$ and ψ has an extremum at α_i.
In each of cases 1, 2, and 3, we may choose a number t_1 such that in a neighborhood of $x = \alpha_i$, the graph of ψ lies on different sides of the horizontal line $y = t_1$. Consider the function

$$f_{t_1}(x) = f(x) - t_1 g(x) = g(x)(\psi(x) - t_1).$$

Then α_i is an anti-node of f_{t_1} since $f_{t_1}(\alpha_i) = 0$ and both $g(x)$ and $\psi(x) - t_1$ change sign as x passes through α_i, thus contradicting the assumption that each f_t has no anti-nodes. In case 4, let $t_1 = \ell_i^-$. Then there exists $t_2 < t_1$ such that $f_{t_2}(x) = g(x)(\psi(x) - t_2)$ has two nodes, one on either side of α_i and retains its node at α_i since $g(\alpha_i) = 0$, thus contradicting the assumption that the number of nodes of both f_{t_1} and f_{t_2} is either p or $p+1$. We conclude that the internal limits $\ell_i^{\pm}$ must be infinite. Since ψ varies monotonically on (α_i, α_{i+1}), we conclude that either $\ell_i^{\pm} = \pm\infty$ or $\ell_i^{\pm} = \mp\infty$. Consequently, since ψ is continuous on (α_i, α_{i+1}), there exists exactly one point $\beta_i \in (\alpha_i, \alpha_{i+1})$ for which $\psi(\beta_i) = 0$. Hence $f(\beta_i) = 0$. This proves the lemma. □

Definition 2.1.6. A sequence of functions $\{\varphi_i\}_{i=0}^{\infty}$ is said to form a *Markov system* on the interval (a,b) if, for all integers $n \ge 1$, the finite collection $\{\varphi_i\}_{i=0}^{n}$ forms a Chebyshev system on (a,b).

Theorem 2.1.7. *If an orthonormal sequence of functions* $\{\varphi_i\}_{i=0}^{\infty}$ *forms a Markov system on the interval* (a,b), *then*

(a) The function φ_0 *has no zeros on* (a,b);
(b) For all $j \ge 0$, φ_j *has* j *nodes in* (a,b) *and no other zeros in* (a,b);

(c) *For all integers k and m with* $0 < k < m$ *and arbitrary constants* $c_k, c_{k+1}, \ldots, c_m$ *(not all zero), the linear combination* $\varphi(x) = \sum_{i=k}^{m} c_i \varphi_i(x)$ *has at least k nodes and at most m zeros in* (a,b) *(where each anti-node is counted as two zeros). In particular, if* φ *has m distinct zeros in* (a,b)*, then all are nodes;*

(d) *The nodes of any consecutive functions* φ_j *and* φ_{j+1} *alternate.*

Proof. For each positive integer m, the functions $\{\varphi_i\}_{i=0}^{m}$ form a Chebyshev system on (a,b). By Corollary 2.1.4, the function $\varphi(x) = \sum_{i=k}^{m} c_i \varphi_i(x)$ has at most m zeros on (a,b) (where we count each anti-node twice). Suppose φ has p nodes $\xi_1, \ldots, \xi_p$ with $a < \xi_1 < \cdots < \xi_p < b$ and consider the function

$$\psi(x) = \Delta \begin{pmatrix} \varphi_0 & \cdots & \varphi_{p-1} & \varphi_p \\ \xi_1 & \cdots & \xi_p & x \end{pmatrix} \quad (a \le x \le b). \tag{2.4}$$

By Lemma 2.1.2, $\psi(x) \neq 0$ if $x \notin \{\xi_1, \ldots, \xi_p\}$ and changes sign as x passes through each of these points, so that ψ and φ have the same nodes $\xi_1, \ldots, \xi_p$. Consequently, $\langle \varphi, \psi \rangle \neq 0$. However, by expanding the determinant (2.4) along the last column we find that ψ may be written in the form $\psi(x) = \sum_{i=0}^{p} d_i \varphi_i(x)$. By the orthogonality of $\{\varphi_i\}_{i=0}^{\infty}$, we see that $\langle \varphi, \psi \rangle = 0$ if $p < k$, and conclude that $p \geq k$. This completes the proof of statement (c).

To prove statement (b), we apply statement (c) with $k = m = j$. From this we see that φ_j has at least j nodes and at most j zeros. Hence each zero is a node and there are precisely j such points in (a,b).

To prove statement (d) observe that from statement (b), φ_j and φ_{j+1} have j and $j+1$ nodes respectively and no other zeros in (a,b). Also, given a real number t, the function $\mu(x) = \varphi_j(x) - t\varphi_{j+1}(x)$ has, by part (c), at least j nodes and its q anti-nodes must satisfy $j + 2q \leq j+1$, from which we conclude that $q = 0$ and that μ has exactly j zeros, all of which are nodes. Hence φ_j and φ_{j+1} satisfy the conditions of Lemma 2.1.5, from which we conclude that the nodes of φ_j and φ_{j+1} alternate. □

2.1.2 Oscillating Matrices and Kernels

We now gather relevant results in matrix algebra which will be essential in what follows. The proofs can be found in [109].

Given an $n \times n$ matrix A, an integer $1 \leq p \leq n$ and two collections of integers $i_1, i_2, \ldots, i_p$ and $k_1, k_2, \ldots, k_p$ satisfying $1 \leq i_1 < i_2 < \cdots < i_p \leq n$, $1 \leq k_1 < k_2 < \cdots < k_p \leq n$, let $A\begin{pmatrix} i_1 & i_2 & \cdots & i_p \\ k_1 & k_2 & \cdots & k_p \end{pmatrix}$ be the determinant of the $p \times p$ sub-matrix of A with (ℓ, m)th entry $a_{i_\ell k_m}$, that is, the determinant of the $p \times p$ submatrix of A obtained by retaining only the rows of A numbered $i_1, i_2, \ldots, i_p$ and the columns numbered $k_1, k_2, \ldots, k_p$.

Definition 2.1.8. A matrix $A = (a_{ij})_{i,j=1}^{n}$ is said to be *completely non-negative* (respectively *completely positive*) if all its minors of any order are non-negative (respectively positive):

$$A\begin{pmatrix} i_1 & i_2 & \cdots & i_p \\ k_1 & k_2 & \cdots & k_p \end{pmatrix} \geq 0 \quad (\text{resp. } > 0)$$

for $1 \leq i_1 < i_2 < \cdots < i_p \leq n$, $1 \leq k_1 < k_2 < \cdots < k_p \leq n$, $p = 1,2,\ldots,n$.

A is said to be *oscillating* if it is completely non-negative and there exists a positive integer κ such that A^κ is completely positive. The smallest such κ is known as the exponent of A. Completely positive matrices are therefore oscillating matrices with exponent 1.

Definition 2.1.9. The square matrix L with (i,k)th entry ℓ_{ik} is a *one-pair matrix* if there are numbers $\{\psi_i\}_{i=1}^n$ and $\{\chi_i\}_{i=1}^n$ such that

$$\ell_{ik} = \begin{cases} \psi_i \chi_k & \text{if } i \leq k \\ \psi_k \chi_i & \text{if } i \geq k. \end{cases}$$

Of particular interest to us will be those one-pair matrices which are also oscillating. The class of such matrices is characterized as follows [109].

Theorem 2.1.10. *In order for the one-pair matrix of Definition 2.1.9 to be oscillating, it is necessary and sufficient that the following conditions hold:*

(i) The $2n$ numbers $\{\psi_i\}_{i=1}^n$, $\{\chi_i\}_{i=1}^n$ are nonzero and have the same sign;
(ii) The quotients ψ_i/χ_i are increasing: $\psi_1/\chi_1 < \psi_2/\chi_2 < \cdots < \psi_n/\chi_n$.

Let I, J be intervals and $K : I \times J \to \mathbb{R}$. Given $x_1, x_2, \ldots, x_n \in I$ and $s_1, s_2, \ldots, s_n \in J$, we define $K\begin{pmatrix} x_1 & x_2 & \cdots & x_n \\ s_1 & s_2 & \cdots & s_n \end{pmatrix}$ by

$$K\begin{pmatrix} x_1 & x_2 & \cdots & x_n \\ s_1 & s_2 & \cdots & s_n \end{pmatrix} = \det_{i,j}(K(x_i, s_j))_{i,j=1}^n.$$

Let $K = K(x,s)$ be defined on $[a,b] \times [a,b]$. Associated with K and $[a,b]$ we define an interval I_K by

$$I_K = \begin{cases} (a,b) & \text{if } K(a,a) = 0 = K(b,b) \\ [a,b) & \text{if } K(a,a) \neq 0 = K(b,b) \\ (a,b] & \text{if } K(a,a) = 0 \neq K(b,b). \end{cases}$$

Definition 2.1.11. The kernel $K(x,s)$ defined on $[a,b] \times [a,b]$ is said to be *oscillating* if the following three conditions are satisfied:

(i) $K(x,s) > 0$ for $x, s \in I_K$, $\{x,s\} \neq \{a,b\}$;

(ii) $K\begin{pmatrix} x_1 & x_2 & \cdots & x_n \\ s_1 & s_2 & \cdots & s_n \end{pmatrix} \geq 0$ for all $a < x_1 < x_2 < \cdots < x_n < b$ and $a < s_1 < s_2 < \cdots < s_n < b$ $(n = 1,2,\ldots)$;

(iii) $K\begin{pmatrix} x_1 & x_2 & \cdots & x_n \\ x_1 & x_2 & \cdots & x_n \end{pmatrix} > 0$ for $a < x_1 < x_2 < \cdots < x_n < b$ $(n = 1,2,\ldots)$.

The condition $\{x,s\} \neq \{a,b\}$ means that x and s cannot simultaneously equal the two ends a and b of the interval. In the case of a symmetric kernel this condition can be written $a < x < b$, $s \in I_K$.

The next result provides an alternative definition of oscillating kernels.

Lemma 2.1.12. *The kernel $K(x,s)$ defined on $[a,b] \times [a,b]$ is oscillating if and only if for any $x_1, x_2, \ldots, x_n \in I_K$, at least one of which is an interior point, the $n \times n$ matrix A with (i,k)th entry $a_{ik} = K(x_i, x_k)$ is oscillating.*

Definition 2.1.13. A *one-pair kernel* K defined on $[a,b] \times [a,b]$ is one of the form

$$K(x,s) = \begin{cases} \psi(x)\chi(s) & \text{if } x \leq s \\ \psi(s)\chi(x) & \text{if } s \leq x \end{cases}$$

where ψ and χ are continuous functions.

As a corollary of Theorem 2.1.10, we have the following characterization of one-pair oscillating kernels.

Theorem 2.1.14. *The one-pair kernel of Definition 2.1.13 is oscillating if and only if the following conditions are satisfied:*

(i) $\psi(x)\chi(x) > 0$ for all $a < x < b$;
(ii) The function $\psi(x)/\chi(x)$ increases monotonically on the interval (a,b).

Consider now the eigenvalue equation

$$\int_a^b K(x,s)\varphi(s)\,\mathrm{d}s = \lambda\varphi(x). \tag{2.5}$$

The main result of this section is the following.

Theorem 2.1.15. *Let K be a continuous oscillating symmetric kernel on $[a,b] \times [a,b]$. Then*

(i) The eigenvalues λ of (2.5) are all positive and simple: $\lambda_0 > \lambda_1 > \cdots > 0$.
(ii) The fundamental eigenfunction φ_0 corresponding to the largest eigenvalue λ_0 has no zeros in the interval I_K.
(iii) For any $j \geq 1$, the eigenfunction φ_j corresponding to the eigenvalue λ_j has exactly j nodes in I_K.
(iv) Given integers m,k with $0 \leq k \leq m$ and real numbers $c_k, c_{k+1}, \ldots, c_m$ not all zero, the linear combination $\varphi(x) = \sum_{i=k}^m c_i\varphi_i(x)$ has no more than m zeros and no fewer than k nodes in the interval I_K.
(v) The nodes of two neighboring eigenfunctions φ_j and φ_{j+1} alternate.

We will show that the prolates are eigenfunctions of a continuous oscillating symmetric kernel and hence that they inherit the properties outlined in Theorem 2.1.15. The proof of Theorem 2.1.15 is postponed until Sect. 2.6.2, as it requires several new concepts and intermediate results.

2.1.3 Sturm–Liouville Theory

Given a constant $c > 0$, we consider here the differential operator $\mathscr{P}$ of (1.6), the eigenfunctions of which are the prolates $\{\bar{\phi}_n^{(c)}\}_{n=0}^{\infty}$. We aim to show that $\mathscr{P}$ is inverted by an integral operator $\mathscr{G}$ associated with a continuous, symmetric, oscillating kernel. The eigenfunctions of $\mathscr{P}$ therefore coincide with those of $\mathscr{G}$, and an application of Theorem 2.1.15 will reveal that the prolates form a Markov system on $[-1,1]$.

The bulk of this section comes from private communication between the authors and Mark Tygert. The main result of this section is as follows.

Theorem 2.1.16. *The differential operator $\mathscr{P}$ of (1.6) has infinitely many eigenvalues $\{\chi_n\}_{n=0}^{\infty}$ with $0 < \chi_0 < \chi_1 < \cdots$. The corresponding eigenfunctions $\bar{\phi}_n = \bar{\phi}_n^{(c)}$ are orthonormal and complete in $L^2[-1,1]$ and form a Markov system on $[-1,1]$.*

The proof of Theorem 2.1.16 is deferred until after the proof of Lemma 2.1.19, at which time we will have developed the required machinery.

The standard form of the equation $\mathscr{P}u = 0$ is $u'' - P(x)u' + Q(x)u = 0$ with coefficient functions $P(x) = -2x/(1-x^2)$ and $Q(x) = c^2x^2/(1-x^2)$. Since $(1-x)P(x)$ and $(1-x)^2 Q(x)$ are analytic at $x = 1$, the point $x = 1$ is a *regular singular* point, implying that there exists a power series solution (Frobenius solution) of the form

$$u(x) = \sum_{n=0}^{\infty} a_n (x-1)^{n+r} \tag{2.6}$$

with $a_0 \neq 0$. Rewriting the equation $\mathscr{P}u = 0$ in the form

$$(1-x)u'' - \frac{2x}{1+x}u' - \frac{c^2x^2}{1+x}u = 0 \tag{2.7}$$

and substituting (2.6) into (2.7) gives

$$\sum_{n=0}^{\infty} a_n(n+r)(n+r-1)(x-1)^{n+r-1} + \frac{2x}{1+x}\sum_{n=0}^{\infty} a_n(n+r)(x-1)^{n+r-1} + \frac{c^2x^2}{1+x}\sum_{n=0}^{\infty} a_n(x-1)^{n+r} = 0.$$

Extracting the lowest order term (the coefficient of $(x-1)^{r-1}$) gives the *indicial equation* $r^2 = 0$, from which we conclude that there is a Frobenius series solution of the form $u_1(x) = \sum_{n=0}^{\infty} a_n(x-1)^n$ centered at $x = 1$ with $a_0 \neq 0$. To obtain another solution centered at $x = 1$, we use the method of *reduction of order* to produce a solution of the form $u_2 = vu_1$ with $v = C\log(x-1) + \sum_{n=0}^{\infty} b_n(x-1)^{n+1}$. From the form of this solution we see that u_2 has at most a logarithmic singularity at $x = 1$. Since the coefficient functions in (2.7) are analytic at $x = 1$ and with power series having radius of convergence at least two, the Frobenius solution u_1 centered at $x = 1$ will also have radius of convergence two. Hence u_1 is also a power series solution for $\mathscr{P}u = 0$ centered at $x = -1$. However, by the previous argument, the equation $\mathscr{P}u = 0$ has a Frobenius solution w_1 centered at $x = -1$ with radius of convergence two and another solution w_2 with a logarithmic singularity at $x = -1$. We conclude that u_1 has at worst a logarithmic singularity at $x = -1$ and w_1 has at worst a logarithmic singularity at $x = 1$. Let

$$C(x) = (1-x^2)(w_1(x)u_1'(x) - u_1(x)w_1'(x)). \tag{2.8}$$

Since $\mathscr{P}u_1 = \mathscr{P}w_1 = 0$, we have

$$\begin{aligned}\frac{\mathrm{d}C}{\mathrm{d}x} &= -2x(w_1u_1' - u_1w_1') + (1-x^2)(w_1u_1'' - u_1w_1'') \\ &= w_1((1-x^2)u_1'' - 2xu_1') - u_1((1-x^2)w_1'' - 2xw_1') \\ &= w_1(c^2x^2u_1) - u_1(c^2x^2w_1) = 0\end{aligned}$$

which shows that $C(x) = C$ is a constant. Suppose that $C = 0$. Then, from (2.8), $w_1'/w_1 = u_1'/u_1$ so that $|u_1|$ and $|w_1|$ have the same logarithmic derivative. This implies that $|u_1| = \lambda\,|w_1|$ for some constant λ (provided $u_1, w_1 \neq 0$). Suppose $w_1(x) = 0$ and $u_1(x) \neq 0$. Then the Wronskian equation $w_1(x)u_1'(x) - u_1(x)w_1'(x) = 0$ shows that $w_1'(x) = 0$. Successively differentiating the Wronskian equation and using the hypothesis $u_1(x) \neq 0$ shows that all the derivatives of w_1 are zero, so that $w_1 \equiv 0$, a contradiction. Hence $w_1(x) = 0$ implies $u_1(x) = 0$. Further, $u_1(x) = \pm\lambda w_1(x)$ on any open interval over which w_1 is non-vanishing. Since $u_1 \in C^2(-1,1]$, $w_1 \in C^2[-1,1)$, and $u_1 = \lambda w_1$, we see that $u_1, w_1 \in C^2[-1,1]$. Since $p(x) = 1 - x^2$ vanishes at $x = \pm 1$ and $\mathscr{P}u_1 = 0$, an integration by parts yields

$$\begin{aligned}0 = \int_{-1}^{1} (\mathscr{P}u_1)(x)u_1(x)\,\mathrm{d}x &= \int_{-1}^{1} (((1-x^2)u_1'(x))' - c^2x^2u_1(x))u_1(x)\,\mathrm{d}x \\ &= (1-x^2)u_1'(x)u_1(x)\Big|_{-1}^{1} - \left(\int_{-1}^{1} (1-x^2)((u_1'(x))^2 + c^2x^2u_1(x)^2)\,\mathrm{d}x\right)\end{aligned}$$

from which we conclude that $\int_{-1}^{1}(1-x^2)u_1'(x)^2\,\mathrm{d}x = c^2\int_{-1}^{1} x^2u_1(x)^2\,\mathrm{d}x = 0$. But the integrands are non-negative, so $u_1 \equiv 0$, a contradiction. Thus, $C \neq 0$.

Let the nonzero constant C be as in (2.8) and define G on the square $Q = [-1,1] \times [-1,1]$ by

$$G(x,s) = \begin{cases} w_1(x)u_1(s)/C & \text{if } x < s \\ w_1(s)u_1(x)/C & \text{if } s < x \\ w_1(x)u_1(x)/C & \text{if } s = x \in (-1,1) \\ \infty & \text{if } x = s = 1 \text{ or } x = s = -1. \end{cases} \tag{2.9}$$

Lemma 2.1.17. *With G as in (2.9), let $\mathscr{G}$ be the integral operator*

$$(\mathscr{G}f)(x) = \int_{-1}^{1} G(x,s)f(s)\,\mathrm{d}s.$$

Then $\mathscr{G} : L^2[-1,1] \to L^2[-1,1]$ is a Hilbert–Schmidt operator.

Proof. Direct calculation gives

$$\mathscr{P}_x G(x,s) = 0 \text{ if } x < s \text{ and } \mathscr{P}_s G(x,s) = 0 \text{ if } s < x$$

and also

$$\begin{aligned}
\lim_{x\to s^+} \frac{\partial G}{\partial x}(x,s) - \lim_{x\to s^-} \frac{\partial G}{\partial x}(x,s) &= \lim_{x\to s^+} \frac{w_1'(x)u_1(s)}{C} - \lim_{x\to s^-} \frac{w_1(s)u_1'(x)}{C} \\
&= \frac{w_1'(s)u_1(s) - w_1(s)u_1'(s)}{C} = \frac{1}{1-s^2},
\end{aligned}$$

the last equality being a consequence of the definition (2.8) of the constant C.

Since u_1, w_1 have singularities at $x = -1$ and $x = 1$ respectively which are at worst logarithmic, there exists a constant B such that

$$|u_1(x)| \le B(1 + \log|1+x|) \text{ and } |w_1(x)| \le B(1 + \log|1-x|).$$

Consequently,

$$\begin{aligned}
\iint_Q |G(x,s)|^2\,\mathrm{d}x\,\mathrm{d}s \le{} & \frac{B^4}{C^2}\int_{-1}^{1} (1+\log|1-x|)^2 \int_x^1 (1+\log|1+s|)^2\,\mathrm{d}s\,\mathrm{d}x \\
&+ \frac{B^4}{C^2}\int_{-1}^{1} (1+\log|1+x|)^2 \int_{-1}^{x} (1+\log|1-s|)^2\,\mathrm{d}s\,\mathrm{d}x < \infty
\end{aligned}$$

from which the result follows. □

As an immediate consequence of Lemma 2.1.17, the eigenfunctions of $\mathscr{G}$ form a complete orthonormal basis for $L^2[-1,1]$.

Lemma 2.1.18. *The integral operator $\mathscr{G}$ in Lemma 2.1.17 is the inverse of the differential operator $\mathscr{P}$ in (1.6).*

Proof. First note that

$$\begin{aligned}
(\mathscr{P}\mathscr{G}f)(x) &= \mathscr{P}_x \int_{-1}^{1} G(x,s)f(s)\,\mathrm{d}s \\
&= \frac{1}{C}\left(\mathscr{P}_x \int_{-1}^{x} w_1(s)u_1(x)f(s)\,\mathrm{d}s + \mathscr{P}_x \int_x^1 w_1(x)u_1(s)f(s)\,\mathrm{d}s\right). \quad (2.10)
\end{aligned}$$

Direct computation, using the fact that $\mathscr{P}u_1 = 0$, gives

$$
\begin{aligned}
&\mathscr{P}_x \int_{-1}^{x} w_1(s)u_1(x)f(s)\,\mathrm{d}s \\
&= (1-x^2)\left(u_1''(x)\int_{-1}^{x} w_1(s)f(s)\,\mathrm{d}s + u_1'(x)w_1(x)f(x) + (u_1w_1f)'(x)\right) \\
&\quad - 2x\left(u_1'(x)\int_{-1}^{x} w_1(s)f(s)\,\mathrm{d}s + u_1(x)w_1(x)f(x)\right) \\
&\qquad + c^2x^2u_1(x)\int_{-1}^{x} w_1(s)f(s)\,\mathrm{d}s \\
&= (1-x^2)(u_1'(x)w_1(x)f(x) + (u_1w_1f)'(x)) - 2xu_1(x)w_1(x)f(x). \qquad (2.11)
\end{aligned}
$$

By a similar calculation,

$$
\begin{aligned}
&\mathscr{P}_x \int_{x}^{1} w_1(x)u_1(s)f(s)\,\mathrm{d}s \\
&= (1-x^2)(-w_1'(x)u_1(x)f(x) - (w_1u_1f)'(x)) + 2xw_1(x)u_1(x)f(x). \qquad (2.12)
\end{aligned}
$$

Combining (2.10), (2.11), and (2.12) and applying (2.8) gives

$$
(\mathscr{P}\mathscr{G}f)(x) = \frac{(1-x^2)}{C}(u_1'(x)w_1(x) - w_1'(x)u_1(x))f(x) = f(x),
$$

that is, $\mathscr{P}\mathscr{G} = I$. Furthermore,

$$
\begin{aligned}
(\mathscr{G}\mathscr{P}f)(x) &= \int_{-1}^{1} G(x,s)(\mathscr{P}f)(s)\,\mathrm{d}s \\
&= \frac{u_1(x)}{C}\int_{-1}^{x} w_1(s)(\mathscr{P}f)(s)\,\mathrm{d}s + \frac{w_1(x)}{C}\int_{x}^{1} u_1(s)(\mathscr{P}f)(s)\,\mathrm{d}s. \qquad (2.13)
\end{aligned}
$$

Integrating the first integral on the right-hand side of (2.13) by parts and using the fact that $\mathscr{P}u_1 = 0$ gives

$$
\int_{-1}^{x} w_1(s)\mathscr{P}f(s)\,ds = (1-x^2)(w_1(x)f'(x) - w_1'(x)f(x)). \qquad (2.14)
$$

Similarly, for the second integral on the right-hand side of (2.13), we have

$$
\int_{x}^{1} u_1(s)\mathscr{P}f(s)\,ds = (1-x^2)(u_1'(x)f(x) - u_1(x)f'(x)). \qquad (2.15)
$$

Combining (2.13), (2.14), and (2.15) gives

$$
(\mathscr{G}\mathscr{P}f)(x) = \frac{(1-x^2)}{C}(u_1'(x)w_1(x) - w_1'(x)u_1(x))f(x) = f(x),
$$

that is, $\mathscr{G}\mathscr{P} = I$. Hence $\mathscr{G} = \mathscr{P}^{-1}$. □

As an immediate consequence of Lemma 2.1.18 we see that $\mathscr{G}$ and $\mathscr{P}$ have the same collection of eigenfunctions, namely the prolates $\{\bar{\phi}_n^{(c)}\}_{n=0}^{\infty}$.

Lemma 2.1.19. *The one-pair kernel G is oscillating.*

Proof. We first show that u_1, w_1 have no zeros on $[-1,1]$. The proof of Lemma 2.6.8 shows that $u_1(1) \neq 0$. Suppose now that $u_1(x_0) = 0$ for some $x_0 \in (-1,1)$. Since $\mathscr{P} u_1 = 0$, an integration by parts gives

$$0 = \int_{x_0}^{1} (\mathscr{P} u_1)(x) u_1(x)\,\mathrm{d}x = \int_{x_0}^{1} [(1-x^2)u_1'(x)^2 + c^2x^2u_1(x)^2]\,\mathrm{d}x,$$

so that $u_1(x) = 0$ on $[x_0, 1]$ which in turn implies that $u_1 \equiv 0$, a contradiction. We conclude that $u_1(x) \neq 0$ for all $x \in (-1,1]$. Similarly, $w_1(x) \neq 0$ for all $x \in [-1,1)$, that is,

$$|u_1(x)w_1(x)| > 0 \text{ for all } x \in (-1,1).$$

Furthermore,

$$\left|\frac{\mathrm{d}}{\mathrm{d}x}\left(\frac{u_1(x)}{w_1(x)}\right)\right| = \left|\frac{w_1(x)u_1'(x) - u_1(x)w_1'(x)}{w_1(x)^2}\right| = \frac{C}{(1-x^2)w_1(x)^2} > 0.$$

Now we apply Theorem 2.1.14 to conclude that the kernel G is oscillating. □

Proof (of Theorem 2.1.16). Since the eigenfunctions of $\mathscr{P}$ coincide with those of $\mathscr{G}$, the orthonormality and completeness of the eigenfunctions follows from the fact that $\mathscr{G}$ is self-adjoint and Hilbert–Schmidt (hence compact). The simplicity of the eigenvalues and the fact that they form a Markov system now follow by appealing to Lemma 2.1.19 and Theorem 2.1.15. □

2.2 Quadratures

Quadrature involves the estimation of integrals $\int_a^b \varphi(x)\omega(x)\,\mathrm{d}x$ for a given function (or class of functions) φ and fixed integrable non-negative weight function ω by sums of the form $\sum_{k=1}^{n} w_k \varphi(x_k)$. The points $\{x_k\}_{k=1}^{n}$ are the *nodes* of the quadrature while the coefficients $\{w_k\}_{k=1}^{n}$ are its *weights*. Classically, quadrature rules consist of n nodes and weights and integrate the monomials $\{x^k\}_{k=0}^{2n-1}$ exactly. By the linearity of such schemes, they integrate polynomials of degree up to $2n-1$ exactly.

Definition 2.2.1. A quadrature formula is said to be *generalized Gaussian* with respect to a collection of $2N$ functions $\varphi_j : [a,b] \to \mathbb{R}$ $(j = 1,2,\ldots,2N)$ and non-negative weight function ω if it consists of N weights and nodes and integrates the functions $\{\varphi_j\}_{j=1}^{2N}$ exactly, that is,

$$\int_a^b \varphi_j(x)\omega(x)\,\mathrm{d}x = \sum_{k=1}^{N} w_k \varphi_j(x_k) \quad (j = 1,2,\ldots,2N). \tag{2.16}$$

The w_k and x_k are called generalized Gaussian weights and nodes respectively.

The following general result about Chebyshev systems is due to Markov [222, 223]. Other proofs can be found in [170, 186].

Theorem 2.2.2. *Suppose $\{\varphi_i\}_{i=1}^{2n}$ is a Chebyshev system on $[a,b]$ and ω is non-negative and integrable on $[a,b]$. Then there exists a unique generalized Gaussian quadrature for $\{\varphi_i\}_{i=1}^{2n}$ on $[a,b]$ with respect to ω. The weights of the quadrature are positive.*

By Theorems 2.2.2 and 2.1.16, there exists a unique N-point generalized Gaussian quadrature for the first $2N$ prolates $\{\bar{\phi}_n\}_{n=0}^{2N-1}$. Computing the weights and nodes of a generalized Gaussian quadrature involves solving the $2N$ equations (2.16). While these equations are linear in the weights $\{w_k\}_{k=1}^{N}$, they are nonlinear in the nodes $\{x_k\}_{k=1}^{N}$. In [72] a Newton iteration for the approximate solution of these equations for arbitrary Chebyshev systems is given. The algorithm is very similar to an algorithm for the computation of nodes and weights of *generalized Gauss–Lobatto quadratures* which we discuss in Sect. 2.4. In the case of the prolates, estimates of the accuracy of such quadratures when applied to arbitrary functions with band limit c typically rely on the eigenfunction property of Corollary 1.2.6, which may be written in the form

$$\mu_n(2c/\pi)\bar{\phi}_n^{(c)}(x) = \int_{-1}^{1} \mathrm{e}^{-\mathrm{i}cxt}\,\bar{\phi}_n^{(c)}(t)\,\mathrm{d}t \tag{2.17}$$

where $\mu_n(a) = 2\mathrm{i}^n\sqrt{\lambda_n(a)/a}$. From now on we abbreviate

$$\mu_n(2c/\pi) = \mathrm{i}^n\sqrt{2\pi\lambda_n(2c/\pi)/c} = \mu_n.$$

Since classical Gaussian quadrature schemes integrate polynomials of degree up to $2N-1$ exactly, they may be used to estimate integrals of functions that are effectively approximated by polynomials. Similarly, generalized Gaussian quadratures for the prolates $\{\bar{\phi}_n^{(c)}\}_{n=0}^{2N-1}$ may be used to approximate integrals of functions that are effectively approximated by this collection. As the following result [364] shows, when N is sufficiently large, the class of functions that are effectively approximated by $\{\bar{\phi}_n^{(c)}\}_{n=0}^{2N-1}$ includes functions of band limit less than or equal to c.

Theorem 2.2.3. *Let $\{x_k\}_{k=1}^{N}$ be nodes and $\{w_k\}_{k=1}^{N}$ be non-negative weights for the generalized Gaussian quadrature for the prolates $\{\bar{\phi}_n^{(c)}\}_{n=0}^{2N-1}$ with $N > c/\pi$. Then for any function f on the line with band limit c,*

$$\left|\sum_{k=1}^{N} w_k f(x_k) - \int_{-1}^{1} f(x)\,\mathrm{d}x\right| \le C|\mu_{2N-1}|\|f\|_{L^2(\mathbb{R})} \tag{2.18}$$

with C a constant independent of N.

Proof. Let $\bar{\phi}_n = \bar{\phi}_n^{(c)}$ and $\mu_n = \mu_n(2c/\pi)$. Since $\{\bar{\phi}_n\}_{n=0}^{\infty}$ is an orthonormal basis for $L^2[-1,1]$, for each x, $t \in [-1,1]$ we have the expansion

$$e^{-icxt} = \sum_{n=0}^{\infty} \left(\int_{-1}^{1} e^{-icxu} \bar{\phi}_n(u)\, du \right) \bar{\phi}_n(t) = \sum_{n=0}^{\infty} \mu_n \bar{\phi}_n(x) \bar{\phi}_n(t) \tag{2.19}$$

where we have used the eigenfunction property (2.17). The error incurred when the quadrature is applied to an exponential e^{-ictx} ($|t| < 1$ fixed) is

$$\begin{aligned} E(t) &= \sum_{k=1}^{N} w_k e^{-icx_k t} - \int_{-1}^{1} e^{-icxt}\, dx \\ &= \sum_{k=1}^{N} w_k \left(\sum_{n=0}^{\infty} \mu_n \bar{\phi}_n(x_k) \bar{\phi}_n(t) \right) - \int_{-1}^{1} \sum_{n=0}^{\infty} \mu_n \bar{\phi}_n(x) \bar{\phi}_n(t)\, dx \\ &= \sum_{n=2N}^{\infty} \mu_n \bar{\phi}_n(t) \left(\sum_{k=1}^{N} w_k \bar{\phi}_n(x_k) - \int_{-1}^{1} \bar{\phi}_n(x)\, dx \right). \end{aligned} \tag{2.20}$$

However, as stated in [296],

$$M_N^c = \max_{0 \le n \le N} \max_{|x| \le 1} |\bar{\phi}_n^{(c)}(x)| \le 2\sqrt{N}. \tag{2.21}$$

Furthermore, the weights w_k of a generalized Gaussian quadrature are non-negative with the sums $\sum_{k=1}^{N} w_k$ bounded independent of N. Consequently, (2.20) and the asymptotic formula of Corollary 1.2.11 may be combined to give the estimate

$$|E(t)| \le C \sum_{n=2N}^{\infty} \sqrt{n}\, |\mu_n| |\bar{\phi}_n(t)| \le C \sum_{n=2N}^{\infty} n\, |\mu_n| \le C |\mu_{2N-1}| \tag{2.22}$$

since $N > c/\pi$ is large enough so that $\{\mu_n\}_{n=2N}^{\infty}$ decreases super-exponentially.

Suppose now that f is a function on the line with band limit c, that is,

$$f(x) = \int_{-1}^{1} \sigma(t) e^{icxt}\, dt \tag{2.23}$$

for some $\sigma \in L^2[-1,1]$. Then, applying the quadrature to f yields an error E_f given by

$$\begin{aligned} E_f &= \sum_{k=1}^{N} w_k f(x_k) - \int_{-1}^{1} f(x)\, dx \\ &= \int_{-1}^{1} \sigma(t) \left(\sum_{k=1}^{N} w_k e^{-icx_k t} - \int_{-1}^{1} e^{-icxt}\, dx \right) dt = \int_{-1}^{1} \sigma(t) E_t\, dt, \end{aligned}$$

from which we conclude that $|E_f| \le C |\mu_{2N-1}| \|\sigma\|_{L^2[-1,1]} \le C |\mu_{2N-1}| \|f\|_{L^2(\mathbb{R})}$. □

In order to obtain estimates of the form (2.18) for functions f on the line of band limit c, it is not necessary that the quadrature rule integrate the first N prolates exactly. With this in mind we make the following definition.

Definition 2.2.4. Let N be a positive integer, $c > 0$, and $\varepsilon \geq 0$. The collection of nodes $\{x_k\}_{k=1}^N \subset [-1,1]$ and positive weights $\{w_k\}_{k=1}^N$ are said to integrate functions on $[-1,1]$ with band limit c to precision ε if for all functions f with band limit c of the form (2.23) we have

$$\left|\int_{-1}^{1} f(x)\,\mathrm{d}x - \sum_{k=1}^{N} w_k f(x_k)\right| \leq \varepsilon \|\sigma\|_{L^1[-1,1]}.$$

A collection of such weights and nodes will be referred to as an *approximate generalized Gaussian quadrature* of band limit c and precision ε. The class of such quadratures is denoted $\mathrm{GGQ}(c,\varepsilon)$.

This definition weakens the condition on the nodes and weights of the quadrature and allows for more flexible constructions. The question then becomes: What are appropriate weights and nodes? In Theorem 2.2.6 below, we show that the nodes may be chosen to be the zeros of a prolate $\bar{\phi}_N^{(c)}$ with N sufficiently large. Before that, however, we require the following generalization of the Euclidean division algorithm due to Xiao, Rokhlin, and Yarvin [364].

Theorem 2.2.5. *Let $\sigma,\rho \in C^1[-1,1]$ with $\sigma(\pm 1) = 0 \neq \rho(\pm 1)$. Let $f = F_{-4c/\pi}(\sigma)$ and $p = F_{-2c/\pi}(\rho)$. Then there exist unique η,ξ such that with $q = F_{-2c/\pi}(\eta)$ and $r = F_{-2c/\pi}(\xi)$ we have $f = pq + r$.*

The role of the degrees of the polynomials in the classical Euclidean algorithm is replaced in Theorem 2.2.5 by the band limits of the functions f, p, q, and r.

Proof. The proof is constructive. Let p, q be as in the theorem. Then

$$\begin{aligned}
p(x)q(x) &= \int_{-1}^{1}\int_{-1}^{1} \rho(t)\eta(s)\mathrm{e}^{-\mathrm{i}cx(t+s)}\,\mathrm{d}t\,\mathrm{d}s \\
&= \int_{0}^{2} \mathrm{e}^{-\mathrm{i}cxu} \int_{u-1}^{1} \rho(u-s)\eta(s)\,\mathrm{d}s\,\mathrm{d}u + \int_{-2}^{0} \mathrm{e}^{-\mathrm{i}cxu} \int_{-1}^{u+1} \rho(u-s)\eta(s)\,\mathrm{d}s\,\mathrm{d}u \\
&= \int_{-2}^{2} \mathrm{e}^{-\mathrm{i}cxu} \int_{\max(-1,u-1)}^{\min(1,u+1)} \rho(u-s)\eta(s)\,\mathrm{d}s\,\mathrm{d}u.
\end{aligned}$$

Consequently,

$$\begin{aligned}
&f(x) - p(x)q(x) \\
&\quad = \frac{1}{2}\int_{-2}^{2} \sigma\left(\frac{u}{2}\right) \mathrm{e}^{-\mathrm{i}cxu}\,\mathrm{d}u - \int_{-2}^{2} \mathrm{e}^{-\mathrm{i}cxu} \int_{\max(-1,u-1)}^{\min(1,u+1)} \rho(u-s)\eta(s)\,\mathrm{d}s\,\mathrm{d}u \\
&\quad = \frac{1}{2}\int_{|u|\le 1} \sigma\left(\frac{u}{2}\right) \mathrm{e}^{-\mathrm{i}cxu}\,\mathrm{d}u - \int_{|u|\le 1} \mathrm{e}^{-\mathrm{i}cxu} \int_{\max(-1,u-1)}^{\min(1,u+1)} \rho(u-s)\eta(s)\,\mathrm{d}s\,\mathrm{d}u \\
&\quad + \frac{1}{2}\int_{1\le|u|\le 2} \sigma\left(\frac{u}{2}\right) \mathrm{e}^{-\mathrm{i}cxu}\,\mathrm{d}u - \int_{1\le|u|\le 2} \mathrm{e}^{-\mathrm{i}cxu} \int_{\max(-1,u-1)}^{\min(1,u+1)} \rho(u-s)\eta(s)\,\mathrm{d}s\,\mathrm{d}u.
\end{aligned}$$

We now want to choose η so that

$$\frac{1}{2}\sigma\left(\frac{u}{2}\right) = \int_{\max(-1,u-1)}^{\min(1,u+1)} \rho(u-s)\eta(s)\,\mathrm{d}s, \quad 1 \le |u| \le 2.$$

The two cases $1 \le u \le 2$ and $-2 \le u \le -1$ give the two equations:

$$\frac{1}{2}\sigma\left(\frac{u}{2}\right) = \begin{cases} \int_{u-1}^{1} \rho(u-s)\eta(s)\,\mathrm{d}s & (1 \le u \le 2) \\ \int_{-1}^{u+1} \rho(u-s)\eta(s)\,\mathrm{d}s & (1 \le -u \le 2). \end{cases} \tag{2.24}$$

Differentiating both of these equations with respect to u gives

$$\frac{1}{4}\sigma'\left(\frac{u}{2}\right) = \begin{cases} -\rho(1)\eta(u-1) + \int_{u-1}^{1} \rho'(u-s)\eta(s)\,\mathrm{d}s & (1 \le u \le 2) \\ \rho(-1)\eta(u+1) + \int_{-1}^{u+1} \rho'(u-s)\eta(s)\,\mathrm{d}s & (1 \le -u \le 2). \end{cases} \tag{2.25}$$

Defining integral operators $V_1 : L^2[0,1] \to L^2[0,1]$ and $V_2 : L^2[-1,0] \to L^2[-1,0]$ by

$$V_1 f(v) = \int_{v}^{1} \rho'(v+1-s) f(s)\,\mathrm{d}s; \quad V_2 f(w) = \int_{-1}^{w} \rho'(w-1-s) f(s)\,\mathrm{d}s,$$

equation (2.25) above becomes

$$(V_1 - \rho(1)I)\eta(v) = \frac{1}{4}\sigma'\left(\frac{v+1}{2}\right); \; (V_2 + \rho(-1)I)\eta(w) = \frac{1}{4}\sigma'\left(\frac{w-1}{2}\right) \tag{2.26}$$

for $v \in [0,1]$ and $w \in [-1,0]$. V_1 and V_2 are Volterra integral operators and $V_1 + \lambda I$, $V_2 + \lambda I$ have bounded inverses if and only if $\lambda \neq 0$. Therefore, to solve the equations (2.26) we require $\rho(1)\rho(-1) \neq 0$. With this proviso, (2.26) provides the definition of η on $[-1,1]$. Integrating (2.25) with respect to u gives

$$\frac{1}{2}\sigma\left(\frac{u}{2}\right) = \begin{cases} \int_{u-1}^{1} \rho(u-s)\eta(s)\,\mathrm{d}s + C_1 & (1 \le u \le 2) \\ \int_{-1}^{u+1} \rho(u-s)\eta(s)\,\mathrm{d}s + C_2 & (1 \le -u \le 2) \end{cases}$$

with C_1, C_2 arbitrary constants. However, by evaluating the first of these equations at $u = 2$ and the second at $u = -2$ we find that $C_1 = \sigma(1)/2$ and $C_2 = \sigma(-1)/2$. Hence, with η so defined, we have a solution of (2.24) only when $\sigma(1) = \sigma(-1) = 0$. Once this is assumed we define ξ by

$$\xi(u) = \frac{1}{2}\sigma\left(\frac{u}{2}\right) - \int_{\max(-1,u-1)}^{\min(1,u+1)} \rho(u-s)\eta(s)\,\mathrm{d}s \qquad (|u| \leq 1).$$

Then $f = pq + r$ as required. □

We now apply the theorem to the construction of approximate generalized Gaussian quadratures for band-limited functions. Recall that by Theorems 2.1.7 and 2.1.16, the Nth prolate $\bar{\phi}_N^{(c)}$ has precisely N nodes in $(-1,1)$. Given a sequence $-1 < x_0 < x_1 < \cdots < x_{N-1} < 1$, the invertibility of the $N \times N$ matrix $(\bar{\phi}_j(x_k))_{j,k=0}^{N-1}$ ensures the existence of real weights $\{w_k\}_{k=0}^{N-1}$ for which

$$\sum_{k=0}^{N-1} w_k \bar{\phi}_j(x_k) = \int_{-1}^{1} \bar{\phi}_j(x)\,\mathrm{d}x \qquad (0 \leq j \leq N-1). \tag{2.27}$$

When the nodes $\{x_k\}_{k=0}^{N-1}$ are the N zeros of $\bar{\phi}_N$ in $(-1,1)$, then the weights are positive and their sum is bounded independently of N and c. To see this, observe that the constant function 1 has prolate expansion $1 = \sum_{j=0}^{\infty} \bar{\phi}_j(x) \int_{-1}^{1} \bar{\phi}_j(t)\,\mathrm{d}t = \sum_{j=0}^{\infty} \mu_j \bar{\phi}_j(0)\bar{\phi}_j(x)$. Evaluating this equation at $x = x_k$, multiplying both sides by w_k, and summing over k gives

$$\begin{aligned}
\sum_{k=0}^{N-1} w_k &= \sum_{k=0}^{N-1}\sum_{j=0}^{\infty} \mu_j \bar{\phi}_j(0) w_k \bar{\phi}_j(x_k) \\
&= \sum_{j=0}^{N-1} \mu_j \bar{\phi}_j(0) \sum_{k=0}^{N-1} w_k \bar{\phi}_j(x_k) + \sum_{j=N}^{\infty} \mu_j \bar{\phi}_j(0) \sum_{k=0}^{N-1} w_k \bar{\phi}_j(x_k) \\
&\leq \sum_{j=0}^{\infty} |\mu_j|^2 \bar{\phi}_j(0)^2 + \sum_{k=0}^{N-1} w_k \sum_{j=N}^{\infty} \mu_j \bar{\phi}_j(0)\bar{\phi}_j(x_k) \\
&\leq 2 + \sum_{k=0}^{N-1} w_k \sum_{j=N}^{\infty} 4j|\mu_j| \leq 2 + C|\mu_{N-1}| \sum_{k=0}^{N-1} w_k,
\end{aligned}$$

where we have used the Plancherel theorem for prolate expansions ($2 = \int_{-1}^{1} \mathrm{d}t = \sum_{j=0}^{\infty} |\langle 1, \bar{\phi}_j\rangle|^2 = \sum_{j=0}^{\infty} |\mu_j|^2 \bar{\phi}_j(0)^2$) and the estimate (2.21). Rearranging this inequality gives $\sum_{k=0}^{N-1} w_k \leq 2/(1 - C|\mu_{N-1}|) \approx 2$ when N is significantly larger than $2c/\pi$. These ingredients allow for the following approximate generalized quadrature result [364].

Theorem 2.2.6. *Let* $\{x_k\}_{k=0}^{N-1} \subset (-1,1)$ *be the nodes of the Nth prolate function* $\bar{\phi}_N = \bar{\phi}_N^{(c)}$ *with bandwidth c, and let* $\{w_k\}_{k=0}^{N-1}$ *be real weights satisfying (2.27). Then for any function f with band limit 2c satisfying the conditions of Theorem 2.2.5 we have the quadrature error estimate*

$$E_f = \left| \sum_{k=0}^{N-1} w_k f(x_k) - \int_{-1}^{1} f(x)\,\mathrm{d}x \right| \leq C|\mu_{N-1}| \left(\|\eta\|_{L^2[-1,1]} + \|\xi\|_{L^2[-1,1]} \right)$$

with η, ξ *as in Theorem 2.2.5.*

Proof. Since $\bar{\phi}_N(\pm 1) \neq 0$ by Lemma 2.6.8 and $\bar{\phi}_N(x) = \int_{-1}^{1} \rho(t)\mathrm{e}^{-icxt}\,\mathrm{d}t$ with $\rho(t) = \bar{\phi}_N(t)/\mu_N$, we may apply Theorem 2.2.5 to produce functions q and r of band limit c with $f = \bar{\phi}_N q + r$. Let $q = F_{-2c/\pi}(\eta)$ and $r = F_{-2c/\pi}(\xi)$. Then, since $\bar{\phi}_N(x_k) = 0$ for all nodes x_k, we have

$$\begin{aligned} E_f &= \left|\sum_{k=0}^{N-1} w_k f(x_k) - \int_{-1}^{1} f(x)\,\mathrm{d}x\right| \\ &= \left|\sum_{k=0}^{N-1} w_k[\bar{\phi}_N(x_k)q(x_k) + r(x_k)] - \int_{-1}^{1}[\bar{\phi}_N(x)q(x) + r(x)]\,\mathrm{d}x\right| \\ &\leq \left|\sum_{k=0}^{N-1} w_k r(x_k) - \int_{-1}^{1} r(x)\,\mathrm{d}x\right| + \left|\int_{-1}^{1}\bar{\phi}_N(x)q(x)\,\mathrm{d}x\right|. \end{aligned} \tag{2.28}$$

By the definition of q and the eigenfunction property (2.17) of the prolates,

$$\begin{aligned} \left|\int_{-1}^{1}\bar{\phi}_N(x)q(x)\,\mathrm{d}x\right| &= \left|\int_{-1}^{1}\eta(t)\int_{-1}^{1}\bar{\phi}_N(x)\mathrm{e}^{icxt}\,\mathrm{d}x\,\mathrm{d}t\right| \\ &= \left|\int_{-1}^{1}\eta(t)\mu_N\bar{\phi}_N(t)\,\mathrm{d}t\right| \leq |\mu_N|\|\eta\|_{L^2[-1,1]}. \end{aligned} \tag{2.29}$$

Furthermore, we may apply (2.22) with $2N$ replaced by N to obtain

$$\begin{aligned} \left|\sum_{k=0}^{N-1} w_k r(x_k) - \int_{-1}^{1} r(x)\,\mathrm{d}x\right| &= \left|\int_{-1}^{1}\xi(t)\left(\sum_{k=0}^{N-1} w_k \mathrm{e}^{-icx_k t} - \int_{-1}^{1}\mathrm{e}^{-icxt}\,\mathrm{d}x\right)\mathrm{d}t\right| \\ &\leq C|\mu_{N-1}|\|\xi\|_{L^2[-1,1]}. \end{aligned} \tag{2.30}$$

Applying (2.29) and (2.30) to (2.28) now gives the required estimate. □

The conditions $\sigma(\pm 1) = 0$ of Theorem 2.2.6 can be removed at the expense of a slower rate of convergence as the following corollary, whose proof can be found in Sect. 2.6.3, shows.

Corollary 2.2.7. *Let* $\{x_k\}_{k=0}^{N-1}$, $\{w_k\}_{k=0}^{N-1}$, $\bar{\phi}_N$ *be as in Theorem 2.2.6. Then for any function* f *with band limit* $2c$ *of the form* $f(x) = \int_{-1}^{1}\sigma(t)\mathrm{e}^{-2icxt}\,\mathrm{d}t$ *with* σ, $\sigma' \in L^2[-1,1]$, *we have*

$$E_f = \left|\sum_{k=0}^{N-1} w_k f(x_k) - \int_{-1}^{1} f(x)\,\mathrm{d}x\right| \leq C\sqrt{|\mu_{N-1}|}(\|\sigma'\|_{L^2[-1,1]} + \|\sigma\|_{L^2[-1,1]}).$$

Discussion

It is important, when applying Theorem 2.2.6, that the norms $\|\eta\|_{L^2[-1,1]}$ and $\|\xi\|_{L^2[-1,1]}$ be not significantly larger than $\|\sigma\|_{L^2[-1,1]} + \|\sigma'\|_{L^2[-1,1]}$, for otherwise the accuracy of the quadrature is compromised. In [364] it is claimed that such estimates have been obtained in the case where $n > 2c/\pi + 10\log(c)$. Numerical evidence of the accuracy of these procedures is also given in [364], where it is observed that the number of quadrature nodes required to accurately integrate band-limited functions is roughly $\pi/2$ times fewer than the number of classical Gaussian nodes needed to attain the same accuracy.

2.3 Local Sampling Formulas

2.3.1 *Uniform Oversampling*

In order to compute a representation of a band-limited function f that is valid on some interval I by means of the classical Shannon formula (Theorem 5.1.3), samples of f are required far from I. This is because of the slow decay of the cardinal sine function. One remedy to this situation is to replace the cardinal sine with an interpolating function having faster decay. This requires *oversampling*. If $f \in \mathrm{PW}_\Omega$ and $\varphi \in L^2(\mathbb{R})$ is chosen so that

$$\hat{\varphi}(\xi) = \begin{cases} 1 & \text{if } |\xi| < \Omega/2 \\ 0 & \text{if } |\xi| > \Omega \end{cases} \tag{2.31}$$

then f can be reconstructed from the samples $\{f(n/(2\Omega)\}_{n=-\infty}^{\infty}$ via

$$f(t) = \frac{1}{2\Omega} \sum_{n=-\infty}^{\infty} f\Big(\frac{n}{2\Omega}\Big)\varphi\Big(t - \frac{n}{2\Omega}\Big).$$

The sampling rate in this case is twice the *Nyquist* rate of the Shannon theorem, but may be reduced to any rate greater than or equal to the Nyquist rate by reducing the support of $\hat{\varphi}$. When the rate is strictly greater than Ω samples per unit time, φ may be chosen to have faster decay in time. How fast can a function φ satisfying (2.31) decay in time? There is no L^2 function satisfying (2.31) with exponential decay in time: the Fourier transform of such a function would be analytic in a strip in the complex plane containing the real line, a condition that would preclude the compact support of $\hat{\varphi}$ in (2.31).

It is simple to construct a function φ with inverse polynomial decay ($|\varphi(t)| \leq C(1+|t|)^{-N}$ for some fixed positive N) whose Fourier transform $\hat{\varphi}$ is a piecewise polynomial satisfying (2.31). In [244], functions having *root exponential decay* in time, that is, $|\varphi(t)| \leq C\mathrm{e}^{-\sqrt{2\pi|t|}}$, and satisfying (2.31) were constructed. In [93] Dziubański and Hernández showed that *sub-exponential decay* is possible: given any $\varepsilon > 0$, one can construct a function φ whose Fourier transform satisfies (2.31) and such that $|\varphi(t)| \leq C_\varepsilon \mathrm{e}^{-\pi|t|^{1-\varepsilon}}$. The constant C_ε blows up as $\varepsilon \to 0$. Finally, in an unpublished manuscript, Güntürk and DeVore gave a sophisticated construction that achieves perhaps the optimal outcome—a function φ with Fourier transform $\hat{\varphi}$ satisfying (2.31) such that $|\varphi(t)| \le C\exp(-\alpha|x|/\log^{\gamma}|x|)$ for $\alpha > 0$ and $\gamma > 1$. This construction is outlined in Sect. 6.3.2.

Unfortunately, apart from the spline-based examples, none of the constructions just mentioned generate sampling functions φ that can be expressed in closed form—a serious disadvantage when compared with the simplicity of the Shannon theorem. In any event, the oversampling results require knowledge of samples taken off (though near) the interval on which an approximation is desired. In summary, if we are prepared to oversample—at any rate higher than that prescribed by the Shannon theorem—then uniform samples can be used to construct local approximations with samples taken on and near the interval of interest. But the absence of closed forms for the sampling functions of such schemes renders them impractical.

More is possible if we are prepared to forego the uniformity of the samples. As we shall see, relatively few nonuniform samples taken only on the interval of interest may be combined using prolates to construct accurate quadrature-based approximate sampling schemes.

2.3.2 Local Approximate Sampling with Prolates

In [296], the authors generated localized sampling formulas for band-limited functions on the line through the use of approximate quadrature schemes. In particular, given a function f with band limit $c > 0$ of the form (2.23), approximations of f via truncations of its prolate series are made, that is, $f(x) \approx \sum_{n=0}^{N} a_n \bar{\phi}_n^{(c)}(x)$. Here the coefficients $\{a_n\}_{n=0}^{N}$ are obtained from samples $\{f(x_k)\}_{k=0}^{M-1}$ where $\{x_k\}_{k=0}^{M-1}$ are quadrature nodes associated with quadrature weights $\{w_k\}_{k=0}^{M-1}$ for which

$$\sum_{k=0}^{M-1} w_k\, g(x_k) \approx \int_{-1}^{1} g(x)\,\mathrm{d}x$$

for all functions on the line with band limit $2c$.

Since $\{\bar{\phi}_n\}_{n=0}^{\infty}$ is a complete orthonormal system in $L^2[-1,1]$, any $f \in L^2[-1,1]$ admits a prolate expansion of the form

$$f(x) = \sum_{n=0}^{\infty} b_n \bar{\phi}_n(x) \tag{2.32}$$

where $b_n = \langle f, \bar{\phi}_n \rangle = \int_{-1}^{1} f(x)\bar{\phi}_n(x)\,\mathrm{d}x$. We now develop error bounds for truncations of the series (2.32) for functions f with band limit $c > 0$. The starting point is the following estimate of the decay of the coefficients in these expansions [296].

Lemma 2.3.1. *Let f be a function on the line with band limit c as in (2.23) and $M_n = M_n^c$ be as in (2.21). Then*

$$\left|\int_{-1}^{1} f(x)\bar{\phi}_n(x)\,\mathrm{d}x\right| \le |\mu_n| M_n \|\sigma\|_{L^1[-1,1]}.$$

Proof. With an application of (2.17) we have

$$\begin{aligned}\left|\int_{-1}^{1} f(x)\bar{\phi}_n(x)\,\mathrm{d}x\right| &= \left|\int_{-1}^{1}\left(\int_{-1}^{1}\sigma(t)\mathrm{e}^{-icxt}\,\mathrm{d}t\right)\bar{\phi}_n(x)\,\mathrm{d}x\right| \\ &= \left|\int_{-1}^{1}\sigma(t)\int_{-1}^{1}\bar{\phi}_n(x)\mathrm{e}^{-icxt}\,\mathrm{d}x\,\mathrm{d}t\right| \\ &= |\mu_n|\left|\int_{-1}^{1}\sigma(t)\bar{\phi}_n(t)\,\mathrm{d}t\right| \le |\mu_n| M_n \|\sigma\|_{L^1[-1,1]} \qquad \square\end{aligned}$$

Convergence of the series (2.32) is now investigated by estimating its tail. The following result is a variation on Lemma 3.2 of [296].

Lemma 2.3.2. *Let $c > 0$ and $N \ge 2c$ be an integer. Then*

$$\sum_{n=N+1}^{\infty} |\mu_n|(M_n)^2 \le C(N+1)^{3/2}(0.386)^N.$$

Proof. By Lemma 2.6.7, for all n we have

$$|\mu_n| \le \frac{\sqrt{\pi}c^n(n!)^2}{(2n)!\Gamma(n+3/2)} \le \frac{4c^n}{2^{2n}(n-1)!}, \tag{2.33}$$

so that by (2.21),

$$\sum_{n=N+1}^{\infty} |\mu_n|(M_n)^2 \le 16 \sum_{n=N+1}^{\infty} \frac{(c/4)^n n}{(n-1)!} \tag{2.34}$$

Let $h(x) = x(x+1)\mathrm{e}^x = \sum_{n=1}^{\infty} nx^n/(n-1)!$. With $S_N(x)$ the Nth partial sum of this series and $R_N(x) = h(x) - S_N(x)$ the Nth remainder, the Taylor remainder theorem gives

$$\begin{aligned}|R_N(x)| &\le \sup_{|t|\le|x|} |h^{(N+1)}(t)| \frac{|x|^{N+1}}{(N+1)!} \\ &\le (|x|^2 + (2N+3)|x| + (N+1)^2)\mathrm{e}^{|x|}\frac{|x|^{N+1}}{(N+1)!}.\end{aligned}$$

Consequently, since $c \le N/2$, the sum in (2.34) may be estimated by

$$\sum_{n=N+1}^{\infty} |\mu_n|(M_n)^2 \leq 16 R_N(c/4)$$
$$\leq 16\left(\left(\frac{c}{4}\right)^2 + (2N+3)\left(\frac{c}{4}\right) + (N+1)^2\right) e^{c/4} \frac{(c/4)^{N+1}}{(N+1)!}$$
$$\leq \left(\frac{81}{4}\right)(N+1) e^{N/8} \frac{(N/8)^{N+1}}{N!}.$$

An application of the Stirling inequality $n^{n+1/2}\sqrt{2\pi} \leq e^n n!$ now gives

$$\sum_{n=N+1}^{\infty} |\mu_n|(M_n)^2 \leq \frac{81}{32\sqrt{2\pi}}(N+1)\sqrt{N}\left(\frac{e^{9/8}}{8}\right)^N$$

from which the result follows. □

Lemma 2.3.2 remains true with the condition $N \geq (0.9)c$ in the sense that $\sum_{n=N+1}^{\infty} |\mu_n|(M_n)^2 \leq C(N+1)^{3/2}\beta^N$ with $0 < \beta < 1$.

These estimates allow us to investigate the nature of the convergence of the prolate series (2.32). The following result appears in [296].

Corollary 2.3.3. *Let $c > 0$ and f be a function on the line with band limit c of the form (2.23) for some $\sigma \in L^2[-1,1]$. Then the prolate series (2.32) converges uniformly to f on $[-1,1]$.*

Proof. Given non-negative integers $M \leq N$, with M_n as in (2.21),

$$\left|\sum_{n=M}^{N}\left(\int_{-1}^{1} f(t)\bar{\phi}_n(t)\,dt\right)\bar{\phi}_n(x)\right| \leq \|\sigma\|_{L^1[-1,1]} \sum_{n=M}^{\infty} |\mu_n|(M_n)^2 \to 0$$

as $M,N \to \infty$ due to Lemma 2.3.2. Hence the series (2.32) satisfies the Cauchy criterion for convergence and consequently converges uniformly. Since it also converges in $L^2[-1,1]$ to f, it must converge uniformly to f on $[-1,1]$. □

As a corollary of Lemma 2.3.1, if $c > 0$, N is a non-negative integer, and f is a band-limited function of the form (2.23), then for all $x \in [-1,1]$,

$$\left|f(x) - \sum_{n=0}^{N}\langle f,\bar{\phi}_n\rangle\bar{\phi}_n(x)\right| \leq \left(\sum_{n=N+1}^{\infty} |\mu_n|(M_n)^2\right)\|\sigma\|_{L^1[-1,1]}. \tag{2.35}$$

Lemma 2.3.2 may be applied to the right-hand side of (2.35) to show that if $N \geq 2c$ then the truncation error that appears on the left-hand side of (2.35) is bounded by $C(N+1)^{3/2}(0.386)^N\|\sigma\|_{L^1[-1,1]}$.

Given a function f with band limit c as in (2.23) and $n \geq 0$, the product $f\bar{\phi}_n^{(c)}$ has band limit $2c$. In fact, $f(x)\bar{\phi}_n^{(c)}(x) = \int_{-1}^{1}\rho_n(r)e^{2icxr}\,dr$ with

$$\rho_n(r) = \frac{2}{\mu_n}\int_{-1}^{1}\bar{\phi}_n^{(c)}(t)\sigma(2r-t)\,dt = \frac{2}{\mu_n}\bar{\phi}_n^{(c)} * \sigma(2r), \tag{2.36}$$

from which we obtain

$$\|\rho_n\|_{L^1[-1,1]} \leq \frac{1}{|\mu_n|}\|\bar{\phi}_n^{(c)}\|_{L^1[-1,1]}\|\sigma\|_{L^1[-1,1]} \leq \frac{\sqrt{2}}{|\mu_n|}\|\sigma\|_{L^1[-1,1]}. \tag{2.37}$$

Given weights $\{w_k\}_{k=0}^{K-1}$ and nodes $\{x_k\}_{k=0}^{K-1}$ which together form a quadrature in $\mathrm{GGQ}(2c,\varepsilon)$ for some $\varepsilon > 0$ (see Definition 2.2.4), and a band-limited function f of the form (2.23), we may accurately estimate the $L^2[-1,1]$ inner products $b_n = \langle f, \bar{\phi}_n\rangle$ that appear in (2.32) by the appropriate quadrature sum as in [296].

Lemma 2.3.4. *Let $c > 0$, N, K be positive integers, and suppose the nodes $\{x_k\}_{k=0}^{K-1}$ and weights $\{w_k\}_{k=0}^{K-1}$ form an element of $GGQ(2c,|\mu_N|^2)$. If f is a band-limited function on the line with band limit c as in (2.23), then*

$$\left|\langle f,\bar{\phi}_n\rangle - \sum_{k=0}^{K-1} w_k f(x_k)\bar{\phi}_n(x_k)\right| \leq \sqrt{2}|\mu_N|\|\sigma\|_{L^1[-1,1]}. \tag{2.38}$$

Proof. Since $f(x)\bar{\phi}_n(x) = \int_{-1}^{1}\rho_n(t)\mathrm{e}^{-2icxt}\,\mathrm{d}t$ with ρ_n as in (2.36), by (2.37) we have

$$\begin{aligned}\left|\langle f,\bar{\phi}_n\rangle - \sum_{k=0}^{K-1} w_k f(x_k)\bar{\phi}_n(x_k)\right| &\leq |\mu_N|^2\|\rho_n\|_{L^1[-1,1]}\\ &\leq \sqrt{2}\frac{|\mu_N|^2}{|\mu_n|}\|\sigma\|_{L^1[-1,1]} \leq \sqrt{2}|\mu_N|\|\sigma\|_{L^1[-1,1]}. \quad \square\end{aligned}$$

As shown in [296], the quadratures outlined above yield accurate localized sampling formulas for functions with band limit c.

Theorem 2.3.5. *Let $c \geq 0$, N, K be positive integers, and suppose the nodes $\{x_k\}_{k=0}^{K-1}$ and weights $\{w_k\}_{k=0}^{K-1}$ form an element of $GGQ(2c,|\mu_N|^2)$. For each integer $0 \leq k \leq K-1$, define the interpolating functions S_k by*

$$S_k(x) = w_k\sum_{n=0}^{N}\bar{\phi}_n(x_k)\bar{\phi}_n(x). \tag{2.39}$$

Then if $N > 2c$, for each band-limited function on the line of the form (2.23) and each $x \in [-1,1]$ we have the estimate

$$\left|f(x) - \sum_{k=0}^{K-1} f(x_k)S_k(x)\right| \leq C(N+1)^{3/2}(0.386)^N\|\sigma\|_{L^1[-1,1]}.$$

Proof. We write

$$\left|f(x)-\sum_{k=0}^{K-1} f(x_k)S_k(x)\right| \le \left|f(x)-\sum_{n=0}^{N}\langle f,\bar{\phi}_n\rangle\bar{\phi}_n(x)\right| + \left|\sum_{n=0}^{N}\langle f,\bar{\phi}_n\rangle\bar{\phi}_n(x)-\sum_{k=0}^{K-1} f(x_k)S_k(x)\right| = A+B. \tag{2.40}$$

With an application of (2.35) and Lemma 2.3.2, A can be estimated by

$$A \le \sum_{n=N+1}^{\infty} |\mu_n|(M_n)^2\|\sigma\|_{L^1[-1,1]} \le C(N+1)^{3/2}(0.386)^N\|\sigma\|_{L^1[-1,1]}. \tag{2.41}$$

Also, by Lemma 2.3.4 and the estimates (2.33) and (2.21),

$$\begin{aligned} B &\le \sum_{n=0}^{N}\left|\langle f,\bar{\phi}_n\rangle - \sum_{k=0}^{K-1} f(x_k)w_k\bar{\phi}_n(x_k)\right||\bar{\phi}_n(x)| \\ &\le 2\sqrt{2}|\mu_N|\sum_{n=0}^{N}\sqrt{n}\|\sigma\|_{L^1[-1,1]} \\ &\le C\frac{c^N(N+1)^{3/2}}{2^{2N}(N-1)!}\|\sigma\|_{L^1[-1,1]} \le C(N+1)^2(e/8)^N\|\sigma\|_{L^1[-1,1]} \end{aligned} \tag{2.42}$$

where we have used the Stirling inequality. Combining (2.40), (2.41), and (2.42) now gives the result. □

2.4 Spectral Elements for Differential Equations

Boyd [44] considered the possibility of improving the efficiency of spectral element methods by replacing the Legendre polynomials in such algorithms by prolates. The unevenly spaced collocation grids associated with the Legendre polynomials translate into suboptimal efficiency of Legendre-based spectral elements. For appropriate values of the band-limit parameter c, the spacing of the prolate grid points is more uniform, allowing for improved accuracy and longer time steps when prolate elements are combined with time marching.

2.4.1 Modal and Nodal Bases

In either the Legendre or prolate formulations, *nodal* or *cardinal* functions C_ℓ satisfy $C_\ell(x_k)=\delta_{k\ell}$. Here $\{x_k\}_{k=1}^N$ is the collection of collocation points. The basis-dependent information in an Nth order spectral method includes:

(i) N collocation points $\{x_k\}_{k=1}^N$;
(ii) N quadrature weights $\{w_k\}_{k=1}^N$;

(iii) matrices $\alpha_{k\ell} = \mathrm{d}C_\ell/\mathrm{d}x\big|_{x=x_k}$ (for first-order equations) and $\beta_{k\ell} = \mathrm{d}^2C_\ell/\mathrm{d}x^2\big|_{x=x_k}$ (for second-order equations).

Changing the basis from Legendre polynomials to prolates or vice versa is equivalent to changing the collocation points, quadrature weights, and the matrices $(\alpha_{k\ell})$ and $(\beta_{k\ell})$.

Quadrature weights and nodes for the Legendre polynomials are well known, but computing them in the prolate case is far more complicated and requires solving a system of nonlinear equations via Newton iteration. For small values of c, it is appropriate to initialize the iteration at the Legendre weights and nodes. However, when N and c are both large, these nodes are not sufficiently accurate for the iteration to converge.

It is desirable, for ease of implementation, that *cardinal* prolate functions be constructed, that is, linear combinations $C_\ell = \sum_{n=0}^{N-1} a_{\ell n}\bar{\phi}_n^{(c)}$ with the property that $C_\ell(x_k) = \delta_{k\ell}$ where $x_1, x_2, \dots, x_N$ are predetermined nodes. The collection $\{C_\ell\}_{\ell=1}^N$ is known as the *nodal basis*. Returning to (1.18), that is, $\bar{\phi}_n = \sum_{m=0}^{\infty} \beta_{nm}\bar{P}_m$, we find that the cardinality condition is equivalent to the matrix equation

$$A\Phi = I$$

where $A,\ \Phi \in M_{N,N}$ with $A_{jn} = a_{jn}$ the unknown coefficient matrix and $\Phi_{nk} = \bar{\phi}_n^{(c)}(x_k)$, so that $A = \Phi^{-1}$. Hence, simple matrix inversion performs the change of basis from the *modal* basis $\{\bar{\phi}_n\}_{n=0}^{N-1}$ to the nodal basis $\{C_\ell\}_{\ell=1}^N$. On the other hand, by evaluating (1.18) at the nodes $\{x_k\}_{k=1}^N$ we have

$$\Phi = B\mathscr{L}$$

with $B_{nj} = \beta_{nj}$ the coefficients in (1.18) and $\mathscr{L}_{jk} = \bar{P}_j(x_k)$. With $\Phi',\ \Phi'' \in M_{N\times N}$ and $\mathscr{L}',\ \mathscr{L}'' \in M_{N\times N}$ derivative matrices with entries $(\Phi')_{nk} = \mathrm{d}\bar{\phi}_n^{(c)}/\mathrm{d}x\big|_{x=x_k}$ and $(\Phi'')_{nk} = \mathrm{d}^2\bar{\phi}_n^{(c)}/\mathrm{d}x^2\big|_{x=x_k}$, $(\mathscr{L}')_{nk} = \mathrm{d}\bar{P}_n/\mathrm{d}x\big|_{x=x_k}$ and $(\mathscr{L}'')_{nk} = \mathrm{d}^2\bar{P}_n/\mathrm{d}x^2\big|_{x=x_k}$, we also have the relations

$$\Phi' = B\mathscr{L}'; \qquad \Phi'' = B\mathscr{L}''. \tag{2.43}$$

The matrices $\mathscr{L}'$ and $\mathscr{L}''$ are easy to compute given the Legendre recurrence relation (1.13) and its derivatives or, alternatively, the differential recurrence relation (1.16).

When solving second-order differential equations via the spectral method using cardinal prolates, we require the derivative matrices $\mathscr{C}'$ and $\mathscr{C}'' \in M_{N\times N}$ given by $(\mathscr{C}')_{\ell k} = \mathrm{d}C_\ell/\mathrm{d}x\big|_{x=x_k}$ and $(\mathscr{C}'')_{\ell k} = \mathrm{d}^2C_\ell/\mathrm{d}x^2\big|_{x=x_k}$. From (2.43),

$$\mathscr{C}' = A\Phi' = AB\mathscr{L}'; \quad \mathscr{C}'' = A\Phi'' = AB\mathscr{L}''.$$

The matrix Φ whose inverse A transforms the modal basis $\{\bar{\phi}_n\}_{n=0}^{N-1}$ to the nodal basis $\{C_j\}_{j=1}^N$ has condition number less than 10^{-4} for all c and $N \le 30$ [44], but becomes ill conditioned for larger values of N, in which case it is recommended that the modal basis be employed.

2.4.2 Computing Quadratures

A Gauss–Lobatto quadrature is a collection of nodes $\{x_k\}_{k=1}^N$ and weights $\{w_k\}_{k=1}^N$ such that $\int_{-1}^1 p(x)\,\mathrm{d}x = \sum_{k=1}^N w_k\, p(x_k)$ for all polynomials p of degree up to $2N-3$. Here $x_1 = -1$, $x_N = 1$ and for $2 \le i \le N-1$, x_k is the $(i-1)$th root of P'_{N-1} while $w_k = 2/(N(N-1)P_{N-1}(x_k)^2)$ for $1 \le k \le N$. We generalize this notion as follows.

Definition 2.4.1. An N-point *generalized Gauss–Lobatto quadrature* (GGLQ) for a set of $2N-2$ basis functions $\{\varphi_n\}_{n=0}^{2N-3}$ defined on $[-1,1]$ is a collection of nodes $\{x_k\}_{k=1}^N$ and weights $\{w_k\}_{k=1}^N$ with $x_1 = -1$, $x_N = 1$ and $\int_{-1}^1 \varphi_n(x)\,\mathrm{d}x = \sum_{k=1}^n w_k \varphi_n(x_k)$ for $0 \le n \le 2N-3$.

An N-point GGLQ for the prolates therefore has the property that with β_{nj} the Legendre coefficients of $\bar{\phi}_n$ as in (1.18), the $2N-2$ residuals

$$\rho_n = \int_{-1}^1 \bar{\phi}_n(x)\,\mathrm{d}x - \sum_{k=1}^N w_k \bar{\phi}_n(x_k) = \sqrt{2}\beta_{n0} - \sum_{k=1}^N w_k \bar{\phi}_n(x_k) = 0$$

for $0 \le n \le 2N-3$. Since each ρ_n is a nonlinear function of the nodes $\{x_k\}_{k=1}^N$, the solution of these simultaneous equations must be approximated via Newton iteration.

We now assume that the nodes x_k are distributed symmetrically about the origin and the weights are symmetric, that is, if N is even then $x = 0$ is not a node and $x_k = x_{N-k+1}$ and $w_k = w_{N-k+1}$ for $1 \le i \le N/2$, while if N is odd then $x_{(N+1)/2} = 0$, $x_k = x_{N-k+1}$, $w_k = w_{N-k+1}$ for $1 \le i \le (N-1)/2$. Under these assumptions, the observation that $\bar{\phi}_{2j+1}^{(c)}$ is an odd function for all j implies that the odd residuals $\rho_{2j+1} = 0$. Consequently, we need only insist on the vanishing of the $N-1$ even residuals $r_j = \rho_{2j}$ $(0 \le j \le N-2)$.

If N is even, the unknowns are the $N/2-1$ nodes $x_{(N+2)/2}, \dots, x_{N-1}$ and the $(N+1)/2$ weights $w_{(N+2)/2}, \dots, w_N$ while, if N is odd, the unknowns are the $(N-3)/2$ nodes $x_{(N+3)/2}, \dots, x_{N-1}$ and the $(N+1)/2$ weights $w_{(N+1)/2}, \dots, w_N$. In either case there are $N-1$ unknowns to be determined. The number of nodes to be determined is

$$M = \frac{1}{4}(2N-5+(-1)^N) = \begin{cases} (N-2)/2 & \text{if } N \text{ is even} \\ (N-3)/2 & \text{if } N \text{ is odd.} \end{cases}$$

Let $\mathbf{u} = (u_0, u_1, \dots, u_{N-2}) \in \mathbb{R}^{N-1}$ be the vector of unknowns

$$\mathbf{u} = \begin{cases} (x_{N-M}, x_{N-M+1}, \dots, x_{N-1}, w_{N-M}, w_{N-M+1}, \dots, w_N) & \text{if } N \text{ is even} \\ (x_{N-M}, x_{N-M+1}, \dots, x_{N-1}, w_{N-M-1}, w_{N-M}, \dots, w_N) & \text{if } N \text{ is odd,} \end{cases}$$

and $\mathbf{r} = \mathbf{r}(\mathbf{u})$ be the residual vector $\mathbf{r}(\mathbf{u}) = (r_0(\mathbf{u}), r_1(\mathbf{u}), \ldots, r_{N-2}(\mathbf{u}))$. Thus we have a mapping $\mathbf{r} : \mathbb{R}^{N-1} \to \mathbb{R}^{N-1}$. Given an initial estimate $\mathbf{u}^{(0)}$ for the unknown weights and nodes, we expand $\mathbf{r}$ as a multivariate Taylor series about $\mathbf{u}^{(0)}$, truncate the series at the linear term, and solve the resulting matrix equation to obtain a second estimate $\mathbf{u}^{(1)}$. This gives the Newton iteration:

$$J(\mathbf{r}(\mathbf{u}^{(n)}))\mathbf{u}^{(n+1)} = -\mathbf{r}(\mathbf{u}^{(n)}) + J(\mathbf{r}(\mathbf{u}^{(n)}))\mathbf{u}^{(n)} \tag{2.44}$$

with $\mathbf{u}^{(n)}$ the nth iterate and $J(\mathbf{r}(\mathbf{u}))$ the $(N-1) \times (N-1)$ Jacobian matrix with (j,k)th entry $[J(\mathbf{r}(\mathbf{u}))]_{jk} = \dfrac{\partial r_j}{\partial u_k}(\mathbf{u})$ $(0 \le j,k \le N-2)$. Because of the symmetry assumptions, each residual r_j may be written as

$$r_j = \sqrt{2}\beta_{2j,0} - 2\sum_{k=N-M}^{N} w_k \bar{\phi}_{2j}(x_k) + s\, w_{N-M-1}\bar{\phi}_{2j}(0)$$

with $s = (1 + (-1)^{N+1})/2$, so that the entries of the Jacobian matrix become

$$[J(\mathbf{u})]_{jk} = \frac{\partial r_j}{\partial u_k} = \begin{cases} -2w_{N-M+k}(\bar{\phi}_{2j})'(x_{N-M+k}), & \text{if } 0 \le k \le M-1 \\ -(2-s)\bar{\phi}_{2j}(x_{N-M-s}), & \text{if } k = M \\ -2\bar{\phi}_{2j}(x_{N-2M-s+k}), & \text{if } M+1 \le k \le N-2. \end{cases}$$

In [44], Boyd describes a variation of this procedure known as *underrelaxation*, which has been shown to expand the domain of convergence, that is, the set of initializations $\mathbf{u}^{(0)}$ for which the algorithm will converge. The following comments summarize the discussion in [44].

When c is small, the Legendre–Lobatto nodes and weights form a good initialization for the iteration. However, in this situation it is usually sufficient to start with more simply defined nodes and weights $\tilde{x}_k$, $\tilde{w}_k$, $k = 1, \ldots, N$ given by

$$\tilde{x}_k = -\cos\left(\frac{\pi(k-1)}{N-1}\right); \quad \tilde{w}_k = \begin{cases} \dfrac{\pi}{N}\sin\left(\dfrac{\pi(k-1)}{N-1}\right), & (1 < k < N) \\ \dfrac{2}{N(N-1)}, & k = 1 \text{ or } N. \end{cases}$$

With $c^*(N) = (N + 1/2)\pi/2$ the *transition bandwidth*, for $N \le 12$, the above definitions give satisfactory initializations in the range $0 \le c \le c^*(N)$. For $N > 12$, the iteration will converge on a smaller range of values of c (see [44]). For large N and c near $c^*(N)$, a better initialization is required. In [44] it is shown that the variation of nodes with c is well approximated by a parabola symmetric about $c = 0$ for all $0 \le c \le c^*(N)$. This suggests that if $\mu/c^*(N)$ is small enough so that the Legendre–Lobatto points provide a successful initialization of the Newton iteration for the prolates $\{\bar{\phi}_n^{(\mu)}\}$, then a good initialization for a Newton iteration for the prolates $\{\bar{\phi}_n^{(c)}\}$ with $c > \mu$ can be obtained by the quadratic extrapolation $x_k(c) \approx x_k(0) + (x_k(\mu) - x_k(0))^2 c^2/\mu^2$.

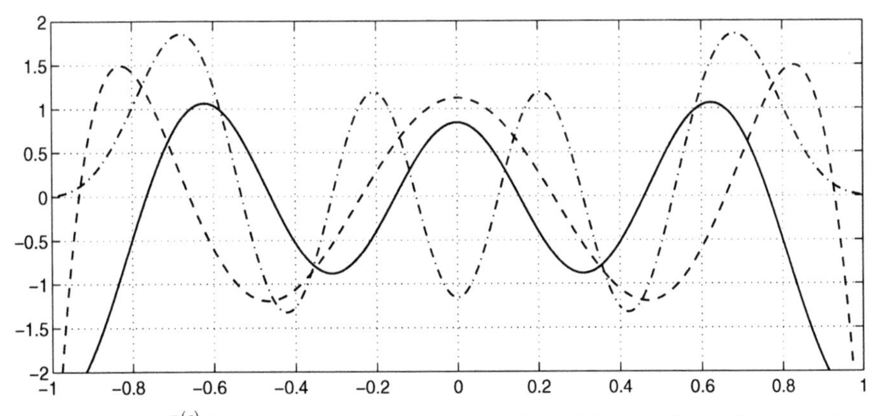

Fig. 2.1 Plots of $\bar{\phi}_6^{(c)}$ for $c = 1$ (dashed), $c = 20$ (dots) and $c = c^*(6) \approx 10.2$ (solid), the transition bandwidth. The zeros are evenly distributed for $c = c^*(6)$

An explanation of the more uniform distribution of the zeros of the prolates relative to that of the Legendre polynomials was given by Boyd in [43], where he observed that if $c < c^*(n)$ then for all $x \in [-1, 1]$,

$$\bar{\phi}_n^{(c)}(x) \sim \frac{\sqrt{\mathscr{E}(x;m)}}{(1-x^2)^{1/4}(1-mx^2)^{1/4}} J_0\left(\frac{c}{\sqrt{m}}\mathscr{E}(x;m)\right)$$

where J_0 is the Bessel function of order zero, $m = c^2/\chi_n$ with χ_n the nth eigenvalue of the prolate operator $\mathscr{P}$ as in (1.6), and $\mathscr{E}(x;m) = \int_x^1 (\sqrt{1-mt^2}/\sqrt{1-t^2})\,\mathrm{d}t$ is the difference of the complete and incomplete elliptic integrals of the second kind. Outside small neighborhoods of ± 1 of width $O(c^{-1/2})$, J_0 can be approximated to obtain

$$\bar{\phi}_n^{(c)}(x) \sim \sqrt{\frac{2}{\pi\chi_n}} \frac{m^{1/4}}{(1-x^2)^{1/4}(1-mx^2)^{1/4}} \cos(cm^{-1/2}\mathscr{E} - \pi/4).$$

Since $\chi_n(c^*(n)) = c^*(n)^2(1+O(n^{-2}))$, we have

$$\bar{\phi}_n^{c^*(n)}(x) \sim \sqrt{\frac{2}{\pi}\frac{1}{c}}(1-x^2)^{-1/2}\cos(n\pi(1-x)/2),$$

from which we see that when $c \sim c^*(n)$, $\bar{\phi}_n^{(c)}$ oscillates like $\cos(n\pi(1-x)/2)$ so that the zeros of these prolates are nearly uniformly distributed.

A plot of the prolates $\bar{\phi}_6^{(c)}$ for $c = 1, 20$ and the transition bandwidth $c = c^*(6) = \pi(6+1/2)/2 \approx 10.2$ appears in Fig. 2.1. Observe the more uniform arrangement of the zeros of $\bar{\phi}_6^{(c^*)}$ over those of $\bar{\phi}_6^{(1)}$ and $\bar{\phi}_6^{(20)}$.

2.4.3 Convergence and Continuously Adjusted Spectral Parameters

Given a series $S = \sum_{n=0}^{\infty} f_n$ with $\{f_n\}_{n=0}^{\infty}$ functions defined on $[a,b]$, we say that the error in the sequence of partial sums $S_N = \sum_{n=0}^{N} f_n$ has asymptotic rate of geometric convergence $\mu > 0$ if

$$\overline{\lim}_{N\to\infty} \frac{\log E(N)}{N} = -\mu$$

where $E(N) = \|S - S_N\|_{L^\infty[a,b]}$. Series with $\|S - S_N\|_{L^\infty[a,b]} \le A(N)\mathrm{e}^{-N\mu}$ with A an algebraic function of N are examples of such series.

When the basis functions contain an internal parameter (such as the band limit parameter c of the prolate basis $\{\bar{\phi}_n^{(c)}\}_{n=0}^{\infty}$) it is useful to generalize the notion of convergence to allow for variations of this internal parameter.

Definition 2.4.2 (Continuously adjusted asymptotic rate of geometric convergence). Suppose that for each τ in an interval I, the collection $\{\varphi_n^{(\tau)}\}_{n=1}^{\infty}$ is an orthonormal basis for $L^2[a,b]$ and for each integer $n \ge 1$, the mapping $\tau \mapsto \varphi_n^{(\tau)}$ from I to $L^2[a,b]$ is continuous. Let $c : \mathbb{N} \to I$. Given $f \in L^2[a,b]$ and an integer $N \ge 1$, let

$$S_N = S_N^{(c)} = \sum_{n=1}^{N} \langle f, \varphi_n^{(c(N))} \rangle \varphi_n^{(c(N))}$$

and $E^{(c)}(N) = \|S - S_N^{(c)}\|_{L^\infty[a,b]}$. We say that the partial sums $\{S_N^{(c)}\}_{N=1}^{\infty}$ have continuously adjusted asymptotic rate of geometric convergence ν if

$$\overline{\lim}_{N\to\infty} \frac{\log E^{(c)}(N)}{N} = -\nu.$$

In [43], Boyd investigated a two-dimensional array of errors $E^{(c)}(N)$ for prolate expansions of particular functions f as N and c vary and argued that c should be chosen smaller than $c^*(N)$ for accurate representation.

2.5 Prolate Expansions of Non-band-limited Functions

The key to approximating non-band-limited functions by prolate series is first to approximate them by band-limited functions. The following result from [41] deals with approximations of analytic functions by band-limited functions.

Theorem 2.5.1 (Constructive band-limited approximation). *Suppose f is analytic in the rectangle $[-1-\delta, 1+\delta] \times [-\delta, \delta] \subset \mathbb{C}$ for some $\delta > 0$. Let $c > 0$ and $f^{(c)}$ be the function on the line defined by*

$$f^{(c)}(x) = \frac{1}{2\pi} \int_{-c}^{c} \left(\int_{-\infty}^{\infty} f(t) b(t) \mathrm{e}^{-\mathrm{i}t\xi}\, \mathrm{d}t \right) \mathrm{e}^{\mathrm{i}\xi x}\, \mathrm{d}\xi \tag{2.45}$$

where $b(x) \geq 0$ is C^∞ and satisfies $b(x) = 1$ if $|x| \leq 1$ and $b(x) = 0$ if $|x| > 1+\delta$. Then

(i) $f^{(c)}$ is band limited with band limit c, and
(ii) for all integers $n > 0$, there exists a constant $A_n > 0$ such that

$$\max_{|x|\leq 1} |f(x) - f^{(c)}(x)| \leq A_n c^{-n}.$$

Proof. Claim (i) is clear from the definition (2.45). To see (ii), note that $f^{(c)}(x) = (\mathbb{1}_{[-c,c]}\widehat{fb}(\cdot/2\pi))^\vee(x/2\pi)/(2\pi)$ and $|x| < 1$. Then

$$\begin{aligned}
|f(x) - f^{(c)}(x)| &= |f(x)b(x) - f^{(c)}(x)| \\
&\leq \int_{-\infty}^{\infty} |\widehat{fb}(\xi) - \widehat{f^{(c)}}(\xi)|\,\mathrm{d}\xi \\
&= \int_{-\infty}^{\infty} |\widehat{bf}(\xi) - \mathbb{1}_{[-\frac{c}{2\pi},\frac{c}{2\pi}]}(\xi)\widehat{bf}(\xi)|\,\mathrm{d}\xi = \int_{|\xi|>\frac{c}{2\pi}} |\widehat{bf}(\xi)|\,\mathrm{d}\xi.
\end{aligned}$$

But $f \in C^\infty$, so $bf \in C_c^\infty \subset \mathcal{S}(\mathbb{R})$, the Schwartz space of rapidly decreasing functions. Since the Fourier transform preserves $\mathcal{S}$, $\widehat{bf} \in \mathcal{S}$ and for all $n \geq 0$ there exists $B_n > 0$ such that $|\widehat{bf}(\xi)| \leq B_n |\xi|^{-n-1}$. Hence $|f(x) - f^{(c)}(x)| \leq \int_{|\xi|>c/2\pi} B_n |\xi|^{-n-1}\,\mathrm{d}\xi = A_n c^{-n}$. □

If f is analytic on the rectangle $[-1-\delta, 1+\delta] \times [-\delta, \delta] \subset \mathbb{C}$ for some $\delta > 0$ then the Legendre and prolate series for f on $[-1,1]$ converge geometrically. In fact, when $f \in H^s[-1,1]$ (the Sobolev space of functions in $L^2[-1,1]$ with square-integrable distributional derivatives up to order s) we have the following variation on Theorem 3.1 of [69].

Theorem 2.5.2. *Let $f \in H^s[-1,1]$, b_n $(n \geq 0)$ be as in (2.32), $\{\chi_n\}_{n=0}^\infty$ be the corresponding eigenvalue of the differential operator $\mathscr{P}$ in (1.6), and suppose $\chi_N \geq 400$. Then there exists a positive constant D such that*

$$|b_N| \leq D\left[\left(\frac{\chi_N}{16}\right)^{-s/2} \|f\|_{H^s[-1,1]} + \left(\frac{(3.75)c}{\sqrt{\chi_N}}\right)^{\sqrt{\chi_N}/4} \|f\|_{L^2[-1,1]}\right]. \tag{2.46}$$

Since $\chi_n \sim n(n+1)[1+O(c^2/2n^2)]$ (see [43]), (2.46) forces the rate of decay of the prolate coefficients of $f \in H^s[-1,1]$ to be geometric. However, (2.46) also forces the rate of decay of the prolate coefficients of $f \in C^\infty[-1,1]$ (in particular for band-limited f) to be roughly of the order of n^{-n} once $\chi_N \geq (3.75)c$.

The result of Chen, Gottlieb, and Hesthaven [69] gives decay of the form

$$|b_N| \leq C(N^{-2s/3}\|f\|_{H^s[-1,1]} + (c/\sqrt{\chi_N})^{\delta N}\|f\|_{L^2[-1,1]})$$

with $\delta > 0$, while Wang and Zhang [349] work in a different Sobolev-type space, showing that

$$|b_N| \le C(N^{-s}\|(1-x^2)^{s/2}D_x^s f\|_{L^2[-1,1]} + (2^{1/6}c/\sqrt{\chi_N})^{\delta N}\|f\|_{L^2[-1,1]}).$$

The proof of Theorem 2.5.2 can be reduced to the following lemmas, whose proofs may be found in Sect. 2.6.4.

Lemma 2.5.3. *Suppose $\bar{\phi}_N^{(c)}$ has the Legendre expansion $\bar{\phi}_N^{(c)} = \sum_{k=0}^{\infty} \beta_k \bar{P}_k$ and $q = q_N = c/\sqrt{\chi_N} < 1$. Let $m = [\sqrt{\chi_N}/4]$ be the largest integer less than or equal to $\sqrt{\chi_N}/4$. Then there exists a positive constant D (independent of k and m) such that for all $k \le 2m$ and with $\alpha = 2.07$,*

$$|\beta_k| \le \begin{cases} D(\alpha/q)^k|\beta_0| & \text{if } k \text{ is even} \\ D(\alpha/q)^k|\beta_1| & \text{if } k \text{ is odd.} \end{cases} \tag{2.47}$$

Let $A_k = \int_{-1}^{1} x^k \bar{\phi}_N^{(c)}(x)\,\mathrm{d}x$. Then $A_0 = \sqrt{2}\beta_0$ and $A_1 = \sqrt{2/3}\beta_1$.

Lemma 2.5.4. *Suppose $\chi_N \ge 400$. Let $m = [\sqrt{\chi_N}/4]$ and $q = c/\sqrt{\chi_N} < 1$. Then there exists a constant C independent of m for which*

$$|A_0| \le Cq^{2m-2}\gamma^m \frac{(3/7)^{\sqrt{\chi_N}/2}}{\sqrt{2m-3/2}} \quad \text{and} \quad |A_1| \le Cq^{2m-2}\gamma^m \frac{(3/7)^{\sqrt{\chi_N}/2}}{\sqrt{2m-1/2}} \tag{2.48}$$

with $\gamma = 4\mathrm{e}^2/3$.

Proof (of Theorem 2.5.2). Let f have Legendre expansion $f = \sum_{k=0}^{\infty} a_k \bar{P}_k$ and prolate expansion (2.32). Then $b_N = \int_{-1}^{1} f(x)\bar{\phi}_N^{(c)}(x)\,\mathrm{d}x = \sum_{k=0}^{\infty} a_k\beta_k$. With $m = [\sqrt{\chi_N}/4]$, let $f_m = \sum_{k=0}^{m} a_k \bar{P}_k$ be the mth partial sum of the Legendre expansion of f. Then

$$b_N = \int_{-1}^{1} f_m(x)\bar{\phi}_N^{(c)}(x)\,\mathrm{d}x + \int_{-1}^{1} [f(x) - f_m(x)]\bar{\phi}_N^{(c)}(x)\,\mathrm{d}x = I + II. \tag{2.49}$$

The truncation error in the Legendre expansion of f satisfies $\|f - f_m\|_{L^2[-1,1]} \le Dm^{-s}\|f\|_{H^s[-1,1]}$ (see [105], [61]), so with an application of the Cauchy–Schwarz inequality, the second integral in (2.49) is bounded by

$$|II| \le \|f - f_m\|_{L^2[-1,1]} \le \frac{D}{m^s}\|f\|_{H^s[-1,1]} \le D\left(\frac{16}{\chi_N}\right)^{s/2}\|f\|_{H^s[-1,1]}. \tag{2.50}$$

Let the constants α and γ be as in Lemma 2.5.3 and Lemma 2.5.4. Application of Cauchy–Schwarz and Lemma 2.5.3 show that the first integral in (2.49) is bounded by

$$|I| = \left|\sum_{k=0}^{m} a_k\beta_k\right| \leq \left(\sum_{k=0}^{m}|a_k|^2\right)^{1/2}\left(\sum_{k=0}^{m}|\beta_k|^2\right)^{1/2}$$
$$\leq D\|f\|_{L^2[-1,1]}\left(\sum_{k=0}^{m}\left(\frac{\alpha}{q}\right)^{2k}\right)^{1/2}\max\{|\beta_0|,|\beta_1|\}$$
$$\leq D\|f\|_{L^2[-1,1]}\left(\frac{\alpha}{q}\right)^{m+1}\max\{|\beta_0|,|\beta_1|\}.$$

Since $\beta_0 = A_0/\sqrt{2}$ and $\beta_1 = \sqrt{3/2}A_1$, Lemma 2.5.4 gives

$$|I| \leq D\|f\|_{L^2[-1,1]}\alpha^{m+1}\gamma^m q^{m-3}\frac{(3/7)^{\sqrt{\chi_N}/2}}{\sqrt{2m-1/2}}. \tag{2.51}$$

Since $\sqrt{\chi_N}/4 - 1 < m \leq \sqrt{\chi_N}/4$ and $\alpha\gamma(3/7)^2 \leq 3.75$, combining (2.49), (2.50), and (2.51) now gives the result. □

In [43], Boyd examined numerically the rate of convergence of prolate series of representative proptotype functions. For fixed $\xi \in \mathbb{R}$, the coefficients of the function $e^{-ix\xi}$ in the Chebyshev, Legendre, and prolate series respectively are

$$b_n^{\mathrm{Cheb}}(\xi) = \int_{-1}^{1} T_n(x)(1-x^2)^{-1/2}e^{-ix\xi}\,dx = \pi i^n J_n(-\xi),$$
$$b_n^{\mathrm{Leg}}(\xi) = \int_{-1}^{1} \bar{P}_n(x)e^{-ix\xi}\,dx = \frac{\pi^{3/2}}{2}i^n\left(\frac{2}{\xi}\right)^{1/2}(n+1/2)J_{n+1/2}(\xi),$$
$$b_n^{\mathrm{pro}}(\xi) = \int_{-1}^{1} \bar{\phi}_n^{(c)}(x)e^{-ix\xi}\,dx = \mu_n\bar{\phi}_n^{(c)}(\xi/c).$$

Coefficients of the Chebyshev and Legendre series can be analyzed through well-known asymptotics of Bessel functions of integer and half-integer order respectively. These oscillate until $n \approx \xi$ and then fall sharply as $n \to \infty$. The prolate eigenvalues μ_n are approximately $\sqrt{2\pi/c}$ for $n < 2c/\pi$ and decay super-exponentially as $n \to \infty$. This leads to the following heuristic:

In the expansions of $e^{-ix\xi}$ ($\xi \in \mathbb{R}$ fixed and $|x| < 1$):

(i) Accurate truncations of Chebyshev and Legendre series require $N \geq \xi$;
(ii) Accurate truncations of prolate series with bandwidth parameter $c = \xi$ require $N \geq 2\xi/\pi$.

Hence, one can approximate a sinusoidal signal accurately using $2/\pi$ as many prolates as with either Legendre or Chebyshev polynomial bases.

We have seen that expansions of complex exponentials lead to expansions of band-limited functions and that by approximating analytic functions by band-limited functions we may write expansions which accurately approximate analytic functions. Boyd [43] quantified the superior performance of prolate expansions over Chebyshev and Legendre expansions, at least in the case of band-limited functions in the following terms.

Let $f \in L^2(\mathbb{R})$ have Fourier transform $\hat{f}$ and $\{g_n\}_{n=1}^{\infty}$ be either the Chebyshev, Legendre, or prolate orthonormal basis for $L^2[-1,1]$. By the Plancherel theorem, the nth coefficient a_n of the series $f(x) = \sum_{n=0}^{\infty} a_n g_n(x)$ ($|x| \leq 1$) may be written as $a_n = \int_{-\infty}^{\infty} \hat{f}(\xi) b_n(\xi/2\pi)\, \mathrm{d}\xi$ with $b_n(\xi) = b_n^{\mathrm{Cheb}}(\xi)$ or $b_n^{\mathrm{Leg}}(\xi)$ or $b_n^{\mathrm{pro}}(\xi)$ the Chebyshev, Legendre, or prolate coefficients respectively of the function $\mathrm{e}^{-\mathrm{i}x\xi}$ (ξ fixed) as above. If f is band limited with band limit c, then

(i) The Chebyshev and Legendre coefficients of f fall super-geometrically for all $|\xi| < c$ when $n > c$, and
(ii) The prolate coefficients with respect to $\{\bar{\phi}_n^{(c)}\}_{n=0}^{\infty}$ fall super-exponentially for $n > 2c/\pi$.

2.6 Notes and Auxiliary Results

2.6.1 Miscellaneous Results on Prolates

Derivative Matrices for Spectral Elements

In Sect. 2.4, formulas are given for first and second derivative matrices for prolate series and cardinal prolate series. This is achieved through prior knowledge of the equivalent matrices for Legendre polynomials and the Legendre expansion of the prolates. However, the derivative matrices for prolate series can also be computed directly without direct reference to the Legendre basis, as the following series of results from [364] shows.

Theorem 2.6.1. *Let $c > 0$ and let m,n be non-negative integers. Then with $\bar{\phi}_n = \bar{\phi}_n^{(c)}$,*

$$\int_{-1}^{1} \bar{\phi}_n'(x)\bar{\phi}_m(x) = \begin{cases} 0 & \text{if } m \equiv n \mod 2 \\ \dfrac{2\mu_m^2}{\mu_m^2 + \mu_n^2}\bar{\phi}_m(1)\bar{\phi}_n(1) & \text{if } m \not\equiv n \mod 2 \end{cases} \tag{2.52}$$

and

$$\int_{-1}^{1} x\bar{\phi}_m(x)\bar{\phi}_n(x)\,\mathrm{d}x = \begin{cases} 0 & \text{if } m \equiv n \mod 2 \\ \dfrac{2}{\mathrm{i}c}\dfrac{\mu_m\mu_n}{\mu_m^2 + \mu_n^2}\bar{\phi}_m(1)\bar{\phi}_n(1) & \text{if } m \not\equiv n \mod 2. \end{cases} \tag{2.53}$$

Proof. Since $\bar{\phi}_n$ is even if n is even and odd if n is odd, the integrals $\int_{-1}^{1} \bar{\phi}_m'(x)\bar{\phi}_n(x)\,\mathrm{d}x$ and $\int_{-1}^{1} x\bar{\phi}_m(x)\bar{\phi}_n(x)\,\mathrm{d}x$ are zero if $m \equiv n \mod 2$ since the integrands are odd. Suppose then that $m \not\equiv n \mod 2$. Differentiating (2.17) with respect to x gives $\mu_n\bar{\phi}_n'(x) = \mathrm{i}c\int_{-1}^{1} t\mathrm{e}^{\mathrm{i}cxt}\bar{\phi}_n(t)\,\mathrm{d}t$. Multiplying both sides of this equation by $\bar{\phi}_m(x)$ and integrating with respect to x over $[-1,1]$ gives

$$\begin{aligned} \mu_n \int_{-1}^{1} \bar{\phi}_n'(x)\,\bar{\phi}_m(x)\,\mathrm{d}x &= \mathrm{i}c\int_{-1}^{1} \bar{\phi}_m(x)\int_{-1}^{1} t\,\mathrm{e}^{\mathrm{i}cxt}\,\bar{\phi}_n(t)\,\mathrm{d}t\,\mathrm{d}x \\ &= \mathrm{i}c\int_{-1}^{1} t\,\bar{\phi}_n(t)\int_{-1}^{1} \mathrm{e}^{\mathrm{i}cxt}\,\bar{\phi}_m(x)\,\mathrm{d}x\,\mathrm{d}t = \mathrm{i}c\mu_m\int_{-1}^{1} t\,\bar{\phi}_n(t)\,\bar{\phi}_m(t)\,\mathrm{d}t. \end{aligned} \tag{2.54}$$

However, by interchanging the roles of m and n in (2.54) we obtain

$$\mu_m \int_{-1}^{1} \bar{\phi}_m'(x)\,\bar{\phi}_n(x)\,\mathrm{d}x = \mathrm{i}c\mu_n \int_{-1}^{1} t\,\bar{\phi}_n(t)\,\bar{\phi}_m(t)\,\mathrm{d}t \tag{2.55}$$

and, since the integrals on the right-hand sides of (2.54) and (2.55) are the same, an integration by parts yields

$$\begin{aligned}\int_{-1}^{1} \bar{\phi}_m'(x)\ \bar{\phi}_n(x)\,\mathrm{d}x &= \frac{\mu_n^2}{\mu_m^2}\int_{-1}^{1} \bar{\phi}_m(x)\bar{\phi}_n'(x)\,\mathrm{d}x \\ &= \frac{\mu_n^2}{\mu_m^2}\left(\bar{\phi}_m(1)\bar{\phi}_n(1) - \bar{\phi}_m(-1)\bar{\phi}_n(-1) - \int_{-1}^{1}\bar{\phi}_m'(x)\bar{\phi}_n(x)\,\mathrm{d}x\right) \\ &= \frac{\mu_n^2}{\mu_m^2}\left(2\bar{\phi}_m(1)\bar{\phi}_n(1) - \int_{-1}^{1}\bar{\phi}_m'(x)\bar{\phi}_n(x)\,\mathrm{d}x\right)\end{aligned} \tag{2.56}$$

since $m \not\equiv n \mod 2$, and thus $\bar{\phi}_m(-1)\bar{\phi}_n(-1) = -\bar{\phi}_m(1)\bar{\phi}_n(1)$. Rearranging (2.56) gives (2.52). Combining (2.52) and (2.54) gives (2.53). □

It is clear from symmetry conditions that if $m \not\equiv n \mod 2$ then

$$\begin{aligned}\int_{-1}^{1} x\bar{\phi}_n'(x)\bar{\phi}_m(x)\,\mathrm{d}x &= \int_{-1}^{1} x^2\bar{\phi}_n''(x)\bar{\phi}_m(x)\,\mathrm{d}x = \int_{-1}^{1} x^2\bar{\phi}_n'(x)\bar{\phi}_m'(x)\,\mathrm{d}x \\ &= \int_{-1}^{1} \bar{\phi}_n''(x)\bar{\phi}_m(x)\,\mathrm{d}x = \int_{-1}^{1} x^2\bar{\phi}_n(x)\bar{\phi}_m(x)\,\mathrm{d}x = 0.\end{aligned}$$

When $m \equiv n \mod 2$, the (nonzero) values of the above integrals are computed in [364] in terms of the values of $\bar{\phi}_n$, $\bar{\phi}_m$ and various derivatives of these functions at $x = 1$, as well as the eigenvalues $\mu_n, \mu_m, \chi_n, \chi_m$. Here is a typical example.

Theorem 2.6.2. *Let $c > 0$ and let m,n be non-negative integers. Then with $\bar{\phi}_n = \bar{\phi}_n^{(c)}$, if $m \equiv n$ mod 2, then*

$$\begin{aligned}\int_{-1}^{1} x^2\bar{\phi}_n''(x)\bar{\phi}_m(x)\,\mathrm{d}x &= \frac{2\mu_n(\bar{\phi}_n'(1)\bar{\phi}_m(1) - \bar{\phi}_m'(1)\bar{\phi}_n(1))}{\mu_m - \mu_n} - \frac{4\mu_n\bar{\phi}_n(1)\bar{\phi}_m(1)}{\mu_m + \mu_n} \\ &= \frac{\mu_n(\chi_n - \chi_m)\bar{\phi}_n(1)\bar{\phi}_m(1)}{\mu_m - \mu_n} - \frac{4\mu_n\bar{\phi}_n(1)\bar{\phi}_m(1)}{\mu_m + \mu_n}.\end{aligned}$$

Numerical Evaluation of μ_n

The methods of Bouwkamp and Boyd for the evaluation of prolates as described in Sect. 1.2.4 also produce the associated eigenvalues $\{\chi_n\}_{n=0}^{\infty}$ of the differential operator $\mathscr{P}$ of (1.6). From this information we may apply (2.17) to estimate the eigenvalues $\{\mu_n\}_{n=0}^{\infty}$ of the integral operator F_c defined in Corollary 1.2.6 and hence the eigenvalues $\{\lambda_n\}_{n=0}^{\infty}$ of the iterated projections $P_{2c/\pi}Q$ of Sect. 1.2. To see this, observe that putting $n = 0$ and $x = 0$ in (2.17) gives

$$\mu_0\bar{\phi}_0(0) = \int_{-1}^{1} \bar{\phi}_0(x)\,\mathrm{d}x.$$

Given that the values of $\bar{\phi}_0$ have already been computed accurately by the Bouwkamp or Boyd methods and that $\bar{\phi}_0(0)$ is not small, estimating the integral on the right-hand side numerically gives μ_0. Putting $n = 0$, $m = 1$ in (2.53) and numerically computing the integral on the left-hand side of that equation now gives μ_1. For $m \geq 2$ odd, μ_m may be obtained from (2.53) with $n = 0$, while if $m \geq 2$ is even, we obtain μ_m by applying (2.53) with $n = 1$.

Lemma 2.6.3. *For all non-negative integers n,*

$$\int_{-1}^{1} t^n P_n(t)\,\mathrm{d}t = \frac{2^{n+1}(n!)^2}{(2n+1)!} = \frac{\sqrt{\pi}n!}{2^n\Gamma(n+3/2)}. \tag{2.57}$$

The proof is by induction, the inductive step being provided by the recurrence relation (1.13) and the fact that $\int_{-1}^{1} t^m P_n(t)\,\mathrm{d}t = 0$ for all integers $0 \le m < n$.

Decay of Eigenvalues μ_n

The following lemma appears in Fuchs's paper [103].

Lemma 2.6.4. *The eigenvalues $\mu_n = \mu_n(2c/\pi)$ associated with the eigenfunctions $\bar{\phi}_n^{(c)}$ as in (2.17) satisfy the differential equation*

$$\frac{\mathrm{d}}{\mathrm{d}c}\mu_n(2c/\pi) = \frac{\mu_n(2c/\pi)}{2c}[2\bar{\phi}_n^{(c)}(1)^2 - 1]. \tag{2.58}$$

Proof. Let $\varphi_n^{(c)}(t) = c^{-1/4}\bar{\phi}_n^{(c)}(t/\sqrt{c})$ $(n \ge 0)$. Then

$$\int_{-\sqrt{c}}^{\sqrt{c}} \varphi_n^{(c)}(t)\varphi_m^{(c)}(t)\,\mathrm{d}t = \int_{-1}^{1} \bar{\phi}_n^{(c)}(t)\bar{\phi}_m^{(c)}(t)\,\mathrm{d}t = \delta_{mn},$$

and, because of (2.17),

$$\int_{-\sqrt{c}}^{\sqrt{c}} \mathrm{e}^{\mathrm{i}st}\varphi_n^{(c)}(s)\,\mathrm{d}s = \mu_n(2c/\pi)\sqrt{c}\,\varphi_n^{(c)}(t). \tag{2.59}$$

Multiplying both sides of (2.59) by $\varphi_n^{(b)}(t)$ (with $b > c$) and integrating both sides from $-\sqrt{b}$ to $\sqrt{b}$ gives

$$\begin{aligned}\sqrt{c}\mu_n(2c/\pi)\int_{-\sqrt{b}}^{\sqrt{b}} \varphi_n^{(c)}(t)\varphi_n^{(b)}(t)\,\mathrm{d}t &= \int_{-\sqrt{c}}^{\sqrt{c}} \varphi_n^{(c)}(s)\int_{-\sqrt{b}}^{\sqrt{b}} \varphi_n^{(b)}(t)\mathrm{e}^{\mathrm{i}st}\,\mathrm{d}t\,\mathrm{d}s\\ &= \sqrt{b}\mu_n(2b/\pi)\int_{-\sqrt{c}}^{\sqrt{c}} \varphi_n^{(c)}(s)\varphi_n^{(b)}(s)\,\mathrm{d}s. \end{aligned} \tag{2.60}$$

Since $b > c$ and the integrands $\varphi_n^{(c)}(s)\varphi_n^{(b)}(s)$ are even functions for all n, (2.60) may be rewritten as

$$(\sqrt{b}\mu_n(2b/\pi) - \sqrt{c}\mu_n(2c/\pi))\int_{-\sqrt{c}}^{\sqrt{c}} \varphi_n^{(c)}(t)\varphi_n^{(b)}(t)\,\mathrm{d}t = 2\sqrt{c}\mu_n(2c/\pi)\int_{\sqrt{c}}^{\sqrt{b}} \varphi_n^{(c)}(t)\varphi_n^{(b)}(t)\,\mathrm{d}t.$$

Dividing both sides of this last equation by $b-c$, allowing $b \to c^+$, and applying the orthonormality of $\{\varphi_n^{(c)}\}_{n=0}^{\infty}$ gives

$$\frac{\mathrm{d}^+}{\mathrm{d}c}(\sqrt{c}\mu_n(2c/\pi)) = \mu_n(2c/\pi)\varphi_n^{(c)}(\sqrt{c})^2 = \frac{\mu_n(2c/\pi)}{\sqrt{c}}\bar{\phi}_n^{(c)}(1)^2. \tag{2.61}$$

Reversing the roles of b and c gives

$$\frac{\mathrm{d}^-}{\mathrm{d}c}(\sqrt{c}\mu_n(2c/\pi)) = \mu_n(2c/\pi)\varphi_n^{(c)}(\sqrt{c})^2 = \frac{\mu_n(2c/\pi)}{\sqrt{c}}\bar{\phi}_n^{(c)}(1)^2$$

so that $\frac{\mathrm{d}}{\mathrm{d}c}(\sqrt{c}\mu_n(2c/\pi))$ exists and is equal to the right-hand side of (2.61). Rearranging the resulting equation gives (2.58). □

The following lemma appears in [280].

Lemma 2.6.5. *Let $\{\mu_n(2c/\pi)\}$ be the collection of eigenvalues appearing in (2.17). Then*

$$\lim_{c\to 0^+}\frac{\mu_n(2c/\pi)}{c^n}=\frac{\mathrm{i}^n\sqrt{\pi}(n!)^2}{\Gamma(n+3/2)(2n)!}. \tag{2.62}$$

Proof. Differentiating both sides of (2.17) n times with respect to x and evaluating at $x=0$ gives

$$\mu_n(2c/\pi)D^n\bar{\phi}_n^{(c)}(0)=(\mathrm{i}c)^n\int_{-1}^{1}t^n\bar{\phi}_n^{(c)}(t)\,\mathrm{d}t. \tag{2.63}$$

Furthermore, $\bar{\phi}_n^{(c)}(x)=\bar{P}_n(x)+O(c^2)$. The uniform convergence of $\bar{\phi}_n^{(c)}$ to $\bar{P}_n$ and the analyticity of $\bar{\phi}_n^{(c)}(x)$ as a function of both x and c now give the uniform convergence of $D_x^n\bar{\phi}_n^{(c)}(x)$ to $D_x^n\bar{P}_n(x)=\dfrac{(2n)!\sqrt{n+1/2}}{2^n n!}$. Hence, allowing $c\to 0^+$ in (2.63) gives $\lim_{c\to 0^+}\dfrac{\mu_n(2c/\pi)}{c^n}=\dfrac{\mathrm{i}^n 2^n n!}{(2n)!}\int_{-1}^{1}t^n P_n(t)\,\mathrm{d}t$. Applying Lemma 2.6.3 now gives the result. □

The following result appears in [280].

Lemma 2.6.6. *Let $c>0$ and let n be a non-negative integer. Then*

$$\mu_n(2c/\pi)=\frac{\mathrm{i}^n\sqrt{\pi}c^n(n!)^2}{(2n)!\Gamma(n+3/2)}\exp\left(\int_0^c\left(\frac{2\bar{\phi}_n^{(\tau)}(1)^2-1}{2\tau}-\frac{n}{\tau}\right)\mathrm{d}\tau\right).$$

Proof. Let $\nu_n=i^{-n}\mu_n$ so that each $\nu_n=\sqrt{2\pi\lambda_n/c}$ is positive. We first rearrange (2.58) so that

$$\frac{\frac{\mathrm{d}}{\mathrm{d}\tau}(\nu_n(2\tau/\pi))}{\nu_n(2\tau/\pi)}=\frac{2\bar{\phi}_n^{(\tau)}(1)^2-1}{2\tau}.$$

With $0<c_0<c$, we integrate both sides of this equation with respect to τ from c_0 to c to obtain

$$\ln(\nu_n(2c/\pi))-\ln(\nu_n(2c_0/\pi))=\int_{c_0}^{c}\frac{2\bar{\phi}_n^{(\tau)}(1)^2-1}{2\tau}\,\mathrm{d}\tau.$$

Since $\bar{\phi}_n^{(c)}(x)=\bar{P}_n(x)+O(c^2)$ and $\bar{P}_n(1)=\sqrt{n+1/2}$ we have $\dfrac{2\bar{\phi}_n^{(\tau)}(1)^2-1}{2\tau}=\dfrac{n}{\tau}+q_n(\tau)$ with q_n smooth. Hence

$$\ln(\nu_n(2c/\pi))=\ln\left(\frac{\nu_n(2c_0/\pi)}{c_0^n}\right)+n\ln c+\int_{c_0}^{c}q_n(\tau)\,\mathrm{d}\tau.$$

Exponentiating both sides of this equation, allowing $c_0\to 0+$, and applying (2.62) now gives the result. □

Rokhlin and Xiao [280] also showed that the prolates satisfy the bound

$$|\bar{\phi}_n^{(c)}(1)|<\sqrt{n+1/2}. \tag{2.64}$$

As a consequence they gave the following estimate of the rate of decay of the eigenvalues $\mu_n(2c/\pi)$.

Lemma 2.6.7. *Let $c > 0$ and let n be a non-negative integer. Then*

$$|\mu_n(2c/\pi)| \le \frac{\sqrt{\pi}(n!)^2 c^n}{\Gamma(n+3/2)(2n)!}.$$

Proof. Define the function F by

$$F(c) = \int_0^c \left(\frac{2\bar{\phi}_n^{(\tau)}(1)^2 - 1}{2\tau} - \frac{n}{\tau} \right) \mathrm{d}\tau.$$

Then, by Lemma 2.6.6, $\frac{\mu_n(2c/\pi)}{\mathrm{i}^n c^n} = \frac{\sqrt{\pi}(n!)^2}{(2n)!\Gamma(n+3/2)} \exp(F(c))$. However, by (2.64), $F'(c) = \frac{2\bar{\phi}_n^{(c)}(1)^2 - 1}{2c} - \frac{n}{c} < 0$, so $\frac{\mu_n(2c/\pi)}{\mathrm{i}^n c^n}$ is decreasing, since

$$\frac{\mathrm{d}}{\mathrm{d}c}\left(\frac{\mu_n(2c/\pi)}{\mathrm{i}^n c^n} \right) = \frac{\sqrt{\pi}(n!)^2}{(2n)!\Gamma(n+3/2)} F'(c)\exp(F(c)) < 0.$$

By (2.62), $\frac{\mu_n(2c/\pi)}{\mathrm{i}^n c^n} \le \lim_{c\to 0^+} \frac{\mu_n(2c/\pi)}{\mathrm{i}^n c^n} = \frac{\sqrt{\pi}(n!)^2}{\Gamma(n+3/2)(2n)!}$ and the result follows. □

Lemma 2.6.7 should be compared to Corollary 1.2.11. Recalling that $\mu_n(a) = 2i^n\sqrt{\lambda_n(a)/a}$, it is clear that both results provide roughly the same upper bound. The difference is that the more sophisticated (and more difficult) Corollary 1.2.11 also proves an asymptotic lower bound.

Values of $\bar{\phi}_n(1)$, $\bar{\phi}_n(0)$

Lemma 2.6.8. *For any $c > 0$ and integer $n \ge 0$ we have $\bar{\phi}_n(1) \ne 0$.*

Proof. Suppose that for some $n \ge 0$, $\bar{\phi}_n(1) = 0$. Each $\bar{\phi}_n$ satisfies the differential equation

$$(1-x^2)\bar{\phi}_n''(x) - 2x\bar{\phi}_n'(x) + (\chi_n - c^2x^2)\bar{\phi}_n(x) = 0, \tag{2.65}$$

and evaluating both sides of (2.65) at $x = 1$ gives

$$-2\bar{\phi}_n'(1) + (\chi_n - c^2)\bar{\phi}_n(1) = 0 \tag{2.66}$$

from which we see that $\bar{\phi}_n'(1) = 0$. Differentiating (2.65) and evaluating both sides at $x = 1$ gives

$$-4\bar{\phi}_n''(1) + (\chi_n - 2 - c^2)\bar{\phi}_n'(1) - 2c^2\bar{\phi}_n(1) = 0$$

from which we conclude that $\bar{\phi}_n''(1) = 0$. For each integer $k \ge 2$, the kth derivative of (2.65) becomes

$$(1-x^2)\bar{\phi}_n^{(k+2)}(x) - 2(k+1)x\bar{\phi}_n^{(k+1)}(x) + (\chi_n - k(k+1) - c^2x^2)\bar{\phi}_n^{(k)}(x) \\ - 2c^2kx\bar{\phi}_n^{(k-1)}(x) - c^2k(k-1)\bar{\phi}_n^{(k-2)}(x) = 0,$$

and evaluating this equation at $x = 1$ gives

$$-2(k+1)\bar{\phi}_m^{(k+1)}(1) + (\chi_m - k(k+1) - c^2)\bar{\phi}_n^{(k)}(1) - 2c^2k\bar{\phi}_n^{(k-1)}(1) - c^2k(k-1)\bar{\phi}_n^{(k-2)}(1) = 0.$$

An inductive argument now gives $\bar{\phi}_n^{(k)}(1) = 0$ for all integers $k \ge 0$. Since each $\bar{\phi}_n$ is analytic on the complex plane, we conclude that $\bar{\phi} \equiv 0$, a contradiction. □

A similar argument can be used to show that if n is even then $\bar{\phi}_n(0) \neq 0$. The following is a consequence of (2.66).

Corollary 2.6.9. *With $c > 0$ and integers $m, n \geq 0$,*

$$\bar{\phi}_m'(1)\bar{\phi}_n(1) - \bar{\phi}_n'(1)\bar{\phi}_m(1) = \frac{1}{2}(\chi_m - \chi_n)\bar{\phi}_n(1)\bar{\phi}_m(1).$$

2.6.2 *Proof of Theorem 2.1.15*

Definition 2.6.10. Given $a < b$, the n-dimensional simplex $M_n = M_n(a,b)$ is defined by

$$M_n = \{x = (x_1, x_2, \ldots, x_n) \in \mathbb{R}^n;\ a \leq x_1 \leq x_2 \leq \cdots \leq x_n \leq b\}.$$

Note that $M^1(a,b) = [a,b]$.

Lemma 2.6.11. *Let $K(X,S)$ $(X, S \in M_n)$ be a continuous symmetric kernel with*

$$K(X,S) \geq 0, \quad K(X,X) > 0 \ \textit{for all}\ X, S \in M_n.$$

Then the eigenvalue λ_0 of the operator $T_K : L^2(M_n) \to L^2(M_n)$ defined by

$$(T_K F)(X) = \int_{M_n} K(X,S) F(S)\, \mathrm{d}S$$

of largest absolute value is positive, simple, and not equal to the remaining eigenvalues. The eigenfunction Φ_0 corresponding to λ_0 has no zeros inside M_n.

Proof. The supremum of $\{\langle T_K \Phi, \Phi\rangle : \|\Phi\| = 1\}$ is attained at an eigenvalue λ_0 of T_K. If $\Phi(X_0) > 0$ at some $X_0 \in M_n$, then the continuity of K and the conditions $K(X_0,S) \geq 0$, $K(X_0,X_0) > 0$ imply that $\lambda_0 > 0$ and λ_0 is the largest eigenvalue of T_K. Let Φ_0 be a corresponding normalized eigenfunction, that is,

$$\lambda_0 = \sup_{\|\Phi\|=1} \langle T_K \Phi, \Phi\rangle = \langle T_K \Phi_0, \Phi_0\rangle;\ \|\Phi_0\| = 1.$$

We claim that Φ_0 has no zeros in M_n. Since $\lambda_0 > 0$ and $K(X,S) \geq 0$,

$$\begin{aligned} \lambda_0 &= |\lambda_0| = |\langle T_K \Phi_0, \Phi_0\rangle| = \left| \int_{M_n} \int_{M^n} K(X,S) \Phi_0(S)\, \mathrm{d}S\, \Phi_0(X)\, \mathrm{d}X \right| \\ &\leq \int_{M_n} \int_{M_n} K(X,S) |\Phi_0(S)|\, \mathrm{d}S\, |\Phi_0(X)|\, \mathrm{d}X = \langle T_K |\Phi_0|, |\Phi_0|\rangle. \end{aligned} \tag{2.67}$$

Since λ_0 is maximal, we must have equality in (2.67). Thus, $|\Phi_0|$ is also an eigenfunction of T_K corresponding to λ_0, that is,

$$\lambda_0 |\Phi_0(X)| = \int_{M_n} K(X,S) |\Phi_0(S)|\, \mathrm{d}S.$$

Now suppose Φ_0 has a zero on M_n and $X_1 \in M_n$ is a zero of Φ_0 such that in any neighborhood U of X_1 there exists $X \in U$ with $\Phi_0(X) \neq 0$. If none of the zeros of Φ_0 has this property then $\Phi_0 \equiv 0$. Since K is continuous and $K(X_1, X_1) > 0$, there is a neighborhood V of X_1 for which $K(X_1, S) > 0$ for all $S \in V$. Hence

$$0 = \lambda_0 |\Phi_0(X_1)| = \int_{M_n} K(X_1,S) |\Phi_0(S)|\, \mathrm{d}S \geq \int_{U \cap V} K(X,S) |\Phi_0(S)|\, \mathrm{d}S > 0,$$

a contradiction. Hence Φ_0 has no zeros in M_n. Also, any eigenfunction with eigenvalue λ_0 has constant sign on M_n and consequently no two such functions can be orthogonal. We conclude that λ_0 is a simple eigenvalue.

By the maximum principle, λ_0 is larger than all positive eigenvalues of T_K. We wish to show that $\lambda_0 > |\lambda'|$ for all negative eigenvalues λ'. Suppose λ' is such a negative eigenvalue, Ψ is an eigenfunction of T_K with eigenvalue λ', and $\|\Psi\| = 1$, that is,

$$\lambda'\Psi(X) = \int_{M_n} K(X,S)\Psi(S)\,\mathrm{d}S. \tag{2.68}$$

Then

$$|\lambda'||\Psi(X)| \leq \int_{M_n} K(X,S)|\Psi(S)|\,\mathrm{d}S, \tag{2.69}$$

and we claim that the inequality (2.69) is strict at every $X \in M_n$ for which $\Psi(X) \neq 0$. To see this, suppose that $\Psi(X_1) \neq 0$ and equality holds in (2.69). Then, as a function of S, $K(X_1,S)\Psi(S)$ must have the same sign on all of M_n. However, from (2.68),

$$\begin{aligned}\operatorname{sgn}(\Psi(X_1)) = -\operatorname{sgn}(\lambda'\Psi(X_1)) &= -\operatorname{sgn}\left(\int_{M_n} K(X_1,S)\Psi(S)\,\mathrm{d}S\right)\\ &= -\operatorname{sgn}(K(X_1,X_1)\Psi(X_1)) = -\operatorname{sgn}(\Psi(X_1)),\end{aligned}$$

a contradiction. Hence the inequality in (2.69) is strict. Multiplying both sides of (2.69) by $|\Psi(X)|$ and integrating over M_n gives

$$\begin{aligned}|\lambda'| = |\lambda'|\|\Psi\|^2 &< \int_{M_n}\int_{M_n} K(X,S)|\Psi(S)|\,\mathrm{d}S\,|\Psi(X)|\,\mathrm{d}X\\ &= \langle T_K|\Psi|,|\Psi|\rangle \leq \langle T_K\Phi_0,\Phi_0\rangle = \lambda\|\Phi_0\|^2 = \lambda_0,\end{aligned}$$

as required. The proof is complete. □

Definition 2.6.12. Given a kernel $K(x,s)$ $(x,s \in (a,b))$, and integers $n, j \geq 1$, the nth *associated kernel* $K_n(X,S)$ $(X = (x_1,x_2,\ldots,x_n),\ S = (s_1,s_2,\ldots,s_n) \in M_n(a,b))$ is

$$K_n(X,S) = K\begin{pmatrix} x_1 & x_2 & \cdots & x_n\\ s_1 & s_2 & \cdots & s_n\end{pmatrix} = \det_{i,j}(K(x_i,s_j))_{i,j=1}^n.$$

The jth iterated kernel $K^j(x,s)$ associated with K is given by $K^1 = K$ and, when $j \geq 2$,

$$K^j(x,s) = \int_a^b \cdots \int_a^b K(x,t_1)K(t_1,t_2)\cdots K(t_{j-1},s)\,\mathrm{d}t_1\,\mathrm{d}t_2\cdots\mathrm{d}t_{j-1},$$

that is, K^j is the kernel associated with the operator $T_{K^j} = (T_K)^j = T_K \circ T_K \circ \cdots \circ T_K$ (j factors).

Lemma 2.6.13. *Let the kernels K, L, and N defined on $(a,b)\times(a,b)$ be related by*

$$N(x,s) = \int_a^b K(x,t)L(t,s)\,\mathrm{d}t,$$

or equivalently, $T_N = T_K \circ T_L$. Then

(i) *The nth associated kernels K_n, L_n, and N_n defined on $M_j(a,b)\times M_j(a,b)$ are related by*

$$N_n(X,S) = \int_{M_n} K_n(X,T)L_n(T,S)\,\mathrm{d}T.$$

(ii) *If K^j is the jth iteration of the kernel K, then $[K_n]^j = [K^j]_n$, that is, the nth associated kernel of the jth iteration of K is the jth iteration of the nth associated kernel of K.*

Proof. First, if f is defined on the n-dimensional cube $Q^n(a,b) = (a,b)^n$, then

$$\int_{Q^n(a,b)} f(t_1,t_2,\dots,t_n)\,\mathrm{d}t_1\,\mathrm{d}t_2\cdots\mathrm{d}t_n = \sum_{\sigma\in\Sigma_n}\int_{M_n(a,b)} f(t_{\sigma(1)},t_{\sigma(2)},\dots,t_{\sigma(n)})\,\mathrm{d}t_1\,\mathrm{d}t_2\cdots\mathrm{d}t_n$$

where Σ_n is the symmetric group of permutations of $\{1,2,\dots,n\}$. Given two collections of functions $\{\varphi_1,\varphi_2,\dots,\varphi_n\}$ and $\{\psi_1,\psi_2,\dots,\psi_n\}$ defined on $[a,b]$, the behavior of determinants under row or column swaps gives

$$\begin{aligned}
&\int_{Q^n}\Delta\begin{pmatrix}\varphi_1 & \varphi_2 & \cdots & \varphi_n\\ t_1 & t_2 & \cdots & t_n\end{pmatrix}\Delta\begin{pmatrix}\psi_1 & \psi_2 & \cdots & \psi_n\\ t_1 & t_2 & \cdots & t_n\end{pmatrix}\mathrm{d}t_1\,\mathrm{d}t_2\cdots\mathrm{d}t_n\\
&\quad=\sum_{\sigma\in\Sigma_n}\int_{M_n}\Delta\begin{pmatrix}\varphi_1 & \varphi_2 & \cdots & \varphi_n\\ t_{\sigma(1)} & t_{\sigma(2)} & \cdots & t_{\sigma(n)}\end{pmatrix}\Delta\begin{pmatrix}\psi_1 & \psi_2 & \cdots & \psi_n\\ t_{\sigma(1)} & t_{\sigma(2)} & \cdots & t_{\sigma(n)}\end{pmatrix}\mathrm{d}t_1\,\mathrm{d}t_2\cdots\mathrm{d}t_n\\
&\quad=n!\int_{M_n}\Delta\begin{pmatrix}\varphi_1 & \varphi_2 & \cdots & \varphi_n\\ t_1 & t_2 & \cdots & t_n\end{pmatrix}\Delta\begin{pmatrix}\psi_1 & \psi_2 & \cdots & \psi_n\\ t_1 & t_2 & \cdots & t_n\end{pmatrix}\mathrm{d}t_1\,\mathrm{d}t_2\cdots\mathrm{d}t_n.
\end{aligned}$$

Given an $n\times n$ matrix A with (i,j)th entry a_{ij}, its determinant may be computed by the Leibniz formula $\det(A) = \sum_{\omega\in\Sigma_n}\mathrm{sgn}(\omega)\prod_{i=1}^n a_{i,\omega(i)}$. Here $\mathrm{sgn}(\omega) = \pm 1$ is the *sign* of the permutation ω. Consequently, by the product formula for determinants,

$$\begin{aligned}
&\int_{M_n}\Delta\begin{pmatrix}\varphi_1 & \varphi_2 & \cdots & \varphi_n\\ t_1 & t_2 & \cdots & t_n\end{pmatrix}\Delta\begin{pmatrix}\psi_1 & \psi_2 & \cdots & \psi_n\\ t_1 & t_2 & \cdots & t_n\end{pmatrix}\mathrm{d}t_1\,\mathrm{d}t_2\cdots\mathrm{d}t_n\\
&\quad=\frac{1}{n!}\int_{Q^n}\det_{i,k}(\varphi_i(t_k))\det_{j,k}(\psi_j(t_k))\,\mathrm{d}t_1\,\mathrm{d}t_2\cdots\mathrm{d}t_n\\
&\quad=\frac{1}{n!}\int_{Q^n}\det_{i,j}\left(\sum_{m=1}^n\varphi_i(t_m)\psi_j(t_m)\right)\mathrm{d}t_1\,\mathrm{d}t_2\cdots\mathrm{d}t_n\\
&\quad=\frac{1}{n!}\sum_{m_1=1}^n\sum_{m_2=1}^n\cdots\sum_{m_n=1}^n\int_{Q^n}\det_{i,j}(\varphi_i(t_{m_i})\psi_j(t_{m_i}))\,\mathrm{d}t_1\,\mathrm{d}t_2\cdots\mathrm{d}t_n\\
&\quad=\frac{1}{n!}\sum_{\sigma\in\Sigma_n}\int_{Q^n}\det_{i,j}(\varphi_i(t_{\sigma(i)})\psi_j(t_{\sigma(i)}))\,\mathrm{d}t_1\,\mathrm{d}t_2\cdots\mathrm{d}t_n\\
&\quad=\frac{1}{n!}\sum_{\sigma\in\Sigma_n}\sum_{\omega\in\Sigma_n}\mathrm{sgn}\,(\omega)\int_{Q^n}\prod_{\ell=1}^n\varphi_\ell(t_{\sigma(\ell)})\psi_{\omega(\ell)}(t_{\sigma(\ell)})\,\mathrm{d}t_1\,\mathrm{d}t_2\cdots\mathrm{d}t_n\\
&\quad=\frac{1}{n!}\sum_{\sigma\in\Sigma_n}\sum_{\omega\in\Sigma_n}\mathrm{sgn}(\omega)\prod_{\ell=1}^n\langle\varphi_\ell,\psi_{\omega(\ell)}\rangle=\det_{i,j}(\langle\varphi_i,\psi_j\rangle).
\end{aligned}\tag{2.70}$$

Putting $\psi_i(t) = K(x_i,t)$ and $\varphi_i(t) = L(t,s_i)$ now gives the result, since if $X = (x_1,x_2,\dots,x_n)$, $T = (t_1,t_2,\dots,t_n)$, $S = (s_1,s_2,\dots,s_n)\in M_n$, then $\Delta\begin{pmatrix}\psi_1 & \psi_2 & \cdots & \psi_n\\ t_1 & t_2 & \cdots & t_n\end{pmatrix} = K_n(X,T)$, $\Delta\begin{pmatrix}\varphi_1 & \varphi_2 & \cdots & \varphi_n\\ t_1 & t_2 & \cdots & t_n\end{pmatrix} = L_n(T,S)$, and by (2.70),

$$\int_{M^n} K_n(X,T)L_n(T,S)\,\mathrm{d}T$$
$$= \int_{M^n} \Delta\begin{pmatrix}\varphi_1 & \varphi_2 & \cdots & \varphi_n\\ t_1 & t_2 & \cdots & t_n\end{pmatrix}\Delta\begin{pmatrix}\psi_1 & \psi_2 & \cdots & \psi_n\\ t_1 & t_2 & \cdots & t_n\end{pmatrix}\mathrm{d}t_1\,\mathrm{d}t_2\cdots\mathrm{d}t_n$$
$$= \det_{i,j}\left(\int_a^b K(x_i,t)L(t,s_j)\,\mathrm{d}t\right) = \det_{i,j}(N(x_i,s_j)) = N_n(X,S).$$

This completes the proof of statement (i), from which statement (ii) immediately follows. □

Let $\Lambda_n = \{I = (i_1,i_2,\ldots,i_n) \in \mathbb{Z}^n : 1 \le i_1 < i_2 < \cdots < i_n\}$.

Lemma 2.6.14. *Let $\{\varphi_i\}_{i=1}^{\infty}$ be a complete orthonormal system of eigenfunctions of an integral operator T_K with symmetric kernel $K(x,s)$ $(x,s \in [a,b])$ with $\{\lambda_i\}_{i=1}^{\infty}$ the associated eigenvalues. Then the functions*

$$\Phi_I(X) = \Delta\begin{pmatrix}\varphi_{i_1} & \varphi_{i_2} & \cdots & \varphi_{i_n}\\ x_1 & x_2 & \cdots & x_n\end{pmatrix}$$

(where $X = (x_1,x_2,\ldots,x_n) \in M_n$ and $I = (i_1,i_2,\ldots,i_n) \in \Lambda_n$) form a complete orthonormal system of eigenfunctions of the operator T_{K_n} acting on $L^2(M_n)$ by $(T_{K_n}F)(X) = \int_{M^n} K_n(X,S)F(S)\,\mathrm{d}S$ with associated eigenvalues $\lambda_I = \lambda_{i_1}\lambda_{i_2}\cdots\lambda_{i_n}$, that is, $T_{K_n}\Phi_I = \lambda_I\Phi_I$.

Proof. As in the proof of the previous lemma, putting $\psi_k(t) = K(x_k,t)$ gives $\Delta\begin{pmatrix}\psi_1 & \psi_2 & \cdots & \psi_n\\ t_1 & t_2 & \cdots & t_n\end{pmatrix} = K_n(X,T)$, and by (2.70),

$$\int_{M_n} K_n(X,S)\Phi_I(S)\,\mathrm{d}S = \int_{M_n} \Delta\begin{pmatrix}\psi_1 & \psi_2 & \cdots & \psi_n\\ s_1 & s_2 & \cdots & s_n\end{pmatrix}\Delta\begin{pmatrix}\varphi_{i_1} & \varphi_{i_2} & \cdots & \varphi_{i_n}\\ s_1 & s_2 & \cdots & s_n\end{pmatrix}\mathrm{d}s_1\,\mathrm{d}s_2\cdots\mathrm{d}s_n$$
$$= \det_{k,j}(\langle\psi_k,\varphi_{i_j}\rangle) = \det_{k,j}\left(\int_a^b K(x_k,t)\varphi_{i_j}(t)\,\mathrm{d}t\right)$$
$$= \det_{k,j}(\lambda_{i_j}\varphi_{i_j}(x_k)) = \lambda_I\det_{i,j}(\varphi_{i_j}(x_i)) = \lambda_I\Phi_I(X),$$

that is, the functions Φ_I are eigenfunctions of T_{K_n} with associated eigenvalues λ_I. The orthogonality of this collection is an immediate consequence of (2.70):

$$\langle\Phi_I,\Phi_J\rangle = \int_{M_n}\Phi_I(X)\Phi_J(X)\,\mathrm{d}X = \det_{\ell,k}(\langle\varphi_{i_\ell},\varphi_{j_k}\rangle) = \det_{\ell,k}(\delta_{i_\ell,j_k}) = \delta_{I,J}.$$

Finally, we show that the collection $\{\Phi_I\}_{I\in\Lambda_n}$ forms a complete set in $L^2(M_n)$. The kernel admits the eigenfunction (Mercer) expansion $K(x,s) = \sum_{\ell=1}^{\infty}\lambda_\ell\,\varphi_\ell(x)\varphi_\ell(s)$, so that the nth associated kernel takes the form

$$
\begin{aligned}
K_n(X,S) = \det_{i,j}(K(x_i,s_j)) &= \det_{i,j}\left(\sum_{\ell=1}^{\infty}\lambda_\ell\varphi_\ell(x_i)\varphi_\ell(s_j)\right)\\
&= \sum_{\ell_0=1}^{\infty}\sum_{\ell_1=1}^{\infty}\cdots\sum_{\ell_{n-1}=1}^{\infty}\prod_{j=0}^{n-1}\lambda_{\ell_j}\varphi_{\ell_j}(x_j)\det_{i,j}(\varphi_{\ell_i}(s_j))\\
&= \sum_{\{\ell_0,\ell_1,\dots,\ell_{n-1}\ \text{distinct}\}}\prod_{j=0}^{n-1}\lambda_{\ell_j}\varphi_{\ell_j}(x_j)\det_{i,j}(\varphi_{\ell_i}(s_j))\\
&= \sum_{\sigma\in\Sigma_n}\sum_{(\ell_0,\ell_1,\dots,\ell_{n-1})\in\Lambda_n}\prod_{j=0}^{n-1}\lambda_{\ell_{\sigma(j)}}\varphi_{\ell_{\sigma(j)}}(x_j)\det_{i,j}(\varphi_{\ell_{\sigma(i)}}(s_j))\\
&= \sum_{\sigma\in\Sigma_n}\sum_{(\ell_0,\ell_1,\dots,\ell_{n-1})\in\Lambda_n}\prod_{j=0}^{n-1}\lambda_{\ell_{\sigma(j)}}\varphi_{\ell_{\sigma(j)}}(x_j)\,\mathrm{sgn}(\sigma)\det_{i,j}(\varphi_{\ell_i}(s_j))\\
&= \sum_{(\ell_0,\ell_1,\dots,\ell_{n-1})\in\Lambda_n}\left(\prod_{k=0}^{n-1}\lambda_{\ell_k}\right)\sum_{\sigma\in\Sigma_n}\mathrm{sgn}(\sigma)\prod_{j=0}^{n-1}\varphi_{\ell_{\sigma(j)}}(x_j)\det_{i,j}(\varphi_{\ell_i}(s_j))\\
&= \sum_{L\in\Lambda_n}\lambda_L\Phi_L(X)\Phi_L(S).
\end{aligned}
$$

Hence, if $F\in L^2(M_n)$ and $\langle F,\Phi_L\rangle = 0$ for all $L\in\Lambda_n$, then $T_{K^n}F = 0$. Since 0 is not an eigenvalue of T_{K_n}, we have arrived at a contradiction and we conclude that $\{\Phi_L\}_{L\in\Lambda_n}$ is a complete orthonormal system for $L^2(M_n)$. □

We are now in a position to prove Theorem 2.1.15.

Proof (of Theorem 2.1.15). We relabel the eigenvalues of T_K so that $|\lambda_0|\geq|\lambda_1|\geq\cdots$ and denote the corresponding eigenfunctions by $\{\varphi_j\}_{j=0}^{\infty}$. Let K_n be the nth associated kernel of K. By Lemma 2.6.14, the eigenvalues are n-fold products of $\{\lambda_j\}_{j=0}^{\infty}$ (with no repetitions). The product with the largest absolute value will be $\lambda_0\lambda_1\cdots\lambda_{n-1}$, which corresponds to the eigenfunction

$$
\Phi_{(0,1,\dots,n-1)}(X) = \Delta\begin{pmatrix}\varphi_0 & \varphi_1 & \cdots & \varphi_{n-1}\\ x_1 & x_2 & \cdots & x_n\end{pmatrix} = \det_{i,j}(\varphi_{i-1}(x_j)) \quad (X\in M_n).
$$

Since K_n satisfies the conditions of Lemma 2.6.11,

$$
\lambda_0\lambda_1\cdots\lambda_{n-1}>0 \quad\text{and}\quad \lambda_0\lambda_1\cdots\lambda_{n-1}>|\lambda_0\lambda_1\cdots\lambda_{n-2}\lambda_n|.
$$

These inequalities hold for all n, so we conclude that $\lambda_0>\lambda_1>\cdots>\lambda_n>\cdots>0$. Also, from Lemma 2.6.11 we conclude that $\Phi_{(0,1,\dots,n-1)}(X)\neq 0$ for all $X\in M^n$. By Lemma 2.1.2, for every n the finite collection $\{\varphi_j\}_{j=0}^{n-1}$ is a Chebyshev system. Equivalently, the infinite collection $\{\varphi_j\}_{j=0}^{\infty}$ is a Markov system on (a,b). The conclusions of Theorem 2.1.15 now follow from Theorem 2.1.7. The self-adjointness of the operator and the simplicity of the eigenvalues now also imply the orthogonality of the eigenfunctions. □

2.6.3 *Proof of Corollary 2.2.7*

Proof (of Corollary 2.2.7). Given $0<\delta<1/2$, choose a real-valued C^∞ *bump function* b on the line with the properties (see, e.g., [8])

(i) $0\leq b(x)\leq 1$ for all x,
(ii) $b(x)=0$ if $|x|\geq 1$ and $b(x)=1$ if $|x|<1-\delta$,

(iii) $|b'(x)| \leq 2/\delta$ for all x.

An application of Wirtinger's inequality (see Appendix) gives

$$\begin{aligned}
\frac{\delta}{2}\|\sigma b'\|_{L^2[0,1]} &\leq \left(\int_{1-\delta}^{1} |\sigma(t)|^2 \,\mathrm{d}t\right)^{1/2} \\
&\leq \left(\int_{1-\delta}^{1} |\sigma(t)-\sigma(1)|^2 \,\mathrm{d}t\right)^{1/2} + |\sigma(1)|\sqrt{\delta} \\
&\leq \frac{2\delta}{\pi}\left(\int_{1-\delta}^{1} |\sigma'(t)|^2 \,\mathrm{d}t\right)^{1/2} + |\sigma(1)|\sqrt{\delta}.
\end{aligned}$$

A similar estimate may be made for $\|\sigma b'\|_{L^2[-1,0]}$ so that we have

$$\begin{aligned}
\|\sigma b'\|_{L^2[-1,1]} &\leq \frac{2}{\sqrt{\delta}}(|\sigma(1)|+|\sigma(-1)|) + \frac{4\sqrt{2}}{\pi}\left(\int_{1-\delta<|t|<1} |\sigma'(t)|^2 \,\mathrm{d}t\right)^{1/2} \\
&\leq \frac{2}{\sqrt{\delta}}(|\sigma(1)|+|\sigma(-1)|+\|\sigma'\|_{L^2[-1,1]}). \qquad (2.71)
\end{aligned}$$

However, $|\sigma(1)|+|\sigma(-1)| \leq 2\sqrt{2}(\|\sigma\|_{L^2[-1,1]}+\|\sigma'\|_{L^2[-1,1]})$, and substituting this inequality into (2.71) gives

$$\|\sigma b'\|_{L^2[-1,1]} \leq \frac{C}{\sqrt{\delta}}(\|\sigma\|_{L^2[-1,1]}+\|\sigma'\|_{L^2[-1,1]}). \qquad (2.72)$$

Now $\tilde{\sigma} = \sigma b$ satisfies the conditions of Theorems 2.2.5 and 2.2.6, so that $\tilde{f}$ defined by $\tilde{f}(t) = \int_{-1}^{1} \tilde{\sigma}(x)\mathrm{e}^{-2icxt}\,\mathrm{d}t$ satisfies the conditions of Theorem 2.2.6. Furthermore,

$$|f(t)-\tilde{f}(t)| \leq \int_{-1}^{1} |\sigma(x)||b(x)-1|\,\mathrm{d}x \leq \int_{1-\delta\leq|t|\leq 1} |\sigma(t)|\,\mathrm{d}t \leq \sqrt{2\delta}\|\sigma\|_{L^2[-1,1]}.$$

By Theorem 2.2.5 there exist functions $\tilde{q}$, $\tilde{r}$ of band limit c such that $\tilde{f} = \phi_N\tilde{q}+\tilde{r}$. Let $\tilde{q}(t) = \int_{-1}^{1}\tilde{\eta}(x)\mathrm{e}^{-icxt}\,\mathrm{d}x$ and $\tilde{r}(t) = \int_{-1}^{1}\tilde{\xi}(x)\mathrm{e}^{-icxt}\,\mathrm{d}x$. Then

$$\begin{aligned}
\|\tilde{\eta}\|_{L^2[-1,1]}+\|\tilde{\xi}\|_{L^2[-1,1]} &\leq C(\|\tilde{\sigma}\|_{L^2[-1,1]}+\|\tilde{\sigma}'\|_{L^2[-1,1]}) \\
&\leq \frac{C}{\sqrt{\delta}}(\|\sigma\|_{L^2[-1,1]}+\|\sigma'\|_{L^2[-1,1]})
\end{aligned}$$

and the error in the quadrature when applied to f may be estimated by

$$\begin{aligned}
E_f = \left|\sum_{k=0}^{N-1} w_k f(x_k) - \int_{-1}^{1} f(x)\,\mathrm{d}x\right| &\leq \sum_{k=0}^{N-1} w_k|f(x_k)-\tilde{f}(x_k)| \\
&\quad + \left|\sum_{k=0}^{N-1} w_k\tilde{f}(x_k) - \int_{-1}^{1}\tilde{f}(x)\,\mathrm{d}x\right| + \int_{-1}^{1}|f(x)-\tilde{f}(x)|\,\mathrm{d}x \\
&\leq C\sqrt{2\delta}\|\sigma\|_{L^2[-1,1]} + C|\mu_{N-1}|(\|\tilde{\eta}\|_{L^2[-1,1]}+\|\tilde{\xi}\|_{L^2[-1,1]}) \\
&\leq C\sqrt{2\delta}\|\sigma\|_{L^2[-1,1]} + \frac{C}{\sqrt{\delta}}|\mu_{N-1}|(\|\sigma\|_{L^2[-1,1]}+\|\sigma'\|_{L^2[-1,1]}).
\end{aligned}$$

Putting $\delta = |\mu_{N-1}|$ now gives the required estimate. □

2.6.4 Proofs of Other Auxiliary Lemmas for Theorem 2.5.2

Proof (of Lemma 2.5.3). Suppose k is even. Since $\|\bar{P}_k\|_{L^2[-1,1]} = \|\bar{\phi}_N\|_{L^2[-1,1]} = 1$, $|\beta_k| \le 1$ for all k. Hence (2.47) is valid for $k = 2,4$ with $D = \max\{|\beta_4|,|\beta_2|\}/(\alpha^2|\beta_0|)$. Assume now that (2.47) is valid for all even integers less than or equal to k. Setting

$$\gamma(x) = \frac{x(x-1)}{(2x-1)\sqrt{(2x-3)(2x+1)}}; \quad \delta(x) = \frac{2x(x+1)-1}{(2x+3)(2x-1)}$$

the recurrence relation (1.13) may be written

$$\beta_{k+2} = \frac{1}{\gamma(k+2)}\left(\frac{1}{q^2}\left(1-\frac{k(k+1)}{\chi_N}\right)-\delta(k)\right)\beta_k - \frac{\gamma(k)}{\gamma(k+2)}\beta_{k-2}. \tag{2.73}$$

For $x \ge 2$, $1/4 \le \gamma(x) \le 2\sqrt{5}/15$ and $1/2 \le \delta(x) \le 11/21$. Thus if $k \le 2m$,

$$\begin{aligned}\frac{1}{q^2}\left(1-\frac{k(k+1)}{\chi_N}\right)-\delta(k) &\ge \left(1-\frac{2m(2m+1)}{\chi_N}\right)-\delta(k)\\ &\ge 1-\frac{(\sqrt{\chi_N}/2)^2}{\chi_N}-\frac{11}{21} = 1-\frac{1}{4}-\frac{11}{21} > 0,\end{aligned}$$

that is, the coefficient of β_k in (2.73) is positive. Consequently, from (2.73), the bounds on $\gamma(x)$, $\delta(x)$, and the inductive hypothesis, we have

$$\begin{aligned}|\beta_{k+2}| &\le \frac{1}{\gamma(k+2)}\left(\frac{1}{q^2}\left(1-\frac{k(k+1)}{\chi_N}\right)-\delta(k)\right)|\beta_k| + \frac{\gamma(k)}{\gamma(k+2)}|\beta_{k-2}|\\ &\le \frac{4}{q^2}|\beta_k| + \frac{8\sqrt{5}}{15}|\beta_{k-2}| \le D\left(\frac{\alpha}{q}\right)^{k+2}|\beta_0|\left(\frac{4}{\alpha^2}+\frac{1}{\alpha^4}\right) < D\left(\frac{\alpha}{q}\right)^{k+2}|\beta_0|.\end{aligned}$$

The result follows by induction. □

Proof (of Lemma 2.5.4). The Nth prolate $\bar{\phi}_N^{(c)}$ satisfies the prolate differential equation

$$((1-x^2)\bar{\phi}')' + \chi_N(1-q^2x^2)\bar{\phi} = 0. \tag{2.74}$$

Let $0 \le \ell \le m$ be an integer. Multiplying both sides of (2.74) by $x^{2\ell}$ and integrating over $[-1,1]$ with respect to x gives

$$2\ell(2\ell-1)A_{2\ell-2} + (\chi_N - 2\ell(2\ell+1))A_{2\ell} - \chi_N q^2 A_{2\ell+2} = 0. \tag{2.75}$$

In particular, $A_0 = q^2A_2$. Since $2m(2m+1) \le \chi_N$, each of $A_0, A_2, \ldots, A_{2m+2}$ has the same sign, so by (2.75)

$$A_{2\ell} = \frac{\chi_N q^2 A_{2\ell+2} - 2\ell(2\ell-1)A_{2\ell-2}}{\chi_N - 2\ell(2\ell+1)} \le \frac{\chi_N q^2 A_{2\ell+2}}{\chi_N - 2\ell(2\ell+1)} = \frac{q^2 A_{2\ell+2}}{1-\dfrac{2\ell(2\ell+1)}{\chi_N}}.$$

Iterating this formula gives

$$A_0 \le q^{2m-2}A_{2m-2}X_m; \quad X_m = \prod_{\ell=1}^{m-1}\left(1-\frac{2\ell(2\ell+1)}{\chi_N}\right)^{-1}. \tag{2.76}$$

Now

$$
\begin{aligned}
-\log X_m &= \sum_{\ell=1}^{m-1} \log\Big(1-\frac{2\ell(2\ell+1)}{\chi_N}\Big) \geq \int_0^m \log\Big(1-\frac{4x^2}{\chi_N}\Big)\,\mathrm{d}x \\
&= \Big[x\log\Big(1-\frac{4x^2}{\chi_N}\Big) - 2x + \sqrt{\chi_N}\tanh^{-1}\Big(\frac{2x}{\sqrt{\chi_N}}\Big)\Big]_{x=1}^{x=m}.
\end{aligned} \tag{2.77}
$$

Since $\sqrt{\chi_N}/4-1 \leq m \leq \sqrt{\chi_N}/4$, we have $1-4m^2/\chi_N \geq 3/4$ and $2m/\sqrt{\chi_N} \geq 1/2-2/\sqrt{\chi_N} \geq 2/5$ provided $\chi_N \geq 400$. Consequently,

$$
\sqrt{\chi_N}\tanh^{-1}(2m/\sqrt{\chi_N}) \geq \sqrt{\chi_N}\tanh^{-1}(2/5) = \frac{\sqrt{\chi_N}}{2}\log(7/3).
$$

Substituting this into (2.77) yields

$$
-\log X_m \geq m\log(3/4) - 2m + \frac{\sqrt{\chi_N}}{2}\log(7/3)
$$

so that after exponentiating both sides, we have

$$
X_m \leq \Big(\frac{4\mathrm{e}^2}{3}\Big)^m \Big(\frac{3}{7}\Big)^{\sqrt{\chi_N}/2}. \tag{2.78}
$$

Combining (2.76) with (2.78) and the inequality $|A_m| \leq (m+1/2)^{-1/2}$ (which follows directly from the Cauchy–Schwarz inequality) gives the first of the inequalities in (2.48). The second inequality may be established in a similar manner. □

Chapter 3
Thomson's Multitaper Method and Applications to Channel Modeling

3.1 Estimation of Quasi-stationary Processes

3.1.1 Thomson's DPSWF-Based Multitaper Method

Periodogram Estimation of a Stationary Process

One of the most basic applications of Fourier analysis is *power spectrum estimation*. Some historical comments on this age-old problem can be found in review articles by Robinson [279] and Benedetto [23], and in Percival and Walden's book [264]. We refer to Appendix A for basic definitions and properties, and references for stochastic processes. In the problem of estimating a wide-sense stationary random process from data, one has only finitely many samples, often of a single realization of the process, with which to work.

Suppose that X is a discrete-time, zero mean wide-sense stationary random process with autocorrelation function R: $R_{xx}(k) = \mathbf{E}\{x(k+\ell)\bar{x}(\ell)\}$ and power spectral density S: $S(\omega) = \sum_k R_{xx}(k)\,\mathrm{e}^{2\pi i k\omega}$. Given values $\{x(0),\dots,x(N-1)\}$, one can divide the data into K blocks of length $L = N/K$ by setting $y_k(\ell) = x(\ell + kL)$. In order to obtain an estimate of the spectrum of X, one then computes the periodograms

$$P_X(\omega;k) = \frac{1}{L}\left|\sum_{\ell=0}^{L-1} y_k(\ell)\mathrm{e}^{-2\pi i\omega\ell}\right|^2$$

and sets

$$P_X(\omega) = \frac{1}{K}\sum_{k=0}^{K-1} P_X(\omega;k)\,.$$

This method of approximating the spectrum of X is justified if X satisfies the *ergodic* hypothesis that the long time temporal averages of any particular realization of X will equiconverge with the ensemble averages of $x_k(n)$ over a large number of realizations x_k of X, for any n. Even then, one needs to know that the given samples of a single realization are sufficient to provide suitable approximations. While the periodogram is an *asymptotically unbiased* estimator of the power spectral density $S_X(\omega)$ (see, e.g., [264]), it is *inconsistent* in the sense that variance does not decrease with the number of samples. In [323], Thomson proposed the use of prolate spheroidal wave functions (PSWFs) as window functions for estimating spectra non-parametrically, that is, in the absence of assumptions on the probability distribution from which the data is drawn. Thomson was motivated by the desire to minimize variance without increasing bias, when one is forced to compute direct spectral estimates from very limited data.

One assumes that N discrete-time samples of data $x(k)$ can be expressed in terms of the *spectral representation* [48, 50]

$$x(k) = \int_{-1/2}^{1/2} \mathrm{e}^{\pi \mathrm{i}(2k+1-N)\xi}\, \mathrm{d}Z(\xi), \quad k = 0, 1, \ldots, N-1. \tag{3.1}$$

Here, dZ is a zero-mean process (i.e., $\mathbf{E}\{\mathrm{d}Z(\xi)\} = 0$) with orthogonal increments, meaning that if $\xi \neq \omega$ then $\mathrm{d}Z(\xi)$ and $\overline{\mathrm{d}Z}(\omega)$ are uncorrelated. The index in the exponential serves to center time on the observation epoch. In turn, the measurements $x(k)$ produce an estimate of dZ by taking their time-centered discrete Fourier transform (DFT),

$$z(\omega) = \sum_{k=0}^{N-1} \mathrm{e}^{-\pi \mathrm{i}(2k+1-N)\omega} x(k). \tag{3.2}$$

Replacing the data by their spectral representation (3.1) and expressing the exponential series in terms of the Dirichlet kernel,

$$D_N(\omega - \xi) = \frac{\sin N\pi(\omega - \xi)}{\sin \pi(\omega - \xi)} = \sum_{k=0}^{N-1} \mathrm{e}^{-\pi \mathrm{i}(\omega-\xi)(2k+1-N)}, \tag{3.3}$$

gives the representation

$$z(\omega) = \int_{-1/2}^{1/2} D_N(\omega - \xi)\, \mathrm{d}Z(\xi). \tag{3.4}$$

The problem posed by Thomson is that of how to use (3.4) to estimate the spectrum dZ of the process that generated $\mathbf{x}$ defining $z(\omega)$ via (3.2).

The convolution with the Dirichlet kernel expresses the spectral leakage inherent in estimating an infinitesimal spectrum directly from finite data. Thomson's solution to the spectrum estimation problem involves two key steps, the first of which is to use discrete prolate spheroidal sequences (DPSSs) as data tapers to estimate the spectrum directly, *locally in frequency*. The second key step is an adaptive weighting of these direct estimates that seeks to reduce bias introduced by tapering. Bias is particularly problematic in regions in which the true spectrum is small.

DPSS Spectral Coefficients

As above, one assumes that dZ is 1-periodic. Fix $N \in \mathbb{N}$ and $0 < W < 1/2$, and for $n = 0, 1, \ldots, N-1$, let $\mathrm{s}_n = \mathrm{s}_n(N, W)$ be the Slepian DPS sequence defined as in Sect. 1.4, cf. [307, p. 1381], by

$$\begin{aligned} \mathrm{s}_n(k) &= \frac{1}{\varepsilon_n} \int_{-1/2}^{1/2} \sigma_n(\xi) \mathrm{e}^{-\pi \mathrm{i} \xi (2k+1-N)} \, \mathrm{d}\xi \\ &= \frac{1}{\lambda_n \varepsilon_n} \int_{-W}^{W} \sigma_n(\xi) \, \mathrm{e}^{-\pi \mathrm{i} \xi (2k+1-N)} \, \mathrm{d}\xi, \quad (k \in \mathbb{Z}_N) \end{aligned} \tag{3.5}$$

because of double orthogonality. As in Chap. 1, $\varepsilon_n = 1$ if n is even and $\varepsilon_n = \mathrm{i}$ if n is odd, and $\boldsymbol{\sigma}_n = \boldsymbol{\sigma}_n(N, W)$ is the nth discrete prolate spheroidal wave function (DPSWF) solution of (cf. (1.35))

$$\sigma_n(\omega) = \frac{1}{\lambda_n} \int_{-W}^{W} D_N(\omega - \xi) \sigma_n(\xi) \, \mathrm{d}\xi = \varepsilon_n \sum_{k=0}^{N-1} s_n(k) \, \mathrm{e}^{\pi \mathrm{i} \omega (2k+1-N)} \,. \tag{3.6}$$

Finally, one has the Toeplitz sinc matrix eigenvalue property

$$\sum_{\ell=0}^{N-1} \frac{\sin 2\pi W (k-\ell)}{\pi (k-\ell)} \mathrm{s}_n(k) = \lambda_n \mathrm{s}_n(k) \,.$$

The first step in approximately solving for dZ in (3.1) from the data z is to consider, for N, W, and ω fixed, the *unobservable* coefficients

$$Z_n(\omega) = \frac{1}{\sqrt{\lambda_n}} \int_{-W}^{W} \sigma_n(\xi) \, \mathrm{d}Z(\omega + \xi) \,. \tag{3.7}$$

One can think of the *observable* coefficients

$$z_n(\omega) = \frac{1}{\lambda_n} \int_{-W}^{W} \sigma_n(\xi) \, z(\omega + \xi) \, \mathrm{d}\xi \tag{3.8}$$

as *eigencoefficients* of the sample Fourier transform z in (3.2). Using (3.4) and (3.6) allows one to write

$$z_n(\omega) = \int_{-1/2}^{1/2} \sigma_n(\xi) \, \mathrm{d}Z(\omega + \xi), \tag{3.9}$$

the integration now taken over the full interval $[-1/2, 1/2]$. Expressing $\boldsymbol{\sigma}_n$ in terms of s_n as in (3.5), and $z(\xi)$ in terms of the samples $\mathbf{x}$ as in (3.1), one can also write

$$z_n(\omega) = \frac{1}{\varepsilon_n} \sum_{k=0}^{N-1} x(k) \, \mathrm{s}_n(k) \, \mathrm{e}^{-\pi \mathrm{i} \omega (2k+1-N)} \,. \tag{3.10}$$

That is, *the nth eigencoefficient is the DFT of the data tapered by the nth DPSS*. The expression $S_n^{\mathrm{est}}(\omega) = |z_n(\omega)|^2$ amounts to a direct estimate of the power spectrum $S(\omega)$ in which the rectangular window is replaced by $\mathbf{s}_n$.

In order to make effective use of the eigencoefficients in obtaining local estimates of dZ, one needs an effective weighting of them that takes advantage of the hypothesis that dZ has orthogonal increments. First, one forms a local representation that Thomson [323] called the *high resolution estimate*,

$$\zeta(\omega;\omega_0) = \sum_{n=0}^{D-1} \sigma_n(\omega-\omega_0)\frac{z_n(\omega_0)}{\lambda_n}\,.$$

It can be regarded as a local extrapolation from $z(\omega_0)$. It is standard to take $D = \lfloor 2NW \rfloor$, the dimension of the space of time- and index-limited DPSSs. Weighting each of the first D eigencoefficient terms by $2W/D$, one has a corresponding high resolution power spectral estimate,

$$S(\omega;\omega_0) = \frac{2W}{D}\left|\sum_{n=0}^{D-1} \sigma_n(\omega-\omega_0)\frac{z_n(\omega_0)}{\lambda_n}\right|^2.$$

Its local frequency average is just

$$\begin{aligned}\bar{S}(\omega_0) &= \frac{1}{2W}\int_{\omega_0-W}^{\omega_0+W} S(\omega;\omega_0)\,\mathrm{d}\omega\\ &= \frac{1}{D}\int_{\omega_0-W}^{\omega_0+W}\left|\sum_{n=0}^{D-1} \sigma_n(\omega-\omega_0)\frac{z_n(\omega_0)}{\lambda_n}\right|^2\mathrm{d}\omega\\ &= \frac{1}{D}\sum_{n=0}^{D-1}\frac{|z_n(\omega_0)|^2}{\lambda_n}\end{aligned}$$

because of the orthogonality property of the DPSWFs.

Low Order Moments of Eigencoefficients

Since $\mathbf{E}\{\mathrm{dZ}\} = 0$, it follows from *linearity of expectation* that $\mathbf{E}\{z_n(\omega)\} = 0$ as well. Since dZ has orthogonal increments,

$$\begin{aligned}\mathbf{E}\{z_n(\omega)\bar{z}_m(\xi)\} &= \mathbf{E}\left\{\int_{-1/2}^{1/2}\int_{-1/2}^{1/2}\sigma_n(\zeta)\bar{\sigma}_m(\eta)\,\mathrm{d}Z(\omega+\zeta)\right\}\\ &= \int_{-1/2}^{1/2}\sigma_n(\omega+\eta)\sigma_m(\xi+\eta)S(\eta)\,\mathrm{d}\eta\,.\end{aligned}$$

In particular,

$$\mathbf{E}\{|z_n(\omega)|^2\} = |\sigma_n(\omega)|^2 * S(\omega)\,.$$

That is, the expected value of the nth eigenspectrum is the convolution of the true power spectrum with the nth DPSWF spectral window $|\sigma_n(\omega)|^2$. Since $\lambda_n \approx 1$ for $n \ll [2NW]$, each low order eigenspectrum is a good estimate from the point of view of bias and the spectral window corresponding to $\bar{S}(\omega_0)$, namely $\frac{1}{D}\sum_{n=0}^{D-1}|\sigma_n(\omega)|^2$, is close to $\mathbb{1}_{[\omega_0-W,\,\omega_0+W]}$, see Fig. 3.1.

Bias introduced by the convolution defining $\mathbf{E}\{|z_n(\omega)|^2\}$ can be expressed as a *local bias* term

$$L_n(\omega) = \int_{-W}^{W} |\sigma_n(\xi)|^2 S(\omega+\xi)\,\mathrm{d}\xi \tag{3.11}$$

plus the expected value of a *broadband bias* term

$$B_n(\omega) = \Big(\int_{-1/2}^{1/2} - \int_{-W}^{W}\Big)|\sigma_n(\xi)|^2\, S(\omega+\xi)\,\mathrm{d}\xi\,. \tag{3.12}$$

The value of B_n here is the expectation of the one defined in Thomson [323].

By the Cauchy–Schwarz inequality, B_n satisfies

$$B_n(\omega) \leq (1-\lambda_n)\int_{-1/2}^{1/2} S(\omega)\,\mathrm{d}\omega\,. \tag{3.13}$$

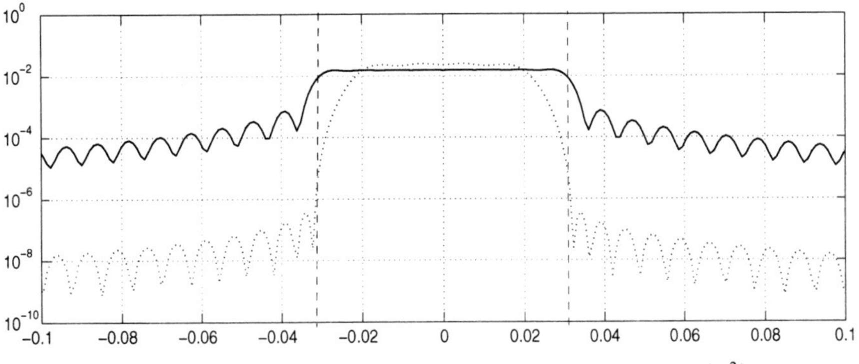

Fig. 3.1 *DPSWF spectral averages*. The solid curve is the arithmetic average of $|\sigma_n^2|$, $n = 0,\ldots,7$ for $N = 128$ and $NW = 4$ so that $W \approx 0.0312$. This average is close to being an ideal low pass filter. The dashed curve is the corresponding average over $n = 0,\ldots,3$

Iterative Spectrum Estimates

Bias characteristics of the nth DPSWF direct estimate will degrade with increasing n since σ_n becomes less localized. This leads to decreased reliability of $\bar{S}(\omega)$ where the power spectrum is relatively weak. Thomson proposed regarding the contributions to $S_n(\omega)$ coming from $[\omega - W, \omega + W]$ as *signal* and from outside as *noise*, and thinking of the order n as analogous to *frequency* in Wiener filtering, which is justified by the sign change properties of the DPSSs. The corresponding *filter* weights, denoted $d_n(\omega)$, are defined so as to minimize the expected mean-squared error (MSE) between $Z_n(\omega)$ in (3.7) and $d_n(\omega)z_n(\omega)$ with $z_n(\omega)$ as in (3.9). The difference can be expressed as

$$
\begin{aligned}
Z_n(\omega) - d_n(\omega)z_n(\omega) \\
&= \frac{1}{\sqrt{\lambda_n}} \int_{-W}^{W} \sigma_n(\xi)\,\mathrm{d}Z(\omega+\xi) - d_n(\omega)\int_{-1/2}^{1/2} \sigma_n(\xi)\,\mathrm{d}Z(\omega+\xi) \\
&= \left(\left(\frac{1}{\sqrt{\lambda_n}} - d_n(\omega)\right)\int_{-W}^{W} - d_n(\omega)\left(\int_{-1/2}^{1/2} - \int_{-W}^{W}\right)\right)\sigma_n(\xi)\,\mathrm{d}Z(\omega+\xi).
\end{aligned}
$$

Since the regions of integration of the two terms are disjoint and $\mathrm{d}Z$ has orthogonal increments, the MSE of the sum is the sum of the squares of the terms.

If the spectrum varies slowly over an interval of length $2W$ then the MSE coming from the first term is

$$
\mathbf{E}\left\{\left|\int_{-W}^{W} \sigma_n(\xi)\,\mathrm{d}Z(\xi+\omega)\right|^2\right\} \approx \lambda_n S_n^{\mathrm{est}}(\omega).
$$

The expectation of the second term is the broadband bias of the nth eigenspectrum and must be estimated from the data. An initial estimate is given by (3.13). Minimizing the MSE from the sum of the terms with respect to $d_n(\omega)$ leads to the weight estimate

$$
d_n(\omega) = \frac{\sqrt{\lambda_n}\,S(\omega)}{\lambda_n S(\omega) + B_n(\omega)} \tag{3.14}
$$

which, in turn, provides a weight averaged spectral density estimate

$$
S^{\mathrm{est}}(\omega) = \sum_{n=0}^{D-1} |d_n(\omega)|^2 S_n^{\mathrm{est}}(\omega) \Big/ \sum_{n=0}^{D-1} |d_n(\omega)|^2. \tag{3.15}
$$

Starting with initial estimates for S and the broadband bias, one estimates the weights in (3.14) and uses them to update the spectrum estimation (3.15) and the broadband bias estimate. Since the definition is recursive, the resulting spectrum estimation is a solution of

$$
\sum_{n=0}^{D-1} \frac{\lambda_n(S^{\mathrm{est}}(\omega) - S_n^{\mathrm{est}})}{|\lambda_n S^{\mathrm{est}} - B_n^{\mathrm{est}}|^2} = 0 \tag{3.16}
$$

with B_n^{est} defined as in (3.12), but with S replaced by S^{est}. One can use convergence of weights, or of the sum of their squares, as a stopping criterion; see [323].

3.1.2 *Bias, Variance, and Bandwidth Estimation*

Variance of Thomson's Estimator

In choosing W for the spectrum estimation just outlined, one requires W to be small enough so that S does not vary too much over any circular interval of length $2W$. But W cannot be too small. One should require, for example, that the time–bandwidth product $[2NW]$ be at least three in order for multitapering to be meaningful. In addition, W should be large enough so that the hypothesis that $\mathrm{d}Z$ has orthogonal increments is justified. Thomson argued that *if the spectrum* $S(\omega)$ *varies sufficiently slowly* then, because of orthogonality of the DPSSs, and since $S_n^{\mathrm{est}} = |z_n(\omega)|^2$ are direct estimates of the power spectrum S for each n using orthogonal tapers, the covariance $\operatorname{cov}\{S_n^{\mathrm{est}}(\omega), S_m^{\mathrm{est}}(\omega)\}$ will be negligible when $m \neq n$. Therefore, if S_D^{est} denotes the unweighted average of S_n^{est}, $n = 0, \ldots, D-1$, then its variance will satisfy

$$\operatorname{var}\{S_D^{\mathrm{est}}\}(\omega) = \frac{1}{D^2}\sum_{n=0}^{D-1}\sum_{m=0}^{D-1}\operatorname{cov}\{S_n^{\mathrm{est}}(\omega), S_m^{\mathrm{est}}(\omega)\} \approx \frac{1}{D^2}\sum_{n=0}^{D-1}\operatorname{var}\{S_n^{\mathrm{est}}(\omega)\}.$$

Walden et al. [341] argued that, *asymptotically*, $\operatorname{var}\{S_n^{\mathrm{est}}(\omega)\} \approx S^2(\omega)$ so that, with each term in the average contributing two degrees of freedom, $S_D^{\mathrm{est}}(\omega)$ is approximately distributed as $(1/2D)S(\omega)\chi^2_{2D}$ and, consequently, that

$$\operatorname{var}\{S_D^{\mathrm{est}}(\omega)/S(\omega)\} \approx 1/D.$$

However, the multitaper method was developed precisely for situations in which the spectrum varies rapidly or exhibits a large dynamic range. For these situations, Walden et al. proposed a method to compute the more precise expression for $\operatorname{var}\{S^{\mathrm{est}}\}(\omega)$ that does not disregard the covariance terms. For complex random variables,

$$\operatorname{cov}\{Z, W\} = \mathbf{E}\{\bar{Z}W\} - \mathbf{E}\{\bar{Z}\}\mathbf{E}\{W\},$$

and the complex form of the Isserlis theorem [184, p. 27] is

$$\operatorname{cov}\{Z_1Z_2, Z_3Z_4\} = \operatorname{cov}\{Z_2, Z_3\}\operatorname{cov}\{Z_1, Z_4\} + \operatorname{cov}\{Z_2, Z_4\}\operatorname{cov}\{Z_1, Z_3\}.$$

With z_n defined as in (3.10) and satisfying $E\{z_n(\omega)\} = 0$, the direct spectral estimates $S_n^{\mathrm{est}}(\omega) = |z_n(\omega)|^2$ satisfy

$$\begin{aligned}
&\operatorname{cov}\{S_n^{\mathrm{est}}(\omega), S_m^{\mathrm{est}}(\omega)\} = \operatorname{cov}\{|z_n(\omega)|^2, |z_m(\omega)|^2\} = \\
&\operatorname{cov}\{z_n(\omega), \bar{z}_m(\omega)\}\operatorname{cov}\{\bar{z}_n(\omega), z_m(\omega)\} + \operatorname{cov}\{z_n(\omega), z_m(\omega)\}\operatorname{cov}\{\bar{z}_n(\omega), \bar{z}_m(\omega)\} \\
&\qquad = \left|\mathbf{E}\{z_n(\omega), \bar{z}_m(\omega)\}\right|^2 + \left|\mathbf{E}\{z_n(\omega), z_m(\omega)\}\right|^2.
\end{aligned}$$

In terms of the autocorrelation sequence R of the power spectral density (PSD) S,

$$\left|\mathbf{E}\{z_n(\omega),\bar{z}_m(\omega)\}\right|^2 = \left|\sum_{k=0}^{N-1}\sum_{\ell=0}^{N-1} s_n(k)\,s_m(\ell)R(k-\ell)\mathrm{e}^{-2\pi\mathrm{i}\omega(k-\ell)}\right|.$$

Altogether, one has the computable expression

$$\begin{aligned}\mathrm{var}\{S^{\mathrm{est}}(\omega)\} = \frac{1}{D^2}\Bigg\{&\sum_{n=0}^{D-1}\sum_{m=0}^{D-1}\left|\sum_{k=0}^{N-1}\sum_{\ell=0}^{N-1} s_n(k)\,s_m(\ell)R(k-\ell)\mathrm{e}^{-2\pi\mathrm{i}(k-\ell)\omega}\right|^2\\ &+\sum_{n=0}^{D-1}\sum_{m=0}^{D-1}\left|\sum_{k=0}^{N-1}\sum_{\ell=0}^{N-1} s_n(k)\,s_m(\ell)R(k-\ell)\mathrm{e}^{-2\pi\mathrm{i}(k+\ell)\omega}\right|^2\Bigg\},\end{aligned} \tag{3.17}$$

which can be computed simply as a sum of fast Fourier transforms (FFTs). Additionally, one can consider the number of tapers needed to minimize the variance as a function of the frequency ω. Walden et al. [341] provided numerical examples showing that, for a white noise process, the covariance terms essentially disappear while, for an autoregressive process, the covariance terms can be substantial.

Effective Bandwidth of a Quadratic Spectrum Estimator

The *estimation bandwidth* of a spectral estimator is usually defined as the minimum frequency separation of approximately uncorrelated estimates. Having such a measure is essential in comparing the resolving power of different spectral estimators, since such a comparison is valid only if the different estimators exhibit the same effective bandwidth. The *equivalent bandwidth* measure of the approximate autocorrelation R between a direct spectral estimate $S^{\mathrm{est}}_{\mathrm{w}}(\omega)$ and $S^{\mathrm{est}}_{\mathrm{w}}(\omega+\xi)$ with single discrete-time taper w is defined as the sum of fourth powers of w. As argued in Walden et al. [342, 344], this measure does not lend itself to multitaper analysis in a compelling way. Instead, they proposed using the *autocorrelation width* in the context of multitaper estimates.

Percival and Walden [264] showed that for any quadratic spectral estimator

$$S^{\mathrm{est}}(\omega) = \sum_{k=0}^{N-1}\sum_{\ell=0}^{N-1} x(k)\,\bar{x}(\ell)\,Q(k,\ell)\,\mathrm{e}^{2\pi\mathrm{i}(\ell-k)\omega}$$

of a process $\mathrm{d}Z$ generating the data $\mathbf{x} = [x(0),\ldots,x(N-1)]^T$ in which Q is positive semi-definite of rank D, starting with an orthogonal basis $\{\mathbf{r}_n\}$ for the cokernel of Q, one can define the nth direct spectral estimator

$$S^{\mathrm{est}}_n(\omega) = \left|\sum_{k=0}^{N-1} \mathbf{r}_n(k)x(k)\,\mathrm{e}^{-2\pi\mathrm{i}k\omega}\right|^2$$

and express the multitaper estimate $S^{\mathrm{est}}(\omega)$ as their arithmetic mean,

$$S^{\mathrm{est}} = \frac{1}{D} \sum_{n=0}^{D-1} S_n^{\mathrm{est}}(\omega).$$

In order to define a bandwidth for this estimator, consider first the case of a direct spectrum estimator. The autocorrelation width associated with the spectral window $P = |\varrho|^2$, where ϱ is the Fourier series of a single taper $\mathbf{r} = [r(0), \dots, r(N-1)]^T$, is

$$\mathrm{width}(P) = \frac{1}{\|P\|_{L^2}^2} = \left(\sum_{\ell=1-N}^{N-1} \left(\sum_{k=0}^{N-1-|\ell|} r(k)\, r(k+|\ell|) \right)^2 \right)^{-1}. \tag{3.18}$$

It is maximized when the autocorrelation is a δ sequence if $\sum r^2(k) = 1$.

In the multitaper case one associates with $\mathbf{r}_n$ the spectral window

$$P_n = |\varrho_n(\omega)|^2, \qquad \varrho_n(\omega) = \sum_{k=0}^{N-1} r_n(k)\, \mathrm{e}^{-2\pi i k\omega}$$

and, for D fixed, sets

$$\mathbf{P}(\omega) = \frac{1}{D} \sum_{n=0}^{D-1} P_n(\omega).$$

One simply replaces P in (3.18) by $\mathbf{P}$ to get

$$\mathrm{width}(\mathbf{P}) = \frac{1}{\|\mathbf{P}\|_{L^2}^2} = \left(\sum_{\ell=1-N}^{N-1} \left(\frac{1}{D} \sum_{n=0}^{D-1} \sum_{k=0}^{N-1-|\ell|} r_n(k)\, r_n(k+|\ell|) \right)^2 \right)^{-1}. \tag{3.19}$$

Walden et al. [342] illustrated through numerical examples that, in the case of the Slepian tapers, the autocorrelation width increases to $2W$ as D increases from 0 to $[2NW]-1$.

3.1.3 Maximum Likelihood and the Karhunen–Loève Transform

The Discrete Karhunen–Loève Transform

Suppose that one is given N samples $\mathbf{x} = \{x(k),\ k = 0, \dots, N-1\}$ of a time series generated by a discrete-time stationary process with a known autocorrelation function

$$R(\ell) = \mathbf{E}\{x(k)\,x(k+\ell)\}, \quad (\ell \in \mathbb{Z}).$$

The discrete *Karhunen–Loève transform* (KLT) [169, 208] associated with the restriction of R to $\mathbb{Z}_N$ (also written as R) is

$$\mathbf{c} = \Psi^T \mathbf{x} \tag{3.20}$$

where Ψ is the orthogonal $N \times N$ matrix whose nth column $\psi_n(k)$ is the finite-time eigenvector of θ_n in the singular value decomposition

$$R = \Psi^T \Theta \Psi, \qquad \Theta = \operatorname{diag}(\theta_0, \dots, \theta_{N-1}). \tag{3.21}$$

This decomposition corresponds to an index-limited version of the eigenspace decomposition of the infinite Toeplitz matrix $M_R(k,\ell) = R(k-\ell)$.

If one is given only finitely many samples of an otherwise unknown stationary, zero-mean Gaussian process, then the autocorrelation function and, thereby, the KLT, has to be estimated from the data. As in the case of band limiting, the KLT can be ill conditioned, so estimates of the KLT based directly on estimates of R can be unstable. Maximum likelihood estimation provides the basis for one possible alternative approach. The coefficients $c_n = \langle \mathbf{x}, \psi_n \rangle$ of the expansion $\mathbf{x} = \sum c_n \psi_n$ satisfy the mean and covariance identities

$$\mathbf{E}\{c_n\} = 0; \qquad \mathbf{E}\{c_n c_m\} = \theta_n \delta_{n,m}$$

with respect to averages over realizations. That is, the coefficients $\{c_n\}$ form a sequence of independent random $N(0,\theta_n)$ variables. Thus, given R and (3.21), the likelihood L of an observed sample vector $\mathbf{x}$ is

$$L(\mathbf{x}) = \prod_{n=0}^{N-1} \frac{1}{\sqrt{2\pi\theta_n}} \mathrm{e}^{-c_n^2/(2\theta_n)}.$$

Calculus of variations leads to the maximizer $c_n^2 = \theta_n$. The problem then is to obtain a spectrum such that the observed data is the most likely data. In order to obtain numerical approximations of the KLT one can use an iterative optimization scheme. However, convergence of such a scheme will depend on a good initialization. The role that the DPSSs can play in this regard will be described now.

Double Orthogonality for KLT

The Wiener–Khintchine relation, see, e.g., [269], says that the autocorrelation is the inverse Fourier transform of the power spectral density. In the discrete setting,

$$R(k) = \int_{-1/2}^{1/2} S(\xi)\mathrm{e}^{2\pi i k\xi}\,\mathrm{d}\xi \quad (k \in \mathbb{Z}).$$

The centered Fourier series of $\psi_n(k)$ defined by (3.21) is

$$\widehat{\psi}_n(\omega) = \sum_{k=0}^{N-1} \psi_n(k)\mathrm{e}^{-\pi i\omega(2k+1-N)}$$

and the Fourier transform of the Wiener–Khintchine formula becomes

$$\theta_n\widehat{\psi}_n(\omega) = \int_{-1/2}^{1/2} S(\xi)D_N(\xi-\omega)\widehat{\psi}_n(\xi)\,\mathrm{d}\xi\,. \tag{3.22}$$

The functions $\widehat{\psi}_n$ are sometimes called *eifenfunctions*, see [323], to emphasize that they are functions in the frequency domain.

Theorem 3.1.1. *The functions $\widehat{\psi}_n$ defined by (3.21) and (3.22) satisfy the double orthogonality relationship*

$$\int_{-1/2}^{1/2} \widehat{\psi}_n(\omega)\overline{\widehat{\psi}}_m(\omega)\,\mathrm{d}\omega = \delta_{n,m}$$

and, with respect to the power spectrum as weight,

$$\int_{-1/2}^{1/2} \widehat{\psi}_n(\omega)\overline{\widehat{\psi}}_m(\omega)S(\omega)\,\mathrm{d}\omega = \theta_n\delta_{n,m}\,.$$

Proof. The orthogonality with respect to Lebesgue measure follows from Parseval's theorem and the orthogonality of the ψ_n. The orthogonality relationship with respect to the weight $S(\omega)$ follows from the Fourier transformed Wiener–Khintchine relationship (3.22): upon multiplying both sides of (3.22) by $\overline{\widehat{\psi}}_m$ and integrating, one obtains

$$\begin{aligned}
\theta_n\int_{-1/2}^{1/2} \widehat{\psi}_n(\omega)\overline{\widehat{\psi}_m(\omega)}\,\mathrm{d}\omega &= \int_{-1/2}^{1/2}\int_{-1/2}^{1/2} S(\xi)D_N(\xi-\omega)\widehat{\psi}_n(\xi)\overline{\widehat{\psi}_m(\omega)}\,\mathrm{d}\omega\,\mathrm{d}\xi \\
&= \int_{-1/2}^{1/2} S(\xi)\widehat{\psi}_n(\xi)\int_{-1/2}^{1/2} D_N(\xi-\omega)\overline{\widehat{\psi}_m(\omega)}\,\mathrm{d}\omega\,\mathrm{d}\xi \\
&= \int_{-1/2}^{1/2} S(\xi)\widehat{\psi}_n(\xi)\overline{\widehat{\psi}_m(\omega)}\,\mathrm{d}\omega\,\mathrm{d}\xi
\end{aligned}$$

since convolution with D_N reproduces the space of index-limited signals. □

The DPSSs can be viewed as the KLT basis functions for the special case in which $S(\xi) = \mathbb{1}_{[-W,W]}(\xi)$ for some $W < 1/2$. According to Szegő's theorem in the discrete context, the KLT eigenvalues will be asymptotically equal to the spectrum at frequencies spaced $1/N$ apart; see, e.g., [118, 119]. The corresponding eifenfunctions should be localized around these frequencies.

In order to motivate an iterative numerical estimation scheme, consider a reformulation of the problem of estimating the eigenvalues θ_n from the spectrum S. If $\{\beta_n\}_{n=0}^{N-1}$ form another basis for the span of the eifenfunctions then one can write

$$\widehat{\psi}_n(\omega) = \sum_{m=0}^{N-1} \gamma_{nm}\beta_m(\omega),$$

and (3.22) becomes

$$\theta_n \sum_{m=0}^{N-1} \gamma_{nm}\beta_m(\omega) = \sum_{m=0}^{N-1} \gamma_{nm} \int_{-1/2}^{1/2} S(\xi) D_N(\xi-\omega)\beta_m(\xi)\,\mathrm{d}\xi\,.$$

Multiply by $\bar{\beta}_p(\omega)$ and integrate over ω. Since D_N reproduces β_p,

$$\theta_n \sum_{m=0}^{N-1} \gamma_{nm}\langle\beta_m,\beta_p\rangle = \sum_{m=0}^{N-1} \gamma_{nm} \int_{-1/2}^{1/2} S(\xi)\beta_m(\xi)\bar{\beta}_p(\xi)\,\mathrm{d}\xi\,. \tag{3.23}$$

If we let Γ denote the matrix with entries γ_{nm} and

$$B_{mp} = \langle\beta_m,\beta_p\rangle,\; Q_{mp} = \int_{-1/2}^{1/2} S(\xi)\beta_m(\xi)\bar{\beta}_p(\xi)\,\mathrm{d}\xi,$$

then (3.23) becomes the matrix identity

$$\Theta\Gamma B = \Gamma Q. \tag{3.24}$$

The purpose of an iterative scheme is to steer an initial basis configuration toward the eigenbasis. This is not the actual method that Thomson used, however.

Consider instead starting with shifted versions $\sigma_n(\xi+2Wq)$ of the DPSWFs $\sigma_n = \sigma_n(N,W)$ in (3.23), multiplying these by a second shifted DPSWF, then integrating not over all of $[-1/2,1/2]$, but instead over a principal domain $(2qW-W, 2qW+W)$. To fix notation for this, we assume that $2NW = D \in \mathbb{N}$ (or approximately so), write $m = p+qD$ whenever $m \in \{0,\dots,N-1\}$, and finally set $\sigma^{(m)}(\xi) = \sigma_p(\xi+2qW)$ to express the shifted DPSWF that is *pth most localized* on $[2qW-W, 2qW+W]$. Define the partial inner product

$$b_{m,m'} = \int_{-W}^{W} \sigma_p(\xi+2(q-q')W)\,\sigma_{p'}(\xi)\,\mathrm{d}\xi$$

to obtain the analogue of (3.22):

$$\theta_n \sum_{m=0}^{N-1} \gamma_{nm} b_{m,m'} = \sum_{m=0}^{N-1} \gamma_{nm}\lambda_{p'} \int_{W(2q'-1)}^{W(2q'+1)} \sigma^{(m)}(\xi)\overline{\sigma^{(m')}}(\xi) S(\xi)\,\mathrm{d}\xi\,.$$

Denoting by Λ the repeated block diagonal matrix with entries $\lambda_0,\dots,\lambda_{D-1}$ and using $\mathscr{B} = \{b_{m,m'}\}$ one obtains the analogue of (3.24),

$$\Theta\Gamma\mathscr{B} = \Gamma\Lambda Q.$$

The two eigenvalue equations differ in the domains of integration defining the test function correlations. Since the eigenvalues tend to unity asymptotically in N for the DPSWFs, this says that the KLT eigenvalues are nearly determined by the local DPSWF moments of S. Thomson indicated that an iterative estimate of Θ starting from the shifted DPSWF moments can be much more accurate than one using the Toeplitz matrix generated by direct estimation of R.

3.2 Time and Band Limiting for Communications Channels

3.2.1 Prediction of Fading Envelopes

The role of fading envelopes in mobile communications will be discussed in more detail in the next section. Here the goal is to see how methods similar to the KLT spectrum estimation method can be used to predict a continuous-time envelope in order to overcome delays in feeding channel state information back from receiver to transmitter (cf. [87, 248]). The eigenvalue decomposition method is advantageous when the prediction involves error-corrupted estimates of a narrowband fading process. The *bathtub-shaped* spectrum of such a process exhibits a very steep transition around the maximum Doppler frequency; see, e.g., [163, 300]. This makes spectral factorization approaches based on rational function approximations difficult.

Let $x(t)$ denote the real component of a complex fading envelope. Consider the problem of predicting a future value $x(r)$ using noisy values $x(t)+e(t)$, where $t \in [r-\tau-2T, r-\tau]$. The errors are modeled as uncorrelated, zero-mean white noise with variance σ^2. The linearly predicted estimate is

$$x^{\mathrm{est}}(r) = \int_{r-\tau-2T}^{r-\tau} (x(t)+e(t))\, h(r-t)\, \mathrm{d}t \tag{3.25}$$

in which h is chosen to minimize the mean-squared prediction error

$$\mathrm{MSE} = \mathbf{E}\{(x(r) - x^{\mathrm{est}}(r))^2\}.$$

For present purposes, we will assume that the error due to replacing the observations $x(t)$ by observed discrete samples is negligible. One can choose r as desired. If $\tau - r = T$ then

$$\mathrm{MSE} = \mathbf{E}\left\{\left(x(\tau+T) - \int_{-T}^{T} (x+e)(t)\, h(\tau+T-t)\, \mathrm{d}t\right)^2\right\} \tag{3.26}$$

and, since x and e are uncorrelated, one can write

$$\mathrm{MSE} = \mathrm{MSE}_x + \mathrm{MSE}_e \tag{3.27}$$

in which

$$\mathrm{MSE}_x = \mathbf{E}\left\{\left(x(\tau+T) - \int_{-T}^{T} x(t)\, h(\tau+T-t)\, \mathrm{d}t\right)^2\right\} \quad \text{and} \tag{3.28}$$

$$\mathrm{MSE}_e = \mathbf{E}\left\{\left(\int_{-T}^{T} e(t)\, h(\tau+T-t)\, \mathrm{d}t\right)^2\right\}. \tag{3.29}$$

The total error can be minimized by computing optimal coefficients in a continuous Karhunen–Loève basis expansion. Thus, let ϕ_n be an eigenbasis for the autocorrelation R_{xx} acting by convolution against the image of Q_T where, as before, Q_T

is the time localization to $[-T,T]$. That is,

$$\int_{-T}^{T} R_{xx}(t-s)\,\phi_n(s)\,\mathrm{d}s = \lambda_n \phi_n(t), \tag{3.30}$$

and let

$$h(t) = \sum_{n=0}^{\infty} a_n\,\phi_n(\tau+T-t), \qquad t \in [\tau,\, \tau+2T]. \tag{3.31}$$

That such an expansion of h exists requires the completeness of the ϕ_n over $[-T,T]$, which holds if R_{xx} is positive definite.

Lemma 3.2.1. *With h expanded as in (3.31) and MSE defined as in (3.26),*

$$\mathrm{MSE}_x = R_{xx}(0) - 2\sum_{n=0}^{\infty} a_n\,\lambda_n\,\phi_n(\tau+T) + a_n^2\lambda_n^2$$

while

$$\mathrm{MSE}_e = \sigma_e^2 \sum_{n=0}^{\infty} a_n^2\,\lambda_n\,.$$

Proof. Starting from (3.31) and taking h to be real-valued,

$$\begin{aligned}
\mathrm{MSE}_x &= \mathbf{E}\Big\{x^2(\tau+T) - 2\int_{-T}^{T} x(\tau+T)x(t)\,h(\tau+T-t)\,\mathrm{d}t \\
&\quad + \int_{-T}^{T}\int_{-T}^{T} x(t)x(s)\,\bar{h}(\tau+T-t)\,h(\tau+T-s)\,\mathrm{d}s\,\mathrm{d}t\Big\} \\
&= R_{xx}(0) - 2\int_{-T}^{T} R_{xx}(\tau+T-t)\,h(\tau+T-t)\,\mathrm{d}t \\
&\quad + \int_{-T}^{T}\int_{-T}^{T} R_{xx}(t-s)\,h(\tau+T-t)\,h(\tau+T-s)\,\mathrm{d}s\,\mathrm{d}t \\
&= R_{xx}(0) - 2\sum_{n=0}^{\infty} a_n \int_{-T}^{T} R_{xx}(\tau+T-t)\,\phi_n(t)\,\mathrm{d}t \\
&\quad + \sum_{n=0}^{\infty}\sum_{m=0}^{\infty} \int_{-T}^{T} \phi_n(t) \int_{-T}^{T} R_{xx}(t-s)\,\phi_m(s)\;\mathrm{d}s\,\mathrm{d}t \\
&= R_{xx}(0) - 2\sum_{n=0}^{\infty} a_n\,\lambda_n \phi_n(\tau+T) + \sum_{n=0}^{\infty}\sum_{m=0}^{\infty} a_n\,a_m\,\lambda_m \int_{-T}^{T} \phi_n(t)\,\phi_m(t)\,\mathrm{d}t\,.
\end{aligned}$$

Assuming that $S_{xx}(\xi) = 0$ for $|\xi| > \Omega$, the Fourier transform of (3.30) is

$$S_{xx}(\xi)(Q_T\phi_n)^\wedge(\xi) = \lambda_n\widehat{\phi}_n(\xi) \quad \text{and} \quad \int_{-\Omega}^{\Omega} |(Q_T\phi_n)^\wedge(\omega)|^2\,S_{xx}(\omega)\mathrm{d}\omega = \lambda_n^2\,.$$

These observations give

$$\|Q_T\phi_n\|^2 = \int \widehat{\phi_n}\overline{(Q_T\phi_n)^\wedge}(\xi)\,\mathrm{d}\xi = \frac{1}{\lambda_n}\int S_{xx}(\xi)|(Q_T\phi_n)^\wedge(\xi)|^2\,\mathrm{d}\xi = \lambda_n .$$

It follows that

$$\mathrm{MSE}_x = R_{xx}(0) - 2\sum_{n=0}^{\infty} a_n\lambda_n\phi_n(\tau+T) + \sum_{n=0}^{\infty} a_n^2\lambda_n^2 .$$

The corresponding identity for MSE_e is obtained in a similar way. □

It follows from the lemma that

$$\mathrm{MSE} = R_{xx}(0) + \sum_{n=0}^{\infty} a_n^2\lambda_n(\sigma_e^2+\lambda_n) - 2a_n\lambda_n\phi_n(\tau+T) .$$

The error can be minimized by setting the derivatives with respect to a_n all equal to zero. This leads to the following.

Theorem 3.2.2. *The MSE of the estimate (3.25) is minimized when the coefficients of h in (3.31) are defined as*

$$a_n = \frac{\phi_n(\tau+T)}{\lambda_n+\sigma_e^2} .$$

The minimum MSE is then

$$\sum_{n=0}^{\infty}\phi_n^2(\tau+T)\frac{\sigma_e^2}{\lambda_n+\sigma_e^2} .$$

In the case of a band-limited process in which S_{xx} has finite support, it is possible to bound this error essentially in terms of a finite partial sum (see Lyman and Sikora, [211]) and to determine a suitable choice of the parameter τ. Numerical estimates of the error and of the fading envelope require methods to evaluate the eigenfunctions ϕ_n (loc. cit.). Other numerical methods for estimating eigenfunctions from discrete samples will be discussed in Chap. 6.

3.2.2 *Modeling Multipath Channels*

Because of sparsity or nonstationarity of data, it can be unrealistic to apply maximum likelihood estimators for problems such as detection and estimation. In this case, rather than attempting to identify an *eigenbasis* for the problem, it makes sense to expand data in a pre-defined basis or dictionary chosen to minimize uncertainty with respect to the environment. The choice of such a representation basis and the rationale underlying its use are often called a *basis expansion model* (BEM). Here, we outline the use of finite discrete prolate (FDP) sequences and variations thereof in the modeling of time-varying wireless communications channels. Further closely related work will be outlined in the chapter notes (Sect. 3.3). Discussion of background material pertaining to the extensive engineering literature on channel modeling will be limited to material essential for motivating the application of FDP sequences—a specific example of a "basis expansion model" (BEM)—in this context. Details of concepts just touched on here can be found in standard references such as [271,301]; cf. Giannakis and Tepedelenlioğlu [113].

In wireless communications, the transmitted signal is received across multiple propagation paths due to scattering—reflections across buildings or other physical objects—and refraction through different media. Each path has a different relative delay and amplitude. Different parts of the transmitted signal spectrum will be attenuated differently. This effect is known as frequency-selective fading. Motion of the transmitter or receiver and transient characteristics of the environment also result in time-varying fading of the received signal. A channel displaying both time-varying and frequency-dependent characteristics is called a *doubly selective* channel. Channel equalizers can be used to correct fading characteristics, provided that the channel can be estimated accurately in real time. In the case of a stationary but noisy channel, it is sufficient to estimate the impulse response of the channel by means of a Wiener filtering technique. Deconvolution can then be performed at the receiver.

When a transmitted communication signal consists of *low-frequency* information modulated onto a *high-frequency* carrier, the time variation of a wireless channel depends on the rate of change of the environment with respect to the carrier frequency f_C. Separate paths experience Doppler shifts f_ℓ depending on the velocity v of the transmitter or receiver. This can be quantified in terms of the sampling period T_S, maximum user velocity $v_{\max}$, and the speed of light c, by the normalized Doppler bandwidth

$$\nu_{\max} = \frac{v_{\max} f_C}{c} T_S \geq |f_\ell T_S| .$$

Traditional channel estimation methods are based on Wiener filtering. Specific drawbacks of the use of the Fourier transform in channel estimation are detailed in Zemen's dissertation [370], which is the primary source of much of the material in this section (cf. also [371,372]). Simulations for detection and channel estimation based on an FDPS BEM were developed in detail in Zemen et al. (loc. cit.). Those simulations require additional machinery beyond the BEM so they are not reproduced here. Instead, we restrict our focus to defining an FDPS BEM and providing a rationale for its use in communications modeling.

To keep notation to a minimum, we will index discrete sequences by units of samples. In particular, when thinking of a sequence $\mathbf{s} = \{s_k\}$ as the samples of a continuous-time function $s(t)$, we will abuse notation and write $s_k = s(k)$ rather than $s_k = s(kT_S)$. Equivalently, discrete time will be indexed in units of samples, and normalized frequency in units of cycles per sample, unless otherwise noted. We start with a very brief review of some basic concepts pertaining to wireless channels. The FDPS and generalized Slepian *GPS* basis expansion models will be described in the context of a single user system, even though the benefits of this approach are greater in multi-carrier systems. The latter will be outlined at the end of this section.

Multipath Channels

A multipath wireless channel is one containing physical objects off of which the transmitted waveform reflects or *scatters*. If the environment results in M different paths P_ℓ, $\ell = 0, \ldots, M-1$ with attenuation factors η_ℓ and delays τ_ℓ, one can express the multipath channel impulse response as

$$h_{\mathrm{mp}}(t) = \sum_{\ell=0}^{M-1} \eta_\ell\, \delta(t - \tau_\ell)\,. \tag{3.32}$$

If one associates to the transmitter a filter h_{trans} and to the receiver a filter h_{rec} then the total channel impulse response model becomes

$$h(t) = h_{\mathrm{rec}} * h_{\mathrm{mp}} * h_{\mathrm{trans}}\,. \tag{3.33}$$

It is standard to assume a Rayleigh fading, exponential power decay profile in the form

$$\mathbf{E}\{|h(\ell)|^2\} = \eta^2(\ell) = \frac{1}{\alpha}\mathrm{e}^{-\ell/T_\Delta};\ \alpha = \sum \mathrm{e}^{-\ell/T_\Delta} \tag{3.34}$$

in which T_Δ denotes the delay spread, the second moment of the power delay profile, namely

$$T_\Delta = \Big(\sum \eta_\ell^2 \tau_\ell - (\sum \eta_\ell^2 \tau_\ell^2)^2\Big) \text{ when } \sum \eta_\ell^2 = 1,$$

when expressed in units of sample periods. The delay spread for a typical urban environment is on the order of tens to hundreds of nanoseconds. The *essential support* of h in samples is selected on the basis of the signal-to-noise ratio, expressed in terms of the ratio E_b/N_0 of energy per bit (E_b) over noise spectral density (N_0), at which the system will operate, namely (cf. [174, Chap. 2], [33])

$$L \geq 1 + T_\Delta \ln\Big(\frac{E_b}{N_0}\Big)\,.$$

The parameter L depends on the complexity of the environment and increases linearly with the sample rate.

If the symbol sequence $\mathbf{d} = \{d(k)\}$ is transmitted directly over a stationary channel modeled by h, then the received symbol sequence $\mathbf{x}$ is given by convolution

$$x(k) = \sum_{\ell=0}^{L} h(\ell) d(k-\ell) = h(0)\, d(k) + \sum_{\ell=1}^{L} h(\ell)\, d(k-\ell),$$

expressing the kth received symbol as a fixed multiple of the transmitted symbol, plus a linear combination of prior symbols, called *intersymbol interference* (ISI), whose complexity increases with L. Using a smaller symbol rate can reduce time-domain equalization complexity, which is $O(L^2)$. One way in which to minimize ISI is to replace N symbols having short duration proportional to $1/N$ units of time, but requiring bandwidth proportional to N units of frequency, by N symbols each occupying on the order of one unit of time but only $1/N$ units of bandwidth. Channel nonidealities can then be overcome through equalization of a number of slowly modulated narrowband signals, rather than a single, rapidly modulated wideband signal.

In orthogonal frequency–division multiplexing (OFDM) data is divided into N parallel data streams or channels corresponding to orthogonal subcarriers, spaced closely in frequency, used to transmit the data. Each subcarrier is modulated using a conventional scheme such as quadrature amplitude modulation (QAM) or phase shift keying (PSK). It is desirable to have the receiver sum over an integer number of sinusoid cycles for each of the multipaths when demodulating the OFDM stream using a DFT. To achieve this, a guard interval containing a "cyclic prefix"—the end of the OFDM symbol—can also be inserted between successive symbols. The guard interval is transmitted followed by the OFDM symbol.

Single User, Stationary Channel Model

Assuming a channel filter length L, OFDM operating on a data symbol block of length N consists of an inverse Fourier transform, followed by a cyclic prefix insertion, followed by a circular convolution with the channel filter, cyclic prefix removal, an inverse Fourier transform, and finally division by the channel DFT. In this simplified model for OFDM, it is not the actual data block $\mathbf{d}$, but rather the chip vector $\boldsymbol{\mu}$—the cyclically prefixed inverse DFT of $\mathbf{d}$—that is being *transmitted*. Altogether, the operation has the form

$$\mathbf{d} \mapsto \mathscr{F}_N(R_{\mathrm{cp}}(H(T_{\mathrm{cp}}(\mathscr{F}_N^{-1}(\mathbf{d}))))), \tag{3.35}$$

whose steps will be described now in terms of matrices. The data block $\mathbf{d} \in \mathbb{C}^N$ is mapped into a *chip vector* $\boldsymbol{\mu} \in \mathbb{C}^Q$, $Q = N + P$, defined by $\boldsymbol{\mu} = T_{\mathrm{cp}} \mathscr{F}_N^{-1} \mathbf{d}$ where $\mathscr{F}_N^{-1}$ denotes the N-point unitary inverse DFT and T_{cp} adds the cyclic prefix. In block form, $T_{\mathrm{cp}} = I_{N,P} \oplus I_N$, where the matrix of $I_{N,P}$ consists of the last P rows of the $N \times N$ identity, I_N, $P \geq L$ and $\oplus$ refers to the matrix direct sum. The chip vector $\boldsymbol{\mu}$ is multiplied by the $Q \times Q$ Toeplitz channel matrix H, which can also be expressed as $H = H_{\mathrm{lower}} + H_{\mathrm{upper}}$ where H_{lower} has entry $h(\ell)$ all along the ℓth diagonal below the main diagonal, $\ell = 0, \ldots, L-1$, and zeros elsewhere, while H_{upper} has entry $h(\ell)$ along the $(Q-\ell)$th diagonal above the main diagonal, $\ell = 1, \ldots, L-1$, and zeros elsewhere. Cyclic prefix removal is accomplished by the $N \times Q$ matrix R_{cp} whose first P columns consist of the $N \times P$ zero matrix and whose last N columns consist of I_N. Since $(H_{\mathrm{upper}})_{k\ell} = 0$ if $k > P$, it follows that $R_{\mathrm{cp}} H_{\mathrm{upper}} = 0_{N \times Q}$. On the other hand, conjugation by the Fourier transform diagonalizes H in the sense that

$$R_{\mathrm{cp}} H_{\mathrm{lower}} T_{\mathrm{cp}} = \mathscr{F}_N^{-1} H_{\mathrm{diag}} \mathscr{F}_N$$

in which H_{diag} is the $N \times N$ diagonal matrix whose main diagonal consists of the multiplication of the vector $\mathbf{h} = [h(0), \ldots, h(L-1)]^T$ by the matrix consisting of $\sqrt{N}$

times the first L columns of $\mathscr{F}_N$. The factor $\sqrt{N}$ serves to preserve the Plancherel relationship.

Assuming a perfectly clean channel, the net result is received data

$$\mathbf{y} = H_{\text{diag}}\mathbf{d} \in \mathbb{C}^N .$$

In that event, perfect recovery of $\mathbf{d}$ requires multiplying $\mathbf{y}$ by H_{diag}^{-1}, if such an inverse exists. In the event that the channel adds white Gaussian noise of mean zero and variance σ^2, it can be shown that the received data will have the form

$$\mathbf{y} = H_{\text{diag}}\mathbf{d} + \mathbf{z},$$

in which $\mathbf{z}$ is also white noise with variance σ^2.

3.2.3 Basis Expansion Models for Channel Estimation

Basis Expansion Model

Basis expansion modeling seeks basis elements specially adapted to a given physical or geometrical setting in order to track changes in behavior with respect to a free parameter in a concise manner. Traditionally, Fourier basis elements have been used in modeling nonstationary channels in which the channel changes sufficiently slowly.

For the sole purpose of illustrating potential drawbacks of Fourier BEM, Zemen [370] proposed the simplifying hypothesis that the channel is *frequency-flat*, that is, the sampling period is much larger than the delay spread, $T_S \gg T_\Delta$, which implies effectively no ISI. The hypothesis allows one to write

$$y(m) \approx h(m)d(m) + z(m) .$$

An approximation of $h \in \mathbb{C}^M$ by a projection onto the span of D linearly independent vectors $\boldsymbol{\beta} = \{\beta(\cdot,0),\dots,\beta(\cdot,D-1)\}$ in $\mathbb{C}^D$ $(D \le M)$ can be expressed in standard coordinates by

$$h(k) \approx h^{(\boldsymbol{\beta})}(k) = \sum_{n=0}^{D-1} c_n \beta(k,n) . \tag{3.36}$$

The dimension $D = D(\boldsymbol{\beta})$ of the span of $\boldsymbol{\beta}$ should be large enough to approximate band-limited and index-limited sequences. Thus it is assumed that

$$2\lceil \nu_{\max} M \rceil \le D(\boldsymbol{\beta}) \le M - 1 .$$

As before, $\nu_{\max} \in (0, 1/2)$ is the normalized Doppler half-bandwidth. The quality of approximation is quantified in terms of

$$\mathrm{MSE}\,(M,\boldsymbol{\beta}) = \frac{1}{M}\sum_{k=0}^{M-1}\mathbf{E}\{|h(k)-h^{(\beta)}(k)|^2\}\,. \tag{3.37}$$

If $\boldsymbol{\beta}$ is orthonormal then $c_n = \langle h,\beta_n\rangle_{\mathbb{C}^M}$ in (3.36).

Fourier Basis for Frequency-flat Channel

The D-dimensional centered finite partial Fourier basis has $M \times D$ standard basis matrix $\boldsymbol{\beta} = \{\beta(k,n)\}$ with entries

$$\beta(k,n) = \frac{1}{\sqrt{M}}\mathrm{e}^{\pi\mathrm{i}(2n+1-D)k/M},\; n = 0,\ldots,D-1\,.$$

Potential drawbacks of using a Fourier BEM to model a multipath channel can be illustrated by the *single path* case in which $h(k)$ can be expressed as a single Fourier mode, $h(k) = \mathrm{e}^{2\pi\mathrm{i}\nu k}$ where $0<\nu<\nu_{\max}$. To define the "basis response" one thinks of the nth basis vector as a transmitted symbol sequence whose kth channelized output is

$$\sum_{\ell=0}^{M-1} h(\ell)\,\beta(k-\ell,n) = \sum_{\ell=0}^{M-1} h(k-\ell)\,\beta(\ell,n) = \overline{\langle\overline{h(k-\cdot)},\beta_n\rangle_{\mathbb{C}_M}}$$

when the circulant convolution is defined over $\mathbb{Z}_M$. For $k\in\{0,\ldots,M-1\}$, one can then define the "instantaneous basis response" as

$$\sum_{n=0}^{D-1}\beta(n,k)\langle\overline{h(k-\cdot)},\beta_n\rangle_{\mathbb{C}_M},$$

roughly expressing the extent to which the instantaneous state of the channel is captured by the basis $\boldsymbol{\beta}$. When $h(k) = \mathrm{e}^{2\pi\mathrm{i}\nu k}$, the instantaneous response by the basis expansion $\boldsymbol{\beta}$ is given by (cf. [245])

$$H(k,\nu) = \sum_{n=0}^{D-1}\beta(k,n)\sum_{\ell=0}^{M-1}\bar{\beta}(\ell,n)\mathrm{e}^{-2\pi\mathrm{i}\nu(k-\ell)}\,. \tag{3.38}$$

Ideally, the basis response is identically equal to one. Deviation from this ideality can be quantified by the amplitude and phase of H via the error

$$E(k,\nu) = |1-H(k,\nu)|^2, \tag{3.39}$$

(see Fig. 3.2) and its block average

$$E_M(\nu) = \frac{1}{M}\sum_{k=0}^{M-1}E(k,\nu)\,. \tag{3.40}$$

In [370] it was shown that taking $\boldsymbol{\beta}$ to be the Fourier basis is insufficient for error-free detection.

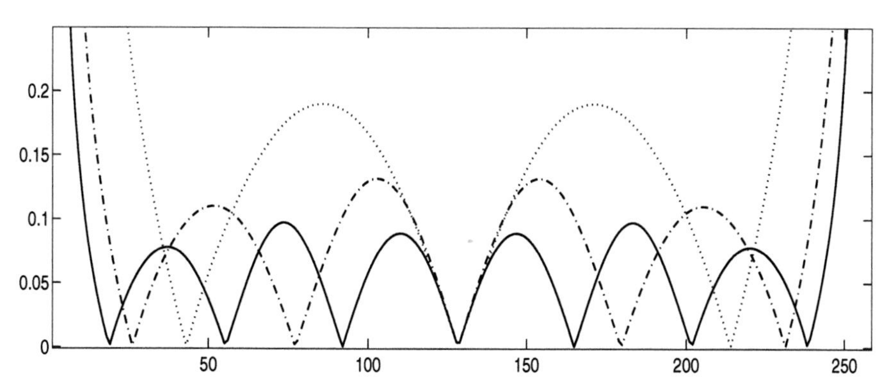

Fig. 3.2 Absolute phase of Fourier error $E(k,\nu)$ (3.39) for $\nu = 1/(2M)$, $M = 256$, and $D = 3$ (dots), $D = 5$ (dash-dot), and $D = 7$ (solid)

Slepian Sequence BEM

The Slepian DPS sequences, $\mathbf{s}_n = \mathbf{s}_n^{(M,\nu_{\max})}$, are index limited to the discrete-time block $\{0,\dots,M-1\}$ and optimally concentrated in frequency inside $[-\nu_{\max}, \nu_{\max}]$. As discussed in Sect. 1.4, the DPSSs are eigenvectors of the Toeplitz sinc matrix $[\mathrm{sinc}_{2\nu_{\max}}(\ell - k)]$, $0 \le k, \ell < M$ (cf. (1.30) with $W = \nu_{\max}$) and they are orthogonal over the finite discrete-time interval $\{0,\dots,M-1\}$, while their centered Fourier series are orthogonal over $[-1/2, 1/2]$ as well as over $[-\nu_{\max}, \nu_{\max}]$. The number of eigenvalues larger than $1/2$ is at most $D = \lceil 2M\nu_{\max} \rceil$, which can be taken as the dimension for an M-index DPSS BEM channel estimation.

Basis Expansion Model Bias

For an ensemble of channel realizations, one can consider the autocorrelation R_h and power spectrum S_h defined by

$$R_h(k) = \mathbf{E}\{h(\cdot)\,\overline{h(k+\cdot)}\} \quad \text{and} \quad S_h(\nu) = \sum_k R_h(k)\,\mathrm{e}^{2\pi i k\nu}.$$

One defines the *square bias* by integrating a single path error (3.39) over the power spectrum, namely

$$\mathrm{bias}^2(k) = \int_{-1/2}^{1/2} E(k,\nu) S_h(\nu)\,\mathrm{d}\nu.$$

For a block of length M, one defines bias_M^2 as the block average or, equivalently, the integration of $E_M(\nu)$ in (3.40) against S_h. The bias part of the BEM MSE (3.37) is independent of noise variance, while the variance expression for the BEM is

$$\text{var}^{(\beta)}(k) \approx \sigma^2 \beta(k,\cdot)\beta^*(\cdot,k),$$

which is exact when $\text{bias}^2(k) = 0$. For orthonormal β, the block averaged variance is $\text{var}_M = \sigma^2 D/M$, which increases linearly with the basis dimension D but otherwise does not depend on β.

Pilot-based Channel Estimation

In mobile wireless communications, multipath channels vary with time due to the motion of transmitters and receivers. In order to maintain an accurate channel model, it is necessary to monitor the channel impulse response. One way in which to do so is by insertion of *pilot symbols* into the data stream. For a frequency-flat channel, information symbols $b(k)$ are interleaved with approximately evenly spaced pilot symbols $p(k)$ to produce the transmit data $d(k)$ and received data $y(k)$ determined by the pilot set $\mathscr{P}$,

$$\begin{cases} d(k) = b(k) + p(k) \\ y(k) = h(k)d(k) + z(k) \end{cases}, \qquad \begin{cases} b(k) = 0, k \in \mathscr{P} \\ p(k) = 0, k \notin \mathscr{P} \\ \mathscr{P} = \{\lfloor \frac{M}{2J} + \frac{qM}{J} \rfloor : q = 0, \ldots, J-1\}\,. \end{cases} \tag{3.41}$$

For example, if $M = 15$ and $J = 3$ then $\mathscr{P} = \{2,7,12\}$ while if $J = 4$ then $\mathscr{P} = \{2,6,10,13\}$. When $2J$ divides M, the pilot indices form a coset of $\mathbb{Z}_M$. Typically, the entries $p(m)$ for $m \in \mathscr{P}$ are i.i.d normalized random elements from the PSK alphabet. Denote by $\mathbb{1}_{\mathscr{P}}$ the indicator function of the pilot indices and by $\mathbf{p}\beta_n$ the componentwise Schur product with kth entry $p(k)\beta_n(k)$. The channel values $h(k)$ have to be estimated from the pilot symbols via the BEM partial correlation

$$\frac{1}{\|\mathbb{1}_{\mathscr{P}}\beta_n\|^2}\langle \mathbf{y}, \mathbf{p}\beta_n \rangle\,. \tag{3.42}$$

Power limitations of mobile devices require taking $\mathscr{P}$ and D as small as possible. To minimize bias, Zemen [370] proposed using *generalized finite Slepian sequences* (GPSs) that are orthogonal over the full block as well as over the pilot set $\mathscr{P}$. Given any subsets S and Σ of $\mathbb{Z}_N$, it is possible to define an operator analogous to $P_\Sigma Q_T$, whose eigenvectors are the finite sequence analogues of the eigenfunctions of $P_\Sigma Q_T$, and it is also possible to give a general criterion for *double orthogonality* as in the continuous case for the operator $P_\Sigma Q_T$ and the finite discrete prolate (FDP) sequences outlined in Sect. 1.5. Before defining the generalized finite DPSSs, we briefly review the FDPS construction.

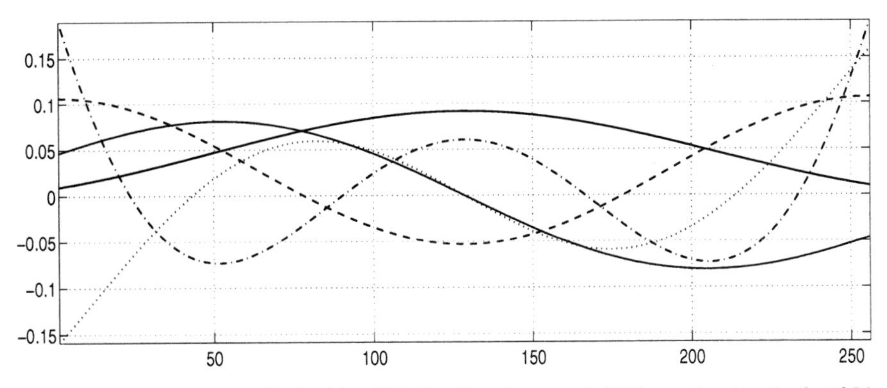

Fig. 3.3 FDP sequences $\mathbf{s}_n/\lambda_n$ for $M=256$, $Q=2$, and $\nu_{\max}=0.0039$, restricted to the first 256 out of 512 indices. Here, $M\nu_{\max}$ rounds up to one. The nth sequence has n sign changes

One can define finite "FDPS" analogues of the DPSSs on $\mathbb{Z}_P$ as in Sect. 1.5 (there we used N for the ambient dimension). The FDP sequences are orthonormal over all of $\mathbb{Z}_P$ and orthogonal over the index subset $\{0,\dots,M-1\}$. For notational convenience, in what follows we will assume that $\nu_{\max}$ is rounded so that $P\nu_{\max}\in\mathbb{N}$ corresponds to the DFT bandwidth K on p. 35. The FDP sequences are taken to be the *left* singular vectors of the $P\times M$ matrix with (q,m)th entry (cf. (1.37))

$$W_{qm}=\frac{\sin(\pi(2\nu_{\max}P+1)(q-m)/P)}{P\sin\pi\left(\frac{q-m}{P}\right)},\ 0\le q<P,0\le m<M.$$

The left singular vectors are the solutions of

$$WW^T\mathbf{s}_q=\lambda_q^2\,\mathbf{s}_q,\quad q=0,\dots,\min\{2P\nu_{\max},M-1\}.$$

One uses the term *finite Slepian vector* or FDPS to refer either to $\mathbf{s}_q$ or to the restriction of $\mathbf{s}_q$ to its first M indices; cf. [370]. Several FDPSs are plotted in Fig. 3.3. As Grünbaum indicated (as noted in Sect. 1.5), in the special case $P=MQ$ one can think of $W=W_Q$ as an approximation of the infinite matrix of the DPS sequences with entries

$$W_{qm}^{\infty}=\frac{\sin 2\pi(q-m)\nu_{\max}}{\pi(q-m)}.$$

The eigenvalues of W_Q converge to those of W^∞ as $Q=P/M\to\infty$ [126, 127].

In analogy with the $P_\Sigma Q_S$ operators, W_Q can be regarded as the composition of an index-limiting and a band-limiting operator, as was proposed in [375], where the index-limiting operator takes the form

$$T=I_M\oplus 0_{P-M};\qquad B=I_{P\nu_{\max}+1}\oplus 0_{P-2P\nu_{\max}-1}\oplus I_{P\nu_{\max}}.$$

Then the matrix

$$W_{P,M}=\left[W\,|\,0_{P\times(P-M)}\right]^T=\mathscr{F}_P^{-1}B\mathscr{F}_PT$$

corresponds to index limiting to $\{0,\dots,M-1\}$, followed by band limiting to $[-\nu_{\max}, \nu_{\max}]$. Applying $W_{P,M}$ to the zero padding of $\mathbf{h} = [h(0),\dots,h(M-1)]^T$ produces an extrapolation

$$W_{P,M}\begin{bmatrix}\mathbf{h}\\0\end{bmatrix} \approx \begin{bmatrix}\mathbf{h}\\\tilde{\mathbf{h}}\end{bmatrix}.$$

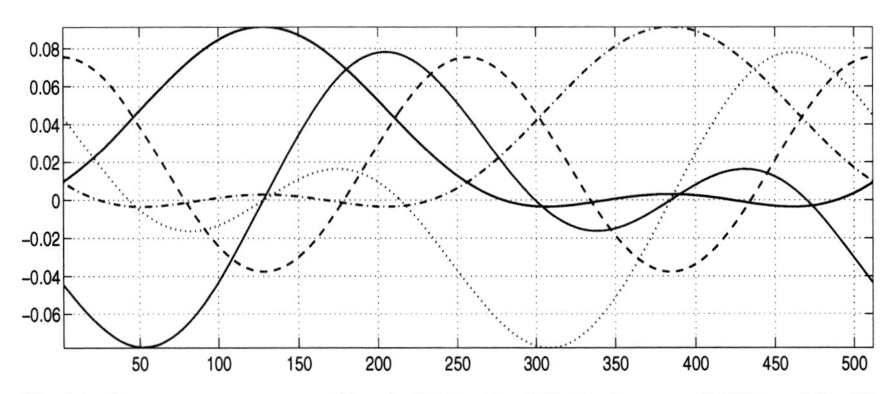

Fig. 3.4 GP sequences $\mathbf{u}_0,\dots,\mathbf{u}_4$ of length 512 for $M = 256$, $Q = 2$, $\nu_{\max} = 0.0039$, and $J = 30$ pilot indices. Here, $M\nu_{\max}$ rounds up to one. The sequence $\mathbf{u}_n$ has n sign changes over the first 256 indices. The first two sequences $\mathbf{u}_0$ and $\mathbf{u}_1$ are index concentrated in the first 256 samples, but $\mathbf{u}_n$, $n \geq 3$ are mostly concentrated in the last 256 samples

Generalized Prolate Sequences

Zemen introduced generalized FDP sequences (i.e., GPSs) as a means of approximating channel impulse response vectors of length M through the channel response at specified pilot positions. The ambient dimension P should be large enough that the DFT on $\mathbb{Z}_P$ has sufficient frequency resolution.

By defining $T = I_M \oplus 0_{P-M}$, one produces a time- and band-limiting operator whose *orthogonality set* consists of the indices $\{0,\dots,M-1\}$. One can change this orthogonality set to the pilot set $\mathscr{P} \subset \mathbb{Z}_M$ by replacing I_M by $\mathbb{1}_{\mathscr{P}}$, the diagonal matrix having ones along the diagonal elements corresponding to pilot indices $p \in \mathscr{P}$, and zeros elsewhere. Thus one sets $T_{\mathscr{P}} = \mathrm{diag}(\mathbb{1}_{\mathscr{P}} \oplus 0_{P-M})$. The corresponding time–frequency localization block matrix is

$$W_{P,\mathscr{P}} = \mathscr{F}_P^{-1} B \mathscr{F}_P T_{\mathscr{P}} = \left[\frac{W_{\mathscr{P}}^T}{0_{(P-M)\times P}}\right]^T.$$

The *GP* sequences $\mathbf{u}_n^{\mathscr{P}}$ are the left singular vectors of $W_{P,\mathscr{P}}$. Several GP sequences are lotted in Fig. 3.4. The *essential* rank of $W_{P,\mathscr{P}}$, which is at most $J = |\mathscr{P}|$, provides an upper bound for the *essential* dimension of the space spanned by the GP sequences having eigenvalues close to one.

From the point of view of channel identification, it is necessary to have a sufficient number of pilot symbols J. Specifically, J should be as large as the dimension of signals band limited to $[-\nu_{\max}, \nu_{\max}]$ that are essentially time supported in $\mathscr{P}$. Denote by $\mathbf{u}_n^{\mathscr{P},M}$ the restriction of $\mathbf{u}_n^{\mathscr{P}}$ to $\mathbb{Z}_M$. The $\mathbf{u}_n^{\mathscr{P},M}$ are nearly orthogonal.

The GP basis approximation is given by

$$h(k) \approx h^{\mathscr{P}}(k) = \sum_{n=0}^{D-1} c_n \mathbf{u}_n^{\mathscr{P},M}(k)$$

in which c_n is estimated via (3.42), that is, with y as in (3.41),

$$c_n = \frac{1}{\|\mathbb{1}_{\mathscr{P}} \mathbf{u}_n^{\mathscr{P}}\|^2} \langle \mathbf{y}, \mathbf{p}\mathbf{u}_n^{\mathscr{P}} \rangle \approx \frac{1}{\lambda_n^2} \sum_{k \in \mathscr{P}} y(k) \overline{p(k)\, u_n^{\mathscr{P}}(k)}\,.$$

Basis Expansion Variance

When $\beta = \{\beta(k,\ell)\}$ has columns that are not necessarily orthogonal, the instantaneous basis frequency response is slightly more complicated than in (3.38). First, define the invertible $D \times D$ Gram matrix $G = \beta^*\beta$. The instantaneous frequency basis response can then be defined as

$$H(k,\nu) = \beta(k,\cdot)\, G^{-1} \sum_{\ell=0}^{M-1} \beta^*(\cdot,\ell) \mathrm{e}^{-2\pi i \nu(k-\ell)}\,. \tag{3.43}$$

The basis expansion instantaneous and mean variance estimators can also be defined by (cf. [245, Sec. 6.1.4])

$$\mathrm{var}(k) = \sigma^2 \beta(k,\cdot)\, G^{-1} \beta^*(\cdot,k) \quad \text{and} \quad \mathrm{var}_M = \frac{\sigma^2}{M} \sum_{k=0}^{M-1} \mathrm{var}(k)$$

where σ^2 is the channel noise variance. In the case of the pilot-based GPSs, these quantities can be defined by restricting the GP sequences to the pilot indices. If one chooses D small enough that the GP eigenvalues $\lambda_n^{\mathscr{P}}$ are close to one for $n = 0,\dots,D-1$, then for this special case the orthogonality of the pilot-restricted GPSs provides the variance estimate

$$\mathrm{var}_M \approx \sigma^2 D/J\,.$$

In particular, the variance decreases as the number J of pilot indices increases.

Zemen also established analytical estimates for the square bias with respect to a GP BEM under the hypothesis of a *Jakes spectrum* and demonstrated a very close correspondence between the theoretical and numerical simulation estimates, which gives errors several orders of magnitude smaller than those produced by a Fourier BEM; see Fig. 3.5.

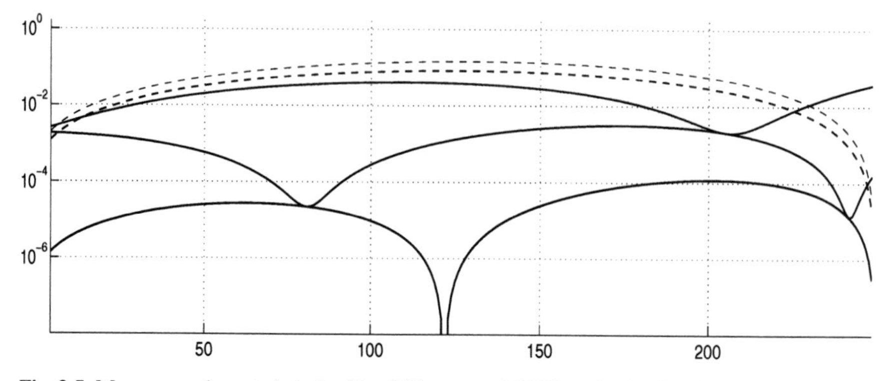

Fig. 3.5 Mean error characteristic for $M = 256$, $\nu_{\max} = 0.0039$, and using Fourier basis expansions with $D = 3$ and $D = 5$ (dashed) and Slepian basis with $D = 3,4,5$. The error $E_M(\nu)$ decreases with D and is much smaller for the Slepian basis

3.2.4 A Multi-carrier, Time-varying Channel Model

Multiple User, Stationary Channel Model

In addition to reducing the complexity of ISI, a broader form of OFDM also allows multiple users to communicate over the same spectrum by spreading data symbols associated with individual users over multiple subcarriers in such a way that different users' data live in orthogonal subspaces. This method is called multi-carrier code division multiple access (MC-CDMA). For a system having U users, each symbol is spread by means of a user-specific *spreading sequence* $\mathbf{s}_{\mathrm{u}} \in \mathbb{C}^N$, $\mathrm{u} = 1,\dots,\mathrm{U}$. Then an information symbol b_{u} associated with user u is pre-encoded as $\mathbf{d}_{\mathrm{u}} = b_{\mathrm{u}}\mathbf{s}_{\mathrm{u}}$. The received data takes the form

$$\mathbf{y}_{\mathrm{u}} = b_{\mathrm{u}} H_{\mathrm{diag}} \mathbf{s}_{\mathrm{u}} + \mathbf{z},$$

with H_{diag} defined as in (3.35). The data vector $\mathbf{d} \in \mathbb{C}^N$ carries only the single bit b rather than N units of information as in the single user case. One refers to the ratio U/N as *load*. In principle, the information associated with $\mathrm{U} \leq N$ users can be transmitted *orthogonally* over the system at the Nyquist rate. In some cases it is possible to recover information in an overloaded system ($\mathrm{U} > N$) that is transmitted via almost orthogonal spreading codes.

When a single channel has to be shared by all users, the operations have to be expressed in a different order. One sets $\mathbf{b}(k) = [b_1(k),\dots,b_{\mathrm{U}}(k)]$ and expresses spreading codes in terms of an $N \times \mathrm{U}$ *spreading matrix* $\mathbf{S}$.

Multiple User Model and Detection

In this brief discussion, we will consider only the case in which all users are transmitted over the same time-varying channel. Effects of path loss and fading will be ignored in order to maintain focus on the role of pilots. A common set of pilot symbols is used to aid in channel estimation. As before, the pilot index set is $\mathscr{P} = \{\lfloor M/(2J) + qM/J \rfloor : q = 0, \dots, J-1\}$. For $m \in \mathscr{P}$, one associates the pilot symbol vector $\mathbf{p}(m) = p(m,n)$, $n = 0, 1, \dots, N-1$. Typically, the entries $p(m,n)$ are i.i.d random elements from the PSK alphabet scaled by the number of users so that the energy per OFDM pilot symbol is the same as per data symbol. If $m \notin \mathscr{P}$ then $\mathbf{p}(m)$ is the zero element of $\mathbb{C}^N$. After OFDM and cyclic prefixing one obtains the chip vector

$$\boldsymbol{\mu}(k) = T_{\mathrm{cp}} \mathscr{F}_N^{-1} \mathbf{d}(k) .$$

If the normalized Doppler bandwidth $\nu_{\max}$ is much smaller than the normalized subcarrier bandwidth P/N, then one is justified in treating the time-variant channel as constant over the duration of an OFDM symbol. This constraint on $\nu_{\max}$ is stronger than the under-spread condition $\nu_{\max} L < 1$. Then one can write $h(k,\ell) = h'(kP,\ell)$ where h' acts on the serialized chip stream μ where $\boldsymbol{\mu}(k) = [\mu(kP), \dots, \mu(kP + P - 1)]^T$ with chip period T_c T_s related to sample period T_s by $T_c = T_s/P$. Setting $\mathbf{h}(k) = [h(k,0), \dots, h(k,L-1)]^T$ and $H_{\mathrm{diag}}(k) = \sqrt{N} \mathrm{diag} \mathscr{F}_{N\times L} \mathbf{h}(k)$, the received signal takes the form

$$\mathbf{y}(k) = H_{\mathrm{diag}}(k)(\mathbf{S}\mathbf{b}(k) + \mathbf{p}(k)) + \mathbf{z}(k) \in \mathbb{C}^N .$$

In the absence of noise, the problem of recovering the information $\mathbf{b}(k)$ boils down to inverting $\mathbf{c}_{\mathrm{u}} = (H_{\mathrm{diag}}(k)\mathbf{S})_{\mathrm{u}}$ on the complement of the pilot indices. In the case of a noisy, unknown time-varying channel, the problem is that of quantifying the most likely information sequences that produced the output $\mathbf{y}(k)$, $k = 0, \dots, M-1$, before applying a left inverse.

It is possible to produce an unbiased minimum MSE filter estimate by iterating the formula [370, p. 23]

$$\mathbf{f}_{\mathrm{u}}^* = \frac{\mathbf{c}_{\mathrm{u}}^* (\sigma^2 I + \mathbf{C} V \mathbf{C}^*)^{-1}}{\mathbf{c}_{\mathrm{u}}^* (\sigma^2 I + \mathbf{C} V \mathbf{C}^*)^{-1} \mathbf{c}_{\mathrm{u}}} \tag{3.44}$$

in which $\mathbf{C}(k) = (H_{\mathrm{diag}}(k)\mathbf{S})$ where $\mathbf{S}$ is the spreading matrix and V denotes the error covariance matrix

$$V = \mathbf{E}\{(\mathbf{b}(k) - \widetilde{\mathbf{b}}(k))(\mathbf{b}(k) - \widetilde{\mathbf{b}}(k))^*\}$$

The matrix V is estimated empirically as the diagonal matrix whose entry $V_{\mathrm{u,u}}$ is taken as the mean value of $|1 - \widetilde{b}_{\mathrm{u}}(k)|$ over each block of length M. Here, $\widetilde{\mathbf{b}}(k)$ contains the *soft* bit estimates that are estimated iteratively using the "BCJR" algorithm [15], based on extrinsic probabilities of occurrence of each symbol. We refer to [370] for details on convergence of this estimation scheme.

3.3 Notes and Auxiliary Results

3.3.1 DPS Sequences and Code Design

Pulse shaping refers to designing specific waveforms to compensate for channel characteristics, thus minimizing degradation resulting from intersymbol (ISI) or intercarrier interference (ICI). The idea of pulse shaping was described already in the works of Nyquist [247] and Hartley [137]. Its specific use for removing ISI was developed further by Gerst and Diamond in 1961 [112], with further aspects described in Dines and Hazony [81] and Papoulis [256]. It has become a standard textbook topic; see, e.g., [271]. DPS sequences and their generalizations can be used as bases for pulse shaping to reduce ISI and ICI.

Orthogonal Pulses

Hua and Sarkar [158] investigated the problem of designing numerically a discrete, finite-duration signal whose spectral energy is maximally concentrated in a given band while being orthogonal to block-shifted copies of itself. Extension of the orthogonality and concentration problem over an ensemble of such signals will be considered below. The problem is phrased in terms of a nonlinear cost function whose minimization nearly gives the desired orthogonality and concentration. Hua and Sarkar worked in units of *baud time* where baud rate refers to the number of symbols per second. Thus, a signal vector extending over N_c baud times corresponding to N_s symbols per baud time will contain $N = N_c N_s$ sample values. In the frequency domain, one assumes N_r samples per baud rate or inverse baud time, with a total of $M = N_r N_s$ DFT points. One thus works with the non-square DFT matrix

$$W_{N,M}(k,\ell) = \frac{1}{\sqrt{M}} \mathrm{e}^{-2\pi i k\ell/M}, \quad k = 0,\dots,N-1,\ \ell = -\frac{M}{2},\dots,\frac{M}{2} \bmod M.$$

The objective is to identify a vector $\mathbf{f}$ having its DFT spectrum energy concentrated within $-N_b \leq \ell \leq N_b$ (mod M) subject to the orthogonality constraint

$$\sum_{k=0}^{N-1} f(k) f(k+qN_s) = \delta_q, \quad q = 0,1,\dots,N_c-1.$$

The corresponding cost function to be minimized over weights $d_0,\dots,d_c$ is

$$J = d_c \sum_{|\ell|>N_b} |(W_{N,M}\mathbf{f})(\ell)|^2 + \sum_{q=0}^{N_c-1} d_q \Big(\sum_{k=0}^{N-1} f(k) f(k+qN_s) - \delta_q \Big).$$

One performs gradient descent on J relative to the values $f(k)$. The weights d_q are updated iteratively in order to steer $\mathbf{f}$ toward the orthogonality constraint.

Spectral Efficiency

Spectral efficiency refers to information rate per unit frequency, often measured in (bits/ sec)/ Hz, although (samples/ sec)/ Hz might also be used. In [131], Hamamura and Hyuga considered this aspect of the use of DPS sequences in MC-CDMA in comparison to *pure OFDM*, in which truncated sinc signals are used for the different carriers. Using the eigenvalue property of the DPS sequences, it is simple to show that the expected *out-of-band* energy of a discrete, finite-time signal expanded in the first U length N DPS sequences band limited to $[-W,W]$ with expected energy per coefficient $\mathscr{E}$, is $\mathrm{U}\mathscr{E}\lambda^{\mathrm{ave}}$, where λ^{ave} is the average of the first U DPSS eigenvalues. The spectral efficiency can be phrased as the dimension U of the subspace of DPS sequences used, versus decay of the average λ^{ave} of the first U eigenvalues for a given discrete time–bandwidth product. Hamamura and Hyuga showed that a 95% spectral efficiency can be achieved using a time–bandwidth product $WT \approx 45$ for DPS sequences, whereas OFDM requires $WT \approx 373$ to achieve the same spectral efficiency. Prior to the work of Hamamura and Hyuga, the use of DPSSs for improved spectral efficiency was proposed by Gruenbacher and Hummels [124] in the context of digital modulation, using a form of phase shift keying called QPSK$''$.

Coding for Temporal Dispersion

Raos et al. [274–276] also considered DPS sequences as possible codes for doubly selective MC-CDMA channels. Given signals of duration T and maximal temporal dispersion T_c (the length of the channel impulse response), one can employ DPS sequences as user codes time limited to $[0,T-T_c]$ in order to avoid ISI. This leads to the problem of identifying matched filters for the multiple user detector/decorrelator at the receiver. Because of time dispersion, a T-time-limited input will produce an output of duration $T+T_c$. If the uth user spreading code is $\mathbf{sp}_{\mathrm{u}} = [\mathrm{sp}_{\mathrm{u}}(0),\dots,\mathrm{sp}_{\mathrm{u}}(M-1)]$, then corresponding matched filter receiver codes $\mathbf{r}_{\mathrm{u}}$ can be defined by

$$r_{\mathrm{u}}(k) = b_{\mathrm{u}}(k) + \sum_{\mathrm{v}\neq\mathrm{u}} b_{\mathrm{v}}(k)\,\mathbf{sp}_{\mathrm{v}}^T\mathbf{H}_{\mathrm{v}}\mathbf{H}_{\mathrm{u}}^{-1}\mathbf{sp}_{\mathrm{u}},$$

where

$$\mathbf{H}_{\mathrm{u}} = \mathrm{diag}(H_{\mathrm{u}}(0),H_{\mathrm{u}}(W),\dots,H_{\mathrm{u}}(\mathrm{U}W)) \quad \text{with}$$

$$H_{\mathrm{u}}(kW,t;T,T_c) = \int_{t-T_c}^{T} h_{\mathrm{u}}(t-\tau)\big(\lfloor\tau\rfloor - \lfloor\tau-T\rfloor\big)\,\mathrm{e}^{-2\pi ikW\tau}\,\mathrm{d}\tau\,.$$

Thus, in order to recover the symbols $b_{\mathrm{u}}(k)$ from the received data $r_{\mathrm{u}}(k)$, one must invert the linear transformation that maps the vectors $\mathbf{b}_1(k),\dots,\mathbf{b}_{\mathrm{U}}(k)$ to $\mathbf{r}_1(k),\dots,\mathbf{r}_{\mathrm{U}}(k)$ or, in the noisy case, apply the minimum MSE inversion.

Coding for Frequency Dispersion

While Raos et al. had the goal of simultaneously minimizing ISI due to time dispersion and multiple access interference, Chen and Chen [71] proposed using DPSS-based pulse shaping in order to protect against frequency dispersion caused by Doppler spread. The optimized pulse-shaping filters maximize the signal energy around the carrier frequencies, resulting in minimal spectral overlap between the adjacent subchannels. Chen and Chen also employed offset QAM in which, in the case of two adjacent OFDM carriers, time offset is introduced in the imaginary part of one of the carriers and the real part of the other. Using the first few DPS sequences enables good spectral localization, but makes producing orthogonal pulse-shaping filters more challenging. Here is a brief outline of the approach taken in [71]. Fix the parameters N and W in the definition of the DPS sequences, see (1.30). Expand the impulse g into even order DPS sequences by taking $g(k) = \sum_{n=0}^{D-1} a_n \mathbf{s}_{2n}(k)$. Suppose now that the OFDM system has M subcarriers (M is even). Fix a modulation parameter α and phase factor $\mathrm{e}^{2\pi ik\alpha/(2M)}$ and define the transformation that maps $\mathbf{x}$ to

$$y(\ell) = \sum_{\mathrm{u}=0}^{M-1}\sum_{k\in\mathbb{Z}}\Big(\mathrm{Re}\,x_{\mathrm{u}}(k)g(\ell-kM) + \mathrm{i}\,\mathrm{Im}\,x_{\mathrm{u}}(k)g\Big(\ell+\frac{M}{2}-kM\Big)\Big)\,\mathrm{e}^{2\pi ik(\frac{2\ell-\alpha}{2M})}\,.$$

The pulse-shaping filter g is said to be *orthogonal* if the separate signals $\mathbf{x}_\mathrm{u}$ can be recovered from the output y. Chen and Chen showed that when g is symmetric about $M/2$, these conditions may be expressed as

$$\sum_{\ell\in\mathbb{Z}} g(\ell-kM)\,g\Big(\ell+\frac{M}{2}\Big)\,\sin\Big(\frac{2\pi m}{M}\Big(\ell-\frac{\alpha}{2}\Big)\Big) = 0 \quad \text{and}$$

$$\sum_{\ell\in\mathbb{Z}} g\Big(\ell+\frac{M}{2}-kM\Big)\,g\Big(\ell+\frac{M}{2}\Big)\,\cos\Big(\frac{2\pi m}{M}\Big(\ell-\frac{\alpha}{2}\Big)\Big) = \delta_k\delta_m\,.$$

In terms of the *discrete Zak transform*,

$$(Zg)(\ell,\theta) = \sum_{\nu=-\infty}^{\infty} g\Big(\ell+\nu\frac{M}{2}\Big)\,\mathrm{e}^{-2\pi i\nu\theta}\,,$$

the orthogonality condition becomes

$$|(Zg)(\ell,\theta)|^2+|(Zg)(\ell,\theta-1/2)|^2 = 2/M, \quad \ell=0,\ldots,(M-2)/2, \quad \theta\in[0,1))\,.$$

Here one makes use of the symmetry of the filter. Chen and Chen used this identity to define an error function for nonlinear minimization on a discrete frequency-sampled grid for the discrete Zak transform. The grid size is defined in terms of the dimension D of DPS sequences used to define the filter g. Then gradient minimization is performed on the filter coefficients $\langle g,\mathbf{s}_n\rangle$.

3.3.2 Further Aspects of DPSSs in Channel Modeling

A variation of the basic DPSS scheme for channel estimation was suggested by Sejdic et al. [291] in which a single Slepian basis is replaced by a dictionary consisting of a finite union of modulated DPS sequences having the ability to zoom in on regions of the spectrum in which the channel is concentrated, as with Thomson's *high resolution estimate*. A matching pursuit scheme provides a *sparse* approximation of the channel that is accurate relative to the number of basis terms used.

Alcocer-Ochoa et al. [3] asked whether the DPSSs might be *universal* in the sense of channel modeling for multiple-input, multiple-output (MIMO) channels. In essence, the question is whether the MIMO channels can be orthogonalized simultaneously *in space* to remove ICI as well as *in time* to remove ISI. They answered this question in the affirmative for the very special case in which the spatial and temporal variations are separable. Channel approximations of both narrowband and broadband stationary channels DPSS BEMs were considered.

One of the most promising approaches to modeling a doubly selective channel using a DPSS BEM was developed by Sigloch et al. [297]. In their approach, the first-order temporal and spectral fluctuations within a single symbol block were quantified. The channel operator expressed in the DPSS basis was shown to be nearly tridiagonal. Sigloch et al. provided a *quantum mechanical* heuristic justification of this observation, based on a comparison with the Hermite functions as limiting case of PSWFs. The multi-carrier extension of the continuous-time channel output

$$y(t)+z(t) = \int_0^\infty h(t,\tau)\,x(t-\tau)\,\mathrm{d}\tau + z(t)$$

with input x and doubly selective channel $h(t,\tau)$ was discretized and expressed in matrix form

$$\mathbf{y} = \mathbf{H}\mathbf{x}\,.$$

Here, $\mathbf{H}$ operates on a symbol block of input samples. If one is willing to replace the discretized channel by its *linearization*—that is, the discretization of a joint time–frequency linearization of

$h(t,\tau)$—then $\mathbf{y} = \mathbf{H}\mathbf{x}$ can be approximated as

$$\mathbf{y} \approx H_{\text{const}}\mathbf{x} + H_{\text{time}}\mathbf{T}\mathbf{x} + H_{\text{freq}}\mathbf{F}\mathbf{x}$$

in which H_{const} is the constant term of the filter linearization and H_{time} and H_{freq} are the respective average complex rates of change of h in the time and frequency directions. Also, $\mathbf{T}$ corresponds to a discretization of the operator of multiplication by t and $\mathbf{F}$ to the multiplication of the Fourier transform by the frequency variable. In the MIMO case, $\mathbf{x}$ and $\mathbf{y}$ are replaced by multiple transmitted and received signals. The approximate form of the signal received at a fixed antenna is then expressed as a mixture of time- and frequency-channelized $\mathbf{x}$'s. Finally, the inputs and outputs were projected onto the space spanned by the DPSSs with eigenvalues close to one. Altogether, this is written

$$\begin{aligned}\mathbf{y}_r &\approx \sum_s \mathbf{H}_{rs}\Psi^T\mathbf{x}_s\\ \mathbf{H}_{rs} &\approx H_{\text{const}}(r,s) + H_{\text{time}}(r,s)\Psi^T\mathbf{T}\Psi + H_{\text{freq}}(r,s)\Psi^T\mathbf{F}\Psi\end{aligned}$$

expressing the rth received signal as a sum of the filtered transmitted signals $\mathbf{x}_s$. Here, Ψ is the matrix whose columns are the corresponding DPSSs.

Sigloch et al. incorporated pilots into this scheme as a means of providing estimates of the cross-carrier slopes $H_{\text{const}}(r,s)$, $H_{\text{time}}(r,s)$, and $H_{\text{freq}}(r,s)$. The number of pilots used per block was based on the desire to obtain effective estimates of the cross-carrier slopes. They reported results of tests both on acoustical data and on simulated wireless data, demonstrating gains in SNR and E_b/N_0 compared to methods that account solely for temporal dispersion or frequency dispersion, but not both.

Similar theoretical frameworks for communications channels based on Gabor bases have been proposed, for example, in the work of Grip and Pfander [120]. Other approaches incorporating DPSSs in modeling doubly selective channels include those of Tugnait and others [142, 312], of Kim, Wang, and Stark [179], and of Liu et al. [205]. Related windowing techniques were used in Peiker et al. [263].

3.3.3 Further Aspects of Multitaper Spectrum Estimation

A Variant of Thomson's Maximum Likelihood (KLT) Method

Fancourt and Principe [97] suggested viewing the eigenvalue problem

$$\int_{-1/2}^{1/2} D_N(\omega-\xi)S(\xi)\psi(\xi) = \theta\psi(\omega) \tag{3.45}$$

for the KLT eifenfunctions (3.22) as a Fredholm equation of the second kind and applying a *separable kernel* approach to expanding a solution in a basis expansion of the reproducing kernel as in Mercer's theorem. In particular, suppose that $\phi_n(\omega)$, $n = 0,\dots,N-1$, form an orthogonal reproducing family for the Dirichlet kernel, that is,

$$D_N(\omega-\xi) = \sum_{n=0}^{N-1} \phi_n(\omega)\bar{\phi}_n(\xi)\,.$$

Substituting into (3.45) gives

$$\sum_{n=0}^{N-1} \phi_n(\omega) \int_{-1/2}^{1/2} \bar{\phi}_n(\xi)\,\psi(\xi)\,S(\xi)\,\mathrm{d}\xi = \theta\psi(\omega) = \sum_{n=0}^{N-1} c_n \phi_n(\omega) \tag{3.46}$$

where

$$c_n = \sum_{n=0}^{N-1} \phi_n(\omega) \int_{-1/2}^{1/2} \phi_n(\xi)\,\psi(\xi)\,S(\xi)\,\mathrm{d}\xi\,.$$

The coefficient vector $\mathbf{c} = [c_0, \ldots, c_{N-1}]^T$ then is an eigenvector, namely

$$A\mathbf{c} = \theta\mathbf{c}, \qquad A_{mn} = \int_{-1/2}^{1/2} \phi_n(\xi)\,\phi_m(\xi)\,S(\xi)\,\mathrm{d}\xi\,. \tag{3.47}$$

The Fourier basis elements $\phi_n(\xi) = \mathrm{e}^{\pi \mathrm{i}(N-1-2n)\xi}$ provide one possible choice for ϕ_n. Thomson [323] used shifted DPSWFs $\sigma_n(\xi - \xi_0)$ for ϕ_n because they can be localized around peaks in the spectrum. An alternative design criterion to impose on the separation basis is that the Karhunen–Loève components have sparse expansions. Fancourt and Principe [97] argued that, for fixed N and $W = 1/(2N)$, the functions $\phi_n(\xi) = (\sigma_n(\xi - \xi_0) + \sigma_n(\xi + \xi_0))/2$ can provide sparse basis expansions. When $W < \xi_0$, the approximation

$$\int_{-1/2}^{1/2} D_N(\xi - \eta)\,\phi_m(\eta)\,\mathrm{d}\eta \approx \theta_m \phi_m$$

is valid. For these ϕ_n's, (3.46) becomes the generalized eigenfunction equation

$$A\mathbf{c} = \theta B\mathbf{c}, \qquad B_{mn} = \left(\int_{-\xi_0-W}^{-\xi_0+W} + \int_{\xi_0-W}^{\xi_0+W}\right) \phi_n(\xi)\bar{\phi}_m(\xi)\,\mathrm{d}\xi\,.$$

Bias and Variance Estimates for Multitapers

Recall that the eigencoefficients of a sample z generated by the wide-sense stationary (WSS) process $\mathrm{d}Z$ with orthogonal increments have the form (cf. (3.8), (3.9), (3.10))

$$z_n(\omega) = \frac{1}{\sqrt{\lambda_n}} \int_{-W}^{W} \sigma_n(\xi)\,z(\omega+\xi)\,\mathrm{d}\xi = \int_{-1/2}^{1/2} \sigma_n(\xi)\,\mathrm{d}Z(\omega+\xi)$$
$$= \frac{1}{\varepsilon_n} \sum_{k=0}^{N-1} x(k)\,\mathbf{s}_n(k)\,\mathrm{e}^{-\pi \mathrm{i}\omega(2k+1-N)}$$

where (see (3.2), (3.4))

$$z(\omega) = \sum_{k=0}^{N-1} \mathrm{e}^{-\pi \mathrm{i}(2k+1-N)\omega}\,x(k) = \int_{-1/2}^{1/2} D_N(\omega - \xi)\,\mathrm{d}Z(\xi)\,. \tag{3.48}$$

Lii and Rosenblatt [202] considered the problem of finding conditions on the index-limiting parameter N, the bandwidth parameter $W = W(N)$, and the dimension parameter $D = D(N, W)$ under which the spectrum estimate

$$S(\omega) \approx \frac{1}{D} \sum_{n=0}^{D-1} |z_n(\omega)|^2 \tag{3.49}$$

of the power spectral density $S(\omega) = \mathbf{E}\{|\mathrm{d}Z(\omega)|^2\}$ is consistent and asymptotically unbiased. Lii and Rosenblatt prove that if S is bounded then

$$\text{bias}\left(\frac{1}{D}\sum_{n=0}^{D-1}|z_n(\omega)|^2\right) \le \sup|S(\omega)|\left(1-\frac{1}{D}\sum_{n=0}^{D-1}\lambda_n\right)+o(1) \tag{3.50}$$

while, if S is uniformly continuous and $W(N) \to 0$ as $N \to \infty$ while D is fixed then

$$\text{var}\left(\frac{1}{D}\sum_{n=0}^{D-1}|z_n(\omega)|^2\right) = \frac{S^2(\omega)}{D^2}\sum_{n=0}^{D-1}\lambda_n^2 + O\left(\frac{1}{D}\sum_{n=0}^{D-1}(1-\lambda_n)\right)+o(1) \tag{3.51}$$

when $\omega \neq 0, \pm 1/2$. The first term in (3.51) is doubled if $\omega = 0$ or $\omega = \pm 1/2$. When S is continuously differentiable, the $o(1)$ term in (3.50) is shown to be essentially W^2. It is shown, consequently, that in order that (3.49) is asymptotically unbiased for a fixed value of D, it is necessary that $\lambda_n(N, W(N)) \to 1$ for $n = 0, \ldots, D-1$, in particular, $NW(N) \to \infty$. One also requires $W(N) \to 0$. On the other hand, it is shown that in order that the estimate (3.49) is asymptotically consistent, it is necessary that $D \to \infty$ as $N \to \infty$, while $W(N)$ must still tend to zero. It is then sufficient that $D \le 2NW(N)$ in order for asymptotic consistency.

In his fundamental paper on combining jackknifing with multitaper analysis [324], Thomson illustrated through a number of concrete examples how jackknifing provides a sensible tool for asking whether a spectrum estimate is *reasonable*. The jackknife variance of an estimator is, in essence, the variance of the ensemble of all estimates obtained from deleting each single data point from the estimate using all data points, and is useful when data is scarce. Thomson stated that,

> *when studying scientific and engineering data where the basic inferences depend on having accurate estimates of spectra or descriptions of the data under study, it is an invaluable diagnostic of possible problems. In such problems, one rarely cares if a particular statistic is significant at the 94% or 96% level, but a variance estimate that is ten or 100 times too large or too small requires serious attention. Thus, when the jackknife variance has a suspicious average or is either extremely low or extremely high at some frequencies, exploratory data analysis is mandatory.*

Multitaper Method in Nonstationary Time Series Analysis

Bayram and Baraniuk [18] extended Thomson's multitaper method to the context of nonstationary time series and time–frequency representations, including algorithms for identification and extraction of chirp components. They used Hermite functions for the windows, since the Hermites are eigenfunctions of localization to a disk in the time–frequency plane (see [76]), whose Wigner distributions are Laguerre functions in two variables. This means that the Hermite eigenspectra can be computed as bivariate convolutions of the Wigner distribution of the signal with the Laguerre functions. Cakrak and Loughlin [55,56] extended Bayram and Baraniuk's work by finding optimal weighting of the tapers in the nonstationary case in a manner similar to Thomson's approach in the stationary case. Later, Xiao and Flandrin [365] combined multitapering with the "reassignment method" for time–frequency distributions as a means of identifying chirps and other nonstationary signals embedded in nonstationary noise. Other uses of multitapering in time–frequency analysis include the work of Lynn and Sha´ameri [212], who compared the ordinary spectrogram of frequency-shift keyed signals to multitaper spectrograms and showed that use of the latter can improve estimation of bit duration.

Cognitive Radio

The term cognitive radio (CR) was coined by Joseph Mitola. In a 1999 paper [237], Mitola and Maguire outlined a technological scenario in which intelligent computer and software radio networks could detect user communications needs and allocate resources automatically as functions of user context. Anticipating evolution toward such a scenario, in 2002 the FCC Spectrum Policy Task Force [220] recommended a shift toward satisfactory adaptive real-time interactions between a transmitter and receiver based on a new metric called the *interference temperature* that would quantify the accumulated sources of interference locally in a radio environment. The interference temperature at a receiver should provide an accurate measure for the acceptable level of RF interference in the frequency band of interest, thus enabling individual use of a frequency band in which the interference temperature would not be exceeded by such use. Citing favorable comparison to standard minimum least squares estimation, e.g., [317], Haykin [138] advocated for combined use of multitapering with the singular value decomposition (SVD), as in Mann and Park's work [219], as an effective procedure for estimating the power spectrum of the noise floor in an RF environment. Haykin suggested that while this "SVD–MTM" method is computationally intensive, it is not prohibitively so.

In contrast, Ariananda et al. [7] expressed concern that severe power limitations might prohibit multitaper analysis as a feasible spectrum sensing tool, though they did not consider its use in tandem with the SVD. They compared multitaper methods with a filter bank spectrum estimation (FBSE) method proposed by Farhang–Boroujeny in [98], as well as with wavelet methods.

Farhang–Boroujeny's FBSE method for multi-carrier modulation-based CR uses banks of matched root Nyquist filters. One treats each point in the frequency spectrum as the output of a single filter, or of multiple filters operating at the same band. In comparison, multitaper spectrum estimation (MTSE) treats each point as the averaged output of different prototype filters at the same frequency band. Ariananda et al. [7] observed the comparative strengths and weaknesses as follows. FBSE exhibits smaller variance than MTSE over a frequency band with low power spectrum density, and also produces less spectral leakage. Generally, FBSE outperforms MTSE given sufficiently many samples, though MTSE is best when sample data is scarce. MTSE is faster than FBSE per iteration, but may take too many iterations to be feasible in practice. Ariananda et al. suggested that wavelet-based edge detection methods provide a better adaptive approach to spectrum sensing than either MTSE or FBSE, cf., [325], but also suggested that better decision-based algorithms are still needed in order to make any of these methods feasible. One such decision tool was proposed by Wang and Zhang [348], who analyzed detection probability versus false alarm rates for MTSE. Other attempts to address the computational complexity of spectrum sensing include Ma et al.'s work [216] on *cooperative detection* and Thao and Guo's work [374].

Other Windows for Multitaper Analysis

DPSSs are not the only window families that can be used for multitaper analysis. Riedel and Sidorenko [278] observed that the DPS sequences $\mathbf{s}_n$ can have large local bias $\int_{-1/2}^{1/2}|\omega\sigma_n(\omega)|^2$ where, as before, σ_n is essentially the Fourier series of $\mathbf{s}_n$. Local bias is particularly problematic when the bandwidth W is large. Riedel and Sidorenko proposed the use of the sinusoidal tapers $\mathbf{v}_n(k) = \sqrt{\frac{2}{N+1}}\sin\frac{k\pi n}{N+1}$ where N is fixed. There is no specific bandwidth associated with these tapers, but $\mathbf{v}_1,\dots,\mathbf{v}_{D-1}$ together provide a multitaper estimator frequency concentrated on $[-D/2(N+1), D/2(N+1)]$. To describe the sense in which these tapers approximately minimize bias, one considers the normalized continuous-time interval $[0,1]$ and $L^2[0,1]$ with the orthonormal sinusoidal basis $\{2\sin\pi nt\}$. Assuming $S''(\Omega)\neq 0$, the local bias of the continuous-time $\mathbf{v}$-taper spectral estimate $S^{\rm est}$ is defined by

$$\mathbf{E}(S^{\rm est}(\omega)-S(\omega)) = \int|\widehat{v}(\omega-\xi)|^2(S(\omega)-S(\xi))\,\mathrm{d}\xi \approx \frac{S''(\omega)}{2}\int|\widehat{v}(\xi)|^2\xi^2\,\mathrm{d}\xi\,.$$

The leading term is simply a multiple of $\int|\widehat{v}(\xi)|^2\xi^2\,\mathrm{d}\xi$ and, upon application of Plancherel's theorem, this term is minimized for the taper $\sqrt{\pi}\sin\pi t$, (see Papoulis [253] and Parzen [262]). Further analysis shows that $\mathbf{v}_n = \sqrt{2}\sin\pi nt$ minimizes the same expression, subject to its being orthogonal to $\mathbf{v}_1,\ldots,\mathbf{v}_{n-1}$. The discrete sinusoidal tapers above are then shown approximately to minimize bias in a similar sense, among all discrete tapers, within a factor of $1/N$. These tapers are then shown to give rise to a minimizer of the local loss, defined as bias^2 + variance of a spectrum estimator. In their study of effective bandwidth estimators, Walden et al. [342] showed that the autocorrelation bandwidth (3.19) of a spectrum estimator based on sinusoidal tapers grows linearly in the number of tapers.

There are other ways in which to reduce the bias of multitaper spectrum estimates in practice. Prieto et al. [270] introduced a quadratic multitaper method that estimates the slope and curvature of the spectrum on $(-W,W)$ by solving a parameter estimation problem relating the derivatives of the spectrum and the D eigencomponents. The curvature estimate enables a corrected spectrum estimate that is unbiased to quadratic structure, but reduces to Thomson's estimate when $S''(\omega)=0$. Examples indicate that this quadratic method reduces bias in comparison with Thomson's method. Somewhat parallel spectrogram-based techniques have been used for time-varying spectrum estimation, e.g., [226].

Another concern with the DPS sequences is their computational complexity for large N. Greenhall [116] considered *sampled trigonometric prolates* as a possible alternative. These are defined by their trigonometric coefficient sequences, which are eigenvectors of the positive definite matrix $B_{k\ell}=\int_{|t|>W}\operatorname{sinc}(t-k)\operatorname{sinc}(t-\ell)\,\mathrm{d}t$, and their eigenvalues are the spectral leakages. The resulting polynomials of degree at most M are *leakage optimal* and orthogonal.

Treating a periodogram, instead of the original time series, as signal plus noise, one can introduce wavelet-based denoising of a log-periodogram as a means of refining spectrum estimation. Walden et al. [343] observed that the same wavelet denoising could be performed on log multitaper spectrum estimates, taking into account correlations among direct spectrum estimates from the different tapers, as a means of combining the virtues of MTSE and of wavelet-based denoising. Wavelet-based log multitaper estimators have the advantage of being essentially Gaussian, with an invertible correlation matrix. Cristán and Walden [74] further observed that wavelet packet-based estimators provide little or no improvement over discrete wavelet transform estimators, for denoising purposes.

Higher Dimensional Multitaper Analysis

Thomson's multitaper method was extended to multidimensional data by Bronez in [51] (cf. [52]). Bronez's approach does not take the route of extending DPSSs to higher dimensions as Slepian [303] did for the PSWFs. Instead, Bronez developed a more general theory that encompasses irregular sampling and produces general DPS sequences that are based on irregular sampling and, hence, fundamentally different from those proposed by Zemen in Sect. 3.2. Bronez introduced the *integrated spectrum*—the average of the power spectrum over a measurable set—as a means of quantifying a priori the resolving power of an estimation method via the following algorithm [51, Table 1, p. 1866].

Algorithm (Bronez's algorithm).

Given: sample points $\mathbf{t}_k$, $k=1,\ldots,N$, samples $\mathbf{x}(\mathbf{t}_k)$, and Fourier support $B=\{\omega: S_{\mathbf{x}}(\omega)\neq 0\}$,

`For` $A\subset B$ `set` $P_{\mathbf{x}}(A)=\int_A S_{\mathbf{x}}(\omega)\,\mathrm{d}\omega$

`Define general sinc matrices`

$$R_A(k,\ell)=\int_A \mathrm{e}^{2\pi i\omega(\mathbf{t}_k-\mathbf{t}_\ell)}\,\mathrm{d}\omega,\quad 1\le k,\ell\le N\,.$$

`Calculate the eigenvalues and eigenvectors` for

$$R_A(k,\ell)\,\mathbf{w}_n=\lambda_n R_B\,\mathbf{w}_n\,.$$

`Select the number of weight sequences` D over which to average. The leakage (or sidelobe energy) of the sequence $\mathbf{w}_n$ is $\gamma_k = 10\log_{10}(1-\lambda_n)$.

`Update the power averaging set` $A \subset B$, decreasing the resolution by choosing a larger A.

`Update` D.

`Normalize the eigenvectors` so that $\mathbf{w}_n^* R_B \mathbf{w}_k = |A|/D$, $1 \leq n \leq D$.

`Update the integrated spectrum estimate` via $P_{\mathbf{x}}^{\mathrm{ave}}(A) = \sum_{n=1}^{D} |\mathbf{w}_n^* \mathbf{x}|^2$.

`Perform same steps on other frequency bands.`

Bronez observed the necessary tradeoffs that have to be made between bias and variance. In particular, the eigenweight sequences $\mathbf{w}_n$, $n = 1,\ldots,D$, are solutions of $R_A \mathbf{w}_n = \lambda_n \mathbf{w}_n$ where $R_{\mathbf{x}}(k,\ell)$ is the process autocorrelation matrix $r_{\mathbf{xx}}(\mathbf{t}_k - \mathbf{t}_\ell)$ and $R_B(k,\ell) = \int_B e^{2\pi i\omega\cdot(\mathbf{t}_k-\mathbf{t}_\ell)}\,\mathrm{d}\omega$. Such eigenweight sequences were shown to satisfy

$$\mathbf{w}_m^* R_A \mathbf{w}_n = \frac{|A|}{D}\delta_{nm},$$

so that, by Plancherel's theorem,

$$\mathrm{var}(\mathbf{w}_1,\ldots,\mathbf{w}_D) = \sum_{n=1}^{D}\sum_{m=1}^{D}\left|\int_B W_n(\omega)\,\bar{W}_m(\omega)\,\mathrm{d}\omega\right|^2 = \frac{|A|^2}{D},$$

while

$$\mathrm{bias}(\mathbf{w}_1,\ldots,\mathbf{w}_D) = \left(\int_B - \int_A\right)\sum_{n=1}^{D}|W_n(\omega)|^2\,\mathrm{d}\omega = \frac{|A|}{D}\sum_{n=1}^{D}(1-\lambda_n).$$

In particular, the variance bound decreases with D while the bias bound increases. In practice, it is suggested that D be chosen as large as possible, provided the choice does not violate an acceptable bound on the leakage γ_D. As a specific two-dimensional application, Bronez considered a simulation of inverse synthetic aperture radar with sampling along a spiral lattice, thought of as a distorted rectangular grid, computing approximations of the eigenweights efficiently.

Besides the extension to multidimensional data done by Bronez, several other approaches to multitaper spectrum of estimation of multidimensional data have been proposed. Hanssen [134] carried out a straightforward extension of Thomson's approach to the case of several spatial variables by means of tensor products. Liu and Van Veen [206] also considered multidimensional data, going beyond a straightforward extension of Thomson's work in that they proposed a *multiple window minimum variance* (MWMV) spectrum estimator based on Capon's work [62], though the "minimum variance" really refers to minimizing out-of-band power or *exterior bias*, as opposed to statistical variance.

Other multidimensional extensions of multitapering take geometry into account. The work of Wieczorek and Simons [355] involves an analogue of multitaper analysis for spherical data (cf. Simons [299]), in order to represent geo-scientific data by means of Slepian functions on two-dimensional Cartesian, and spherical domains. DPS sequences have been used to analyze multidimensional data in several contexts. These include Chen and Vaidyanathan's [68] application to MIMO radar in which the geometry of a "clutter subspace" is described in terms of PSWFs; Ruan, Hanlen and Reed's [281] use of DPS sequences in spatially interpolated beamformer design; and Lindquist et al.'s [204] use of generalized two-dimensional PSWFs for data acquisition in MRI (cf. [295]). Salerno [282] developed a genuinely two-dimensional version of the Papoulis–Gerchberg algorithm for the purpose of image analysis (cf. Larsson et al. [198]). Other applications to image analysis include Wilson and Spann's [357] original application of DPS sequences in image segmentation.

Further Miscellaneous Applications of Multitaper Spectrum Estimation

Multitaper analysis has been used in a number of areas of applications, originating with the work of Park, Lindberg, and Vernon in seismology [219, 259, 260]. Fodor and Stark [102] extended the application of PSWFs in multitaper analysis to analyze helioseismic data with gaps by means of *resampling*, successfully identifying confidence intervals for such data.

Since 2000, biometric applications of multitapering have increased. Examples include the work of Xu, Haykin and Racine [367], who applied Bayram and Baraniuk's [18] ideas on multitaper time–frequency analysis (MW–TFA) to analyze EEG signals. Information in such signals is usually lost in ensemble averaging. The appealing property of MW–TFA here is that it enables a "TF-coherence estimate" from a single time series realization. This initial study of coherence in EEG data using MTSE has been extended by He and Thomson [139–141]. The use of MTSE in brain imaging, particularly using fMRI, has been studied by Mitra and Pesaran [238]. In such data, significant biological artifacts arise from cardiac and respiratory functions of the patient, in addition to other noise sources. Mitra and Pesaran developed an SVD–MTM based algorithm for such artifact removal. Lovett and Myklebust [209] also used sinusoidal tapers, such as those proposed by Riedel and Sidorenko [278], to perform MTSE of heart rate variability.

Other MTSE applications that account for geometrical factors include Wage's multitaper analysis of sonar array data [340] and Das et al.'s use of MTSE in the study of high resolution cosmic microwave background maps [75].

Chapter 4
Time and Band Limiting of Multiband Signals

4.1 The Area Theorem for Unions of Time–Frequency Rectangles

4.1.1 *Landau and Widom's Theorem for Multiple Intervals*

When $a = 2\Omega T$, the operator $P_\Omega Q_T$ corresponding to single time and frequency intervals has an eigenvalue $\lambda_{\lfloor a \rfloor} \approx 1/2$, as Theorem 4.1.2 below will show. The norm $\lambda_0(a=1)$ of the operator $PQ_{1/2}$ satisfies $\lambda_0(a=1) \geq \|\mathrm{sinc}\,\mathbb{1}_{[-1/2,1/2]}\| > 0.88$. The trace of $PQ_{1/2}$ is equal to $a=1$, on the one hand and to $\sum \lambda_n$ on the other, so $\lambda_1(a=1) \leq 1 - \lambda_0(a=1) < 1/2$. Suppose that $T = 1$ and Σ is a finite, pairwise disjoint union of a frequency intervals $I_1, \dots, I_a$ each of unit length. Then $P_\Sigma Q$ should have on the order of a eigenvalues of magnitude *at least* $1/2$. Consider now the limiting case in which the frequency intervals become *separated at infinity*. Any function ψ_j that is concentrated in frequency on I_j will be almost orthogonal over $[-T,T]$, in the separation limit, to any function ψ_k that is frequency-concentrated on I_k when $j \neq k$. To see this, write $\psi_j(t) = \mathrm{e}^{2\pi i m_j t}\varphi_j(t)$ where m_j is the midpoint of I_j and $\widehat{\varphi}_j$ is essentially concentrated on $[-1/2,1/2]$. Then

$$\int_{-1/2}^{1/2} \mathrm{e}^{2\pi i(m_j - m_k)t}\varphi_j(t)\overline{\varphi_k}(t)\,\mathrm{d}t = \widehat{\varphi}_j * \widehat{\overline{\varphi_k}} * \mathrm{sinc}\,(m_j - m_k) = O\Big(\frac{1}{|m_j - m_k|}\Big)$$

as $|m_j - m_k| \to \infty$. This almost orthogonality prevents eigenvalues from the separate interval operators $P_I Q$ from coalescing into significantly larger eigenvalues of $P_\Sigma Q$. Consequently, $P_\Sigma Q$ will have on the order of a eigenvalues of size *approximately equal to* $\lambda_0(a=1)$ in the separation limit, while the remaining eigenvalues will not be significantly larger than $\lambda_1(a=1) < 1/2$. Incidentally, similar reasoning shows that $P_\Sigma Q$ cannot have *any* eigenvalues larger than $1/2$ when Σ is a union of a large number of short, mutually distant intervals, even if $a = |\Sigma| > 1$.

The general result of Landau and Widom [197] provides an estimate of the distribution of eigenvalues for $P_\Sigma Q_S$, asymptotically in the sense of rescaling one of the sets S or Σ, when both sets are finite unions of intervals. Because of the role of rescaling, there will be no loss in generality in assuming that $|S| = |\Sigma| = 1$. The asymptotic formula states that the number of eigenvalues close to one grows like the area of the union of rectangles. However, the transition region over which the eigenvalues pass from being *close to one* to being *close to zero* now has a width proportional to the number of intervals, which is consistent with the observation made in the previous paragraph.

Theorem 4.1.1. *Suppose that S and Σ are finite pairwise disjoint unions of N_S and N_Σ intervals respectively, with $|S| = |\Sigma| = 1$. Set $B_a = B_a(S,\Sigma) = P_{a\Sigma} Q_S P_{a\Sigma}$ where $a\Sigma = \{a\xi : \xi \in \Sigma\}$. Then the number $N(B_a, \alpha)$ of eigenvalues of B_a larger than α satisfies*

$$N(B_a,\alpha) = a + \frac{N_S N_\Sigma}{\pi^2} \log\Big(\frac{1-\alpha}{\alpha}\Big) \log a + o(\log a), \qquad a \to \infty. \tag{4.1}$$

The *single interval* case, Theorem 1.3.1 was proved in Chap. 1, so we will only outline here the main steps required to pass to the multiple interval case. The case of finitely many time and frequency intervals involves a reduction to the single interval case, which also requires asymptotic separation of the intervals. Although the factor $N_S N_\Sigma$ disappears when $\alpha = 1/2$, it appears prominently for other $\alpha \in (0,1)$.

Reduction to Single Intervals

To simplify the reduction to the single interval case, we assume that Σ is a union of N_Σ bounded intervals, but that S is a single time interval I. The general case (S is also a union of intervals) poses no additional obstacles. Let $\Sigma = \cup_j \Gamma_j$ with $\Gamma_j = [\alpha_j, \beta_j]$. Let $\Delta_0 = (-\infty, \alpha_0)$, $\Delta_1 = (\beta_1, \alpha_2)$, ..., $\Delta_{N_\Sigma} = (\beta_{N_\Sigma}, \infty)$ be the complementary intervals. As in the single interval case, the *transition* eigenvalues of B_a correspond to the large eigenvalues of $B_a(I - B_a)$, which can be written

$$B_a(I - B_a) = P_\Sigma Q_{aI} P_{\mathbb{R}\setminus\Sigma} Q_{aI} P_\Sigma = \sum_{i,j,k=1}^{N_\Sigma} P_{\Gamma_i} Q_{aI} P_{\Delta_j} Q_{aI} P_{\Gamma_k}\,.$$

Unless $i = k$ and Γ_k is adjacent to Δ_j, the term $P_{\Gamma_i} Q_{aI} P_{\Delta_j} Q_{aI} P_{\Gamma_k}$ will have a trace norm that is uniformly bounded independently of a. As in Chap. 1, we say that such an operator is "$O(1)$". This follows from case (UB1) of Lemma 1.5.6, p. 41. Such terms contribute only to the $o(\log a)$ term in Theorem 4.1.1.

Consequently, one then can write

$$B_a(I - B_a) = \sum_{j=1}^{N_\Sigma} P_{\Gamma_j} Q_{aI} (P_{(-\infty,\alpha_j)} + P_{(\beta_j,\infty)}) Q_{aI} P_{\Gamma_j} + O(1)\,.$$

Since the result is asymptotic one can rescale so that $I = [0,1]$ and all the frequency intervals have length greater than one. As in the single interval case, $(B_a(I - B_a))^n$

then can be written as

$$[B_a(I-B_a)]^n = \sum_j \big(P_{a[\alpha_j+1,\beta_j]}Q_{[0,\infty)}P_{a(-\infty,\alpha_j]}Q_{[0,\infty)}P_{a[\alpha_j+1,\beta_j]}\big)^n$$
$$+3 \text{ similar sums} + O(1).$$

As in the one interval case, to each fixed j there correspond nth powers of four operators each unitarily equivalent to

$$K_{a(\beta_j-\alpha_j)} = Q_{[1,a(\beta_j-\alpha_j))}P_{(0,\infty)}Q_{(-\infty,0)}P_{(0,\infty)}Q_{[1,a(\beta_j-\alpha_j))}\,.$$

Since each interval in Σ is fixed while a varies, for each j one has $K_{a(\beta_j-\alpha_j)} = K_a + O(1)$. As in the single interval case one then has

$$\operatorname{tr}[B_a(I-B_a)]^n = 4\sum_j \operatorname{tr} K^n_{a(\beta_j-\alpha_j)} + O(1) = 4\sum_j \operatorname{tr} K^n_a + O(1) = 4N_\Sigma \operatorname{tr} K^n_a + O(1)\,.$$

A similar reduction applies to $B_a[B_a(I-B_a)]^n$. The remainder of the argument follows the single interval case.

4.1.2 The Number of Eigenvalues of $P_\Omega Q_T$ Larger than $1/2$

Theorem 1.3.1 gives an asymptotic estimate for the number of eigenvalues of $P_\Omega Q_T$ larger than $1/2$. If one were willing to neglect the "$o(\log a)$" term in that theorem, upon taking $\alpha = 1/2$, one would obtain the statement that the number of eigenvalues of $P_\Omega Q_T$ larger than $1/2$ is exactly equal to $2\Omega T$. In [193], (cf. [187]) Landau showed that, indeed, one can *essentially* neglect the "$o(\log a)$" term when $\alpha = 1/2$. Scaling to the case $\Omega = 1$, his theorem can be stated as follows.

Theorem 4.1.2. *The eigenvalues of PQ_T satisfy*

$$\lambda_{\lfloor 2T\rfloor -1} \ge 1/2 \ge \lambda_{\lceil 2T\rceil}\,.$$

In the theorem, $\lfloor x\rfloor$ and $\lceil x\rceil$ refer to the greatest integer less than or equal to x and the least integer greater than or equal to x respectively. The Weyl–Courant minimax characterization of the singular values $\lambda_0 \ge \lambda_1 \ge \cdots$ of $P_\Sigma Q_S P_\Sigma$ can be stated as

$$\lambda_n = \begin{cases} \min_{\mathscr{S}_n} \max\{\|Q_S f\|^2 : f\in \mathrm{PW}_\Sigma,\quad \|f\|=1,\, f\perp \mathscr{S}_n\} \\ \max_{\mathscr{S}_{n+1}} \min\{\|Q_S f\|^2 : f\in \mathrm{PW}_\Sigma,\, \|f\|=1,\, f\in \mathscr{S}_{n+1}\}. \end{cases}$$

Here, $\mathscr{S}_n$ ranges over all n-dimensional subspaces including, notably, the subspace spanned by the first n eigenfunctions of $P_\Sigma Q_S P_\Sigma$. In his 1965 work [187], Landau identified a convolver h such that if $f\in \mathrm{PW}$ and $f*h(m)$ vanishes at a given set of $n=\lceil 2T\rceil$ integers, then $\|Q_T f\|^2 \le 0.6$. A sharper bound with 0.6 replaced by 0.5

was attributed to B.F. Logan in [187], but the sharp bound was never published until Landau's 1993 work [193]. The theorem can also be viewed as a corollary of the multiple interval case Proposition 4.1.3 below.

Extension of Theorem 4.1.2 to Multiple Intervals

Landau's technique can be extended to the case in which S is a finite union of intervals (but still $\Sigma=[-1/2,1/2]$). Let S_+ denote those points whose distance to S is at most $1/2$ and, for a fixed function h, let $\mathscr{S}_+=\mathrm{span}\{h(\cdot-k):k\in S_+\}$. Similarly, let $\mathscr{S}_-=\mathrm{span}\{h(\cdot-k):k\in S_-\}$, where S_- is obtained from S by shrinking the intervals in S by $1/2$ at each end (and omitting any intervals having length less than one). In [187], Landau showed that if S is a union of m intervals then

$$\begin{aligned}|S|-2m &\le \#\{k:(k-1/2,k+1/2)\subset S\}=\dim\mathscr{S}_-\\ &\le \#\{k:(k-1/2,k+1/2)\cap S\neq\emptyset\}=\dim\mathscr{S}_+\le|S|+2m.\end{aligned}$$

Landau's method concedes two eigenvalues bigger than $1/2$ per interval. Thus $\lambda_{\lfloor|S|\rfloor-2m}\ge 1/2\ge\lambda_{\lceil|S|\rceil+2m}$. However, a modest but significant gain can be achieved when the intervals all have length at least one by examining the structure of Landau's proof of Theorem 4.1.2. In Izu's dissertation [159], one can find a proof of an equivalent version of the following.

Proposition 4.1.3. *Let $\Sigma=[-1/2,1/2]$ and let S be a finite union of m pairwise disjoint intervals. Denote by*

$$\begin{aligned}\nu &= \max_{\alpha}\#\{k\in\mathbb{Z}:(k-1/2,k+1/2)\subset(S+\alpha)\}\quad\text{and}\\ \mu &= \min_{\beta}\#\{\ell\in\mathbb{Z}:(\ell-1/2,\ell+1/2)\cap(S+\beta)\neq\emptyset\}\,.\end{aligned}$$

Then the eigenvalues λ_n of Q_SP satisfy

$$\lambda_{\nu-1}\ge 1/2\ge\lambda_{\mu}. \tag{4.2}$$

In particular, for $|S|\ge 1$, $\lfloor|S|\rfloor-2m+2\le\nu\le\mu\le\lceil|S|\rceil+2m-2$ so that

$$\lambda_{\lfloor|S|\rfloor-2m+1}\ge 1/2\ge\lambda_{\lceil|S|\rceil+2m-2}\,. \tag{4.3}$$

Lemma 4.1.4. *Denote by $h(t)=\sqrt{2}\cos\pi t\mathbb{1}_{[-1/2,1/2]}(t)$. Then $\|h\|^2=1$ and $\widehat{h}(\xi)\ge 1/\sqrt{2}$ whenever $|\xi|\le 1/2$.*

Proof (of Lemma 4.1.4). That $\|h\|^2=1$ follows from integrating $2\cos^2\pi t=1+\cos 2\pi t$ over $[-1/2,1/2]$. By direct calculation, $\sqrt{2}\widehat{h}(\xi)=\mathrm{sinc}\,(\xi+1/2)+\mathrm{sinc}\,(\xi-1/2)$, which is convex on $[-1/2,1/2]$ while $\widehat{h}(-1/2)=\widehat{h}(1/2)=1/\sqrt{2}$. □

Proof (of Proposition 4.1.3). Assume for the moment that $a\notin\mathbb{N}$ and that S has no half-integer endpoints. Since translation by β is unitary on $L^2(\mathbb{R})$, there is no loss in generality in assuming that $\#\{\ell\in\mathbb{Z}:(\ell-1/2,\ell+1/2)\cap(S+\beta)\neq\emptyset\}$ is minimized when $\beta=0$. Denote by S_+ those points whose distance to S is at most $1/2$ and by N_+ the integers in S_+. With h as in Lemma 4.1.4, h is supported in $[-1/2,1/2]$ so

$$|f*h(x)|^2=\left|\int_{x-1/2}^{x+1/2}f(t)h(x-t)\,\mathrm{d}t\right|^2\leq\|h\|^2\int_{x-1/2}^{x+1/2}|f(t)|^2\,\mathrm{d}t\,.$$

Let $\mathscr{S}_+$ denote the closed span of the functions $h(k-\cdot)$, $k\in N_+$. If $f\perp\mathscr{S}_+$, then $f*h(k)=0$ when $k\in N_+$ so

$$\begin{aligned}\sum|f*h(k)|^2&=\sum_{k\notin N_+}|f*h(k)|^2\leq\|h\|^2\sum_{k\notin N_+}\int_{k-1/2}^{k+1/2}|f(t)|^2\mathrm{d}t\\&\leq\int_{t\notin S}|f(t)|^2=(\|f\|^2-\|Q_Sf\|^2)\end{aligned}$$

since $\|h\|=1$. On the other hand, since $f*h\in\mathrm{PW}$ when f is,

$$\sum|f*h(k)|^2=\int_{-1/2}^{1/2}|\hat{f}(\xi)\hat{h}(\xi)|^2\mathrm{d}\xi\geq\frac{1}{2}\int_{-1/2}^{1/2}|\hat{f}(\xi)|^2\mathrm{d}\xi=\frac{1}{2}\|f\|^2,$$

since $|\widehat{h}(\xi)|\geq 1/\sqrt{2}$ on $[-1/2,1/2]$. Thus, if $f\perp\mathscr{S}_+$ then

$$\|Q_Sf\|^2\leq\frac{1}{2}\|f\|^2.$$

The dimension of $\mathscr{S}_+$ is the number of integers interior to S_+, which is μ. By the minimax criterion, then $\lambda_\mu\leq 1/2$. In the limiting case $|S|\in\mathbb{N}$, the same estimate follows from the fact that for each $n=0,1,\dots,\ \lambda_n(S)$ varies continuously with S.

For the other inequality in (4.2), we will assume that $|S|>1$. Here, there is no loss in generality in assuming that $\#\{k\in\mathbb{Z}:(k-1/2,k+1/2)\subset(S+\alpha)\}$ is maximized when $\alpha=0$. Let S_- be the subset of S of points whose distance to $\mathbb{R}\setminus S$ is at least $1/2$. With h as above, let $q\in\mathrm{PW}$ be such that $\widehat{q}(\xi)=1/\bar{\hat{h}}(\xi)$ on $[-1/2,1/2]$. Let $\mathscr{S}_-$ be the span of $q(\cdot-k)$ where k ranges over N_-, the integers in the closure of S_-. If $f\in\mathscr{S}_-$ then one can write

$$\widehat{f}(\xi)=\sum_{k\in N_-}b_k\mathrm{e}^{-2\pi ik\xi}/\bar{\hat{h}}(\xi)$$

so that

$$\widehat{f}(\xi)\bar{\hat{h}}(\xi)=\sum_{k\in N_-}b_k\mathrm{e}^{-2\pi ik\xi},\quad|\xi|\leq 1/2\,.$$

Then

$$\sum_{k\in N_-}|b_k|^2=\int_{-1/2}^{1/2}|\widehat{f}(\xi)\hat{h}(\xi)|^2\,d\xi\geq\frac{1}{2}\int_{-1/2}^{1/2}|\widehat{f}|^2=\frac{1}{2}\|f\|^2\,.$$

On the other hand, as before we get

$$b_k = \int_{-1/2}^{1/2} \widehat{f}(\xi)\bar{\widehat{h}}(\xi)\,\mathrm{e}^{2\pi i k\xi}\,d\xi = \int f(t)\overline{h(t-k)}\,\mathrm{d}t\,.$$

Since $\mathrm{supp}(h) \subset [-1/2,1/2]$ and $(k-1/2,k+1/2) \subset S$ if $k \in N_-$, one has

$$\sum_{k\in N_-} |b_k|^2 \le \|h\|^2 \sum_{k\in N_-} \int_{k-1/2}^{k+1/2} |f(t)|^2\,\mathrm{d}t \le \|Q_S f\|^2,$$

since $\|h\| = 1$. Altogether, this shows that if $f \in \mathscr{S}_-$ then

$$\frac{1}{2}\|f\|^2 \le \|Q_S f\|^2\,.$$

The dimension of $\mathscr{S}_-$ is ν so, by the minimax criterion, $\lambda_{\nu-1} \ge 1/2$. □

Discussion

In the case in which $m = 1$, that is, S is an interval, Proposition 4.1.3 provides the bound $\lambda_{\lfloor |S| \rfloor -1} \ge 1/2 \ge \lambda_{\lceil |S| \rceil}$ which is just Landau's estimate, Theorem 4.1.2. When $a \in \mathbb{N}$ there is a jump in $\lambda_{\lfloor |S| \rfloor}$, even though λ_n depends continuously on $|S|$ for each fixed n. It follows that $\lambda_{|S|} \le 1/2 \le \lambda_{|S|-1}$ when $|S| \in \mathbb{N}$ and $m = 1$. When $m > 1$, the upper and lower bounds of (4.3) are not close to the sharper (4.2) in a lot of cases as the following heuristic argument indicates. Suppose that $I_1,\dots,I_m$ are the m pairwise disjoint intervals comprising S and that each has length at least one. For each j then $\lambda_{\lfloor |I_j| \rfloor -1}(PQ_{I_j}) \ge 1/2$. Since the I_j are pairwise disjoint, the eigenfunctions corresponding to separate intervals in S and having eigenvalues larger than $1/2$ are linearly independent. Let $M = \sum \lfloor |I_j| \rfloor$. Then we have a subspace of $L^2(\mathbb{R})$ of dimension at least $M-m$ such that $\|Q_S P\varphi\| \ge \|\varphi\|/2$ whenever φ is in this space. Thus, $\lambda_{M-m-1}(PQ_S) \ge 1/2$. This inequality improves $\lambda_{\lfloor |S| \rfloor -2m+1} \ge 1/2$, at least when $\sum |I_j| - \lfloor |I_j| \rfloor \ll m$, that is, $|S| \ll M+m$. This is another way of quantifying the idea that S is close to being a union of *grid intervals*. A particularly interesting case (see [159]) is that in which each of the intervals comprising S has the form $[k-1/2,k+1/2)$. Then $\nu = \mu = |S|$ and one recovers the bounds $\lambda_{\lfloor |S| \rfloor -1} \ge 1/2 \ge \lambda_{\lceil |S| \rceil}$, even though S can be disconnected. However, there may not be any eigenvalues *close to one* in this case.

4.2 Probabilistic Norm Bounds for Sparse Time–Frequency Products

One can verify numerically the finite version of Theorem 4.1.2 for the finite discrete prolate (FDP) sequences of order N and bandwidth parameter $W \in \mathbb{Z}$, defined as the eigenvectors of the matrix $M_{N,W}(k,\ell) = \mathrm{sinc}_{2W}(k-\ell)$, $k,\ell = 0,1,\dots,N-1$. In this case the *discrete normalized area* amounts to $2W$ and one observes $\lambda_{2W-1} > 1/2 >$

λ_{2W} when $W \in \{1/2, 1, \ldots, (N-1)/2\}$. However, a fundamental difference between *finite* and *infinite* settings is manifested in support properties of Fourier transforms. No nonzero $f \in L^2(\mathbb{R})$ that vanishes outside of a set of finite measure has a Fourier transform $\hat{f}$ also vanishing off a set of finite measure. The analogous statement for the DFT is a lower bound on the sizes of the supports of $\mathbf{x}$ and $\widehat{\mathbf{x}}$ due to Donoho and Stark [85].

The Donoho–Stark Inequality

As before, $|F|$ denotes the counting measure of a finite set F. The Donoho–Stark uncertainty inequality can be stated as follows.

Theorem 4.2.1. *Any finite signal $\mathbf{x} : \mathbb{Z}_N \to \mathbb{C}$ with N-point discrete Fourier transform $\widehat{\mathbf{x}}$ satisfies*

$$|\{n : x_n \neq 0\}|\,|\{n : \widehat{x}_n \neq 0\}| \geq N, \tag{4.4}$$

$$|\{n : x_n \neq 0\}| + |\{n : \widehat{x}_n \neq 0\}| \geq 2\sqrt{N}. \tag{4.5}$$

In addition, $|\{n : x_n \neq 0\}|\,|\{n : \widehat{x}_n \neq 0\}| = N$ only when, up to appropriate time–frequency shifts, $\mathbf{x}$ is the indicator of a subgroup where $\mathbb{Z}_N$ is regarded as a cyclic group of order N.

The characterization of extremal functions for the Donoho–Stark inequality emphasizes an important (group-theoretic) distinction between the DFT and the Fourier transform on $\mathbb{R}$. The role of group theory is emphasized further in Tao's [322] sharper inequality for $N = P$ prime, namely

$$|\{n : x_n \neq 0\}| + |\{n : \widehat{x}_n \neq 0\}| \geq P + 1\,.$$

Recall that $\mathbb{Z}_P$ has no nontrivial proper subgroups. More information regarding the Donoho–Stark inequality and related finite uncertainty inequalities and proofs can be found in [152].

Quantitative Robust Uncertainty Principles

The finite setting can indicate limitations of the multiband cases of the Bell Labs theory of time and band limiting on $\mathbb{R}$—not to mention precise limitations of the analogous finite theory—when the time or frequency supports become highly disconnected. Candès, Romberg, and Tao (Candès et al.) [57–60] found that norm estimates in such disconnected cases could be useful in signal recovery problems, a situation that has been anticipated in the literature, e.g., [126, p. 137]. Specifically, they considered the problem of finding a bound on the norm—and hence on the largest eigenvalue—of the discrete version of the operator $P_\Sigma Q_S P_\Sigma$ when the time–frequency area is small. In the finite case, the normalized area is $a = |S||\Sigma|/N$, where $|S|$ is the counting measure of S. Denote by $A_{S\Sigma}$ the operator with standard basis matrix $D_S\mathscr{F}_N^{-1}D_\Sigma\mathscr{F}_N$ where $\mathscr{F}_N$ is the matrix of the N-point DFT and $D_S = \operatorname{diag} S$ is the diagonal matrix with $D_S(j,j) = 1$ if $j \in S$ and $D_S(j,k) = 0$ otherwise. Thus D_S is the matrix of multiplication by the index indicator $\mathbb{1}_S$. Then

$$\begin{aligned} A_{S\Sigma}A_{S\Sigma}^* &= D_S\mathscr{F}_N^{-1}D_\Sigma\mathscr{F}_N(D_S\mathscr{F}_N^{-1}D_\Sigma\mathscr{F}_N)^* \\ &= D_S\mathscr{F}_N^{-1}D_\Sigma\mathscr{F}_N\mathscr{F}_N^*D_\Sigma\mathscr{F}_ND_S = D_S\mathscr{F}_N^{-1}D_\Sigma\mathscr{F}_ND_S, \end{aligned} \tag{4.6}$$

since $\mathscr{F}_N^* = \mathscr{F}_N^{-1}$ and D_Σ is idempotent.

Candès et al. were motivated by applications to compressed sensing in which a signal $\mathbf{x}$ could be recovered (using nonlinear optimization techniques) from its values on S provided that its Fourier transform $\widehat{\mathbf{x}}$ vanishes outside Σ. This recovery is contingent upon invertibility of $I - A_{S\Sigma}$, which holds if $\|A_{S\Sigma}\| \ll 1$. Candès et al. were able to obtain such bounds in a probabilistic sense using a combination of probabilistic and combinatorial methods to which we will refer as "CRT methods." We will give a brief outline of their approach here, and a more detailed outline of their technical estimates in Sects. 4.3 and 4.4. In order to quantify the decay of $\operatorname{prob}\|A_{S\Sigma}\| > 1/2$ with respect to relative magnitudes of supports, Candès et al. fixed a parameter $1 \le \beta \le (3/8)\log N$, so β is at most a fraction of $\log N$. Magnitudes of time and frequency supports are then bounded by

$$M(N,\beta) = \frac{N}{\sqrt{(\beta+1)\log N}}\Big(\frac{1}{\sqrt{6}} + o(1)\Big).$$

The "$o(1)$" term arises in Proposition 4.2.5 and Lemma 4.2.6 below.

Theorem 4.2.2. *Fix $S \subset \mathbb{Z}_N$. Let $\Sigma \subset \mathbb{Z}_N$ be randomly generated from the uniform distribution of subsets of $\mathbb{Z}_N$ of given size $|\Sigma|$, where $|S| + |\Sigma| \le M(N,\beta)$. Then, with probability at least $1 - O((\log N)^{1/2}/N^\beta)$, every signal $\mathbf{x}$ supported in S satisfies*

$$\|\widehat{\mathbf{x}}\,\mathbb{1}_\Sigma\|^2 \le \frac{1}{2}\|\mathbf{x}\|^2,$$

while every signal $\mathbf{x}$ frequency-supported in Σ satisfies

$$\|\mathbf{x}\,\mathbb{1}_S\|^2 \le \frac{1}{2}\|\mathbf{x}\|^2.$$

The second inequality says that $\|A_{S\Sigma}\|^2 \le 1/2$. By the arithmetic–geometric inequality, the normalized area satisfies

$$a = \frac{|S||\Sigma|}{N} \le \frac{1}{4N}M(N,\beta)^2 \le \frac{N}{24(\beta+1)\log N}(1+o(1)),$$

suggesting that even *small* N can support time–frequency set pairs of normalized area substantially larger than one on which no signal is mostly localized.

The probability is defined with respect to the uniform distribution among all sets Σ of fixed size $|\Sigma|$. The proof has two main ingredients. The CRT methods require the ability to treat the frequency coordinates independently. For this reason, Candès et al. considered Bernoulli generated frequency sets of random size rather than sets of fixed size. Thus, the first ingredient of the proof involves a random choice of frequency support. In contrast to our earlier usage of the symbol Ω as an interval length, we will use the symbol Ω here to denote a *Bernoulli* generated frequency support set of random size $|\Omega|$, while Σ denotes a *uniformly* generated support set of fixed size $|\Sigma|$. To describe Ω consider, for each $\omega \in \mathbb{Z}_N$, a Bernoulli random variable $I(\omega)$ (that is, a random variable with values in $\{0,1\}$) with $\text{prob}(I(\omega)=1)=\tau$ for each ω. Set $\Omega=\{\omega\in\mathbb{Z}_N : I(\omega)=1\}$. Then $|\Omega|=\sum_{\omega=0}^{N-1} I(\omega)$ has expected value $N\tau$. Take $\tau=|\Sigma|/N$ where, as before, $|\Sigma|$ is fixed. One has to show that $|\Omega|$ does not vary too much from its expected value of $|\Sigma|$. The second ingredient is a probabilistic bound on the norm of $A_{S\Omega}$ with S fixed.

The Main Technical Estimate for Theorem 4.2.2

Consider now the auxiliary matrix

$$H_{S\Omega}(s,t)=\begin{cases}\sum_{\omega\in\Omega} \mathrm{e}^{2\pi i\omega(s-t)/N}, & s,t\in S, s\neq t\\ 0, & \text{else,}\end{cases} \tag{4.7}$$

the finite analogue of the off-diagonal kernel $K(s,t)=\mathbb{1}_S\,\mathbb{1}_{s\neq t}\,\mathbb{1}_\Sigma^\vee(s-t)$. Then $A_{S\Omega}A_{S\Omega}^*=\frac{|\Omega|}{N}D_S+\frac{1}{N}H_{S\Omega}$ and

$$\|A_{S\Omega}\|^2\leq\frac{|\Omega|}{N}+\frac{\|H_{S\Omega}\|}{N}\,. \tag{4.8}$$

The CRT technique for estimating $\|H_{S\Omega}\|$ is to estimate traces of powers of $H_{S\Omega}^2$. It is similar to Landau and Widom's approach to proving Theorem 1.3.1. Section 4.3 contains a detailed outline of the proof of the following.

Proposition 4.2.3 (Main Proposition). *If $\tau\leq \mathrm{e}^{-2}$ then, with $\tau^*=\tau/(1-\tau)$, the powers of $H_{S\Omega}$ satisfy the expectation inequality*

$$\mathbf{E}(\mathrm{tr}(H_{S\Omega}^{2n}))\leq 2\,(n|S|)^{n+1}\left(\frac{4\tau^* N}{\mathrm{e}}\right)^n .$$

Bernoulli Versus Uniform Models

Theorem 4.2.2 states that for $|S|+|\Sigma|$ *small*, $\|A_{S\Sigma}\|^2\leq 1/2$ with overwhelming probability, taken with respect to the uniform distribution of sets of size $|\Sigma|$. But Proposition 4.2.3 applies to Bernoulli generated sets Ω whose sizes are also random. To pass from the proposition to the theorem one needs a probabilistic bound relating the probabilities of the operators $A_{S\Sigma}$ and $A_{S\Omega}$ having large norms, as was obtained by Candès and Romberg [57, 58] as follows.

Lemma 4.2.4. *Let Σ be drawn uniformly at random from the subsets of $\mathbb{Z}_N$ of fixed size $|\Sigma|$, and let $\Omega = \{\omega \in \mathbb{Z}_N : I(\omega) = 1\}$ where, for each ω, $\mathrm{prob}(I(\omega) = 1) = |\Sigma|/N$. Let $A_\Omega = A_{S\Omega}$ and $A_\Sigma = A_{S\Sigma}$ with the same fixed time support S. Then*

$$\mathrm{prob}(\|A_\Omega\|^2 > 1/2) \geq \frac{1}{2}\mathrm{prob}(\|A_\Sigma\|^2 > 1/2).$$

Proof. Conditioning $\|A_\Omega\|^2$ on $|\Omega|$ gives

$$\begin{aligned}\mathrm{prob}(\|A_\Omega\|^2 > 1/2) &= \sum_{m=0}^{N} \mathrm{prob}(\|A_\Omega\|^2 > 1/2 \,|\, |\Omega| = m)\,\mathrm{prob}(|\Omega| = m)\\ &= \sum_{m=0}^{N} \mathrm{prob}(\|A_{\Sigma_m}\|^2 > 1/2)\,\mathrm{prob}(|\Omega| = m).\end{aligned}$$

Here, "Σ_m" indicates a set drawn uniformly at random from all sets of size m. Observe that $\mathrm{prob}(\|A_{\Sigma_m}\|^2 > 1/2)$ is nondecreasing in m. Also, if $\mathbf{E}(|\Omega|) = \tau N \in \mathbb{N}$ then the median size of Ω is also τN; see, e.g., [165]. Therefore,

$$\begin{aligned}\mathrm{prob}(\|A_\Omega\|^2 > 1/2) &\geq \sum_{m=[\tau N]}^{N} \mathrm{prob}(\|A_{\Sigma_m}\|^2 > 1/2)\,\mathrm{prob}(|\Omega| = m)\\ &\geq \mathrm{prob}(\|A_\Sigma\|^2 > 1/2) \sum_{m=[\tau N]}^{N} \mathrm{prob}(|\Omega| = m)\\ &\geq \frac{1}{2}\mathrm{prob}(\|A_\Sigma\|^2 > 1/2).\end{aligned}$$

This proves the lemma. □

Probabilistic Norm Estimates for the Bernoulli Model

Armed with Lemma 4.2.4, Theorem 4.2.2 reduces to a corresponding estimate of $\mathrm{prob}(\|A_{S\Omega}\|^2 > 1/2)$ in which Ω is a *Bernoulli random set*. The following precise estimate uses Proposition 4.2.3.

Proposition 4.2.5. *Let $\rho = \sqrt{2/3}$ and fix $1 \leq \beta \leq (3/8)\log N$ and $q \in (0, 1/2]$. Let $\Omega = \{\omega \in \mathbb{Z}_N : I(\omega) = 1\}$ where $\mathrm{prob}(I(\omega) = 1) = |\Sigma|/N$ and*

$$|S| + |\Sigma| \leq \frac{\rho q N}{\sqrt{(\beta+1)\log N}}.$$

Let $H_{S\Omega}$ be defined as in (4.7). Then

$$\mathrm{prob}(\|H_{S\Omega}\| > qN) \leq \sqrt{6}N^{-\beta}\sqrt{(\beta+1)\log N}.$$

Proof. The moment bound of Proposition 4.2.3 is utilized as follows. Since $H = H_{S\Omega}$ is self-adjoint, the Markov inequality and $\|H^n\| \leq \mathrm{tr}(H^n)$ imply that for $n = 1,2,\ldots,$

$$\mathrm{prob}\,(\|H\| > qN) = \mathrm{prob}\,(\|H^n\|^2 > (qN)^{2n}) \leq \frac{\mathbf{E}\|H^n\|^2}{(qN)^{2n}} \leq \frac{\mathbf{E}(\mathrm{tr}(H^{2n}))}{(qN)^{2n}}\,.$$

Taking $\tau < 1/3$ if necessary (so that $4/(1-\tau) \leq 6$) one obtains from Proposition 4.2.3 and the inequality on arithmetic and geometric means that

$$\mathbf{E}(\mathrm{tr}(H^{2n})) \leq 2n(6/\mathrm{e})^n n^n |S|^{n+1}(\tau N)^n \leq 2\frac{n^{n+1}}{\mathrm{e}^n}|S|\left(\frac{|S|+\tau N}{\rho}\right)^{2n}.$$

Specializing to $n = \lfloor (\beta+1)\log N \rfloor$, the hypothesis $|S| + \tau N \leq \rho q N/\sqrt{(\beta+1)\log N}$ and the last two inequalities imply that

$$\mathrm{prob}\,(\|H\| > qN) \leq 2\mathrm{e}\rho q N^{-\beta}\sqrt{(\beta+1)\log N} \leq \sqrt{6}N^{-\beta}\sqrt{(\beta+1)\log N}\,.$$

This establishes Proposition 4.2.5. □

To pass from a probabilistic bound on $H_{S\Omega}$ to one on $A_{S\Omega}$, see (4.8), one requires that $|\Omega|$ does not deviate too much. When $\tau \geq (\log N)^2/N$, Ω is likely not too large as the following lemma shows.

Lemma 4.2.6. *With ρ, β, q and Σ as in Proposition 4.2.5, let $\Omega = \{\omega \in \mathbb{Z}_N : I(\omega) = 1\}$ where $\mathbf{E}(|\Omega|) = \tau N = |\Sigma|$ such that*

$$\frac{(\log N)^2}{N} \leq \tau \leq \frac{\rho q}{\sqrt{(1+\beta)\log N}}\,.$$

Then, with probability at least $1 - N^{-\beta}$,

$$\frac{|\Omega|}{N} \leq \frac{2\rho q}{\sqrt{(1+\beta)\log N}}\,.$$

Proof. The Bernstein inequality for a random variable (see, e.g., [39]) states that

$$\mathrm{prob}\,(|\Omega| > \mathbf{E}(|\Omega|) + \lambda) \leq \exp\left(\frac{-\lambda^2}{2\mathbf{E}(|\Omega|) + 2\lambda/3}\right) \qquad (\lambda > 0)\,.$$

Taking $\lambda = \mathbf{E}(|\Omega|) = N\tau$, since $\tau \geq (\log N)^2/N \geq 8\beta \log N/(3N)$ (by the hypothesis of Proposition 4.2.5) one obtains $\mathrm{prob}\,(|\Omega| > 2\mathbf{E}(|\Omega|)) \leq N^{-\beta}$. That is,

$$|\Omega| \leq 2\mathbf{E}(|\Omega|) = 2\tau N \leq \frac{2N\rho q}{\sqrt{(1+\beta)\log N}}$$

with probability at least $1 - N^{-\beta}$. This proves the lemma. □

Applying Proposition 4.2.5 and Lemma 4.2.6 to (4.8) one finds that, with probability at least $1 - O(N^{-\beta}\sqrt{\log N})$,

$$\|A_{S\Omega}\|^2 \leq q\left(1 + \frac{2\rho}{\sqrt{(\beta+1)\log N}}\right).$$

With $q = (1 - (2\rho/\sqrt{(\beta+1)\log N}))/2 = 1/2 + o(N)$ one obtains $\|A_{S\Omega}\|^2 < 1/2$ with the desired probability. Theorem 4.2.2 then follows from Lemma 4.2.4.

Discussion

A closer look at the "$o(1)$" term in defining $M(N,\beta)$ in Theorem 4.2.2 led Candès, Romberg, and Tao to replace $M(N,\beta)$ with the specific support bound $|S| + |\Sigma| \leq 0.2971N/\sqrt{(\beta+1)\log N}$ which also gives the norm inequalities of the theorem whenever $N \geq 512$. For concreteness, take $\beta = (13/36)\log N < (3/8)\log N$. The support constraint then is $|S| + |\Sigma| \leq 0.2971N/((7/6)\log N) = 0.2547N/\log N$. For $N = 512$, this requires $|S| + |\Sigma| \leq 20.88$, while $\log(N)^{1/2}/N^{\beta} \approx \sqrt{6.24}/(512)^{13/36} < (2.5)/(2^{13/4})$. In particular, when $N = 512$, if the constant in the $O((\log N)^{1/2}/N^{\beta})$ term in Theorem 4.2.2 is essentially equal to one then, for any fixed time support S, there is less than a one-in-four chance that there is a signal supported in S having at least half of its DFT energy in the randomly generated set Σ, when $|\Sigma| \leq 20 - |S|$.

Table 4.1 gives the results of a simple experiment in which 100 time and frequency support sets, each of the same size ($|S| = |\Sigma|$), were generated for several values of $|S|$. Both S and Σ were generated by taking the indices of the largest values of a vector of dimension N with coordinates generated by a uniform random distribution. With $N = 512$, the maxima and minima of the norms of $A_{S\Sigma} = \mathscr{F}^{-1}D_{\Sigma}\mathscr{F}D_S$ among the 100 instances per each value of S are given. The normalized time–frequency area a needs to be approximately 10 before one begins to find $\mathbf{x}$ frequency-supported in Σ and having at least half of its energy in S.

Table 4.1 Norm of $A = \mathscr{F}^{-1}\mathbb{1}_{\Sigma}\mathscr{F}\mathbb{1}_S$ for randomly generated supports, $N = 512$

$\|S\| = \|\Sigma\|$	a	max $\|A\|$	min $\|A\|$	$\%\|A\|^2 > 1/2$
10	0.20	.27	.22	0
20	0.78	0.39	0.33	0
40	3.12	0.54	0.49	0
80	12.5	0.73	0.69	59
100	19.53	0.80	0.76	100

For visual comparison, Fig. 4.1 illustrates the eigenvalues of $A_{S\Sigma} = \mathscr{F}^{-1}D_{\Sigma}\mathscr{F}D_S$ as the frequency support becomes disconnected. In contrast to Table 4.1, the normalized area is fixed at $a = 16$ in this case, and the time support is an interval. Only the bottom row of Fig. 4.1 corresponds to a typical *randomly generated* frequency support in the sense described for Table 4.1. For larger values of the number N of

DFT points, experimental results indicate a decrease in the *expected* norm of A, as is consistent with the appearance of the factor $1/N^{\beta}$ in Theorem 4.2.2.

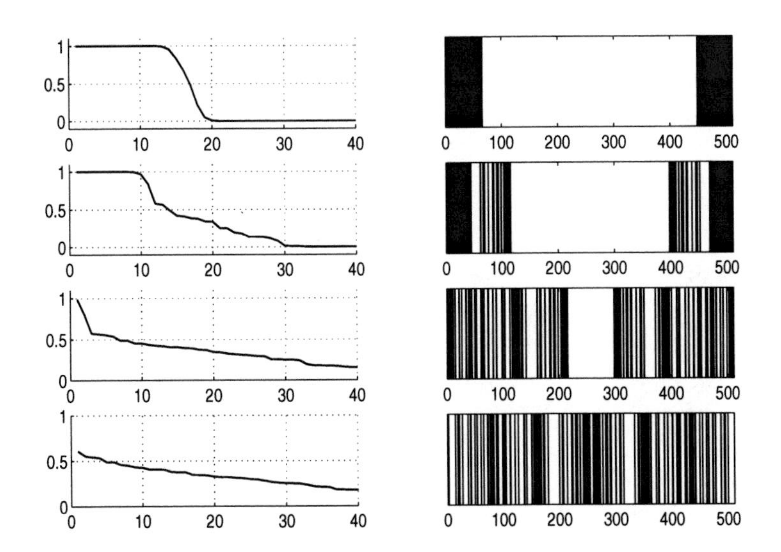

Fig. 4.1 *Eigenvalues for disconnected Fourier supports.* Plots show eigenvalues (left) and Fourier supports (right) for a 512 point DFT. In each case, the time support is a discrete interval of length 64 points and the Fourier support has 128 points. The normalized area is $a = 16$. As the Fourier support becomes more disconnected, the eigenvalues become more evenly spread out

4.3 Proof of the Main Proposition

This section contains a detailed proof of Proposition 4.2.3. First we briefly review a few tools that will be used.

4.3.1 Relational Calculus on $\mathbb{Z}_N^{2n}$.

Partitions of $\{1,\ldots,n\}$

A partition P of $A = \{1,\ldots,n\}$ is a collection $U_1,\ldots,U_k$ of subsets of A that are pairwise disjoint ($U_i \cap U_j = \emptyset$) and such that each element of A is contained in one of the U_k ($\cup U_k = A$). A partition can also be thought of as a decomposition of an equivalence relation into its equivalence classes. This was the approach taken by Candès,

Romberg, and Tao [59]. In the following outline of their arguments, we translate their *relational calculus* into the language of partitions. When $P = \{U_1,\ldots,U_k\}$, we will write $U_j \in P$ and we will denote by $|P|$ the number k of nonempty subsets (classes) into which P partitions A. A partition $Q = \{V_1,\ldots,V_\ell\}$ is called a *refinement* of $P = \{U_1,\ldots,U_k\}$, expressed by $P \leq Q$, if for each $V_i \in Q$, there is a $U_j \in P$ such that $V_i \subset U_j$. In other words, Q repartitions each of the partition elements of P. Finally, we will write $\mathscr{P} = \mathscr{P}(A)$ for the lattice of partitions of A, partially ordered by refinement.

Stirling Numbers

Partitions of finite sets satisfy certain combinatorial properties. For example, the *Stirling numbers of the second kind*, $\mathrm{St}\,(n,k)$, enumerate the number of distinct ways in which a set of n elements can be partitioned into k pairwise disjoint subsets. These numbers satisfy

$$\mathrm{St}\,(n+1,k) = \mathrm{St}\,(n,k-1) + k\,\mathrm{St}\,(n,k) \tag{4.9}$$

since, if a is a distinguished element of A (where $|A| = n+1$), then $\mathrm{St}\,(n,k-1)$ counts all of those k-partitions containing the singleton a as one of its elements, while $k\,\mathrm{St}\,(n,k)$ counts the ways that a can be added to one of the elements of a k-partition of $A \setminus \{a\}$ to obtain a k-partition of A.

Function Partitions

If A and B are finite sets, $B = \{b_1,\ldots,b_k\}$, and $f : A \to B$ is a *surjective* function, then f partitions A by setting $U_j = f^{-1}(b_j)$ and $P_f = \{U_1,\ldots,U_k\}$. If σ is any permutation of the elements of B then $g = \sigma \circ f$ gives rise to the same partition $P = P_f = P_g$ where, now, $U_j = g^{-1}(\sigma(b_j))$. The enumeration of the partition sets might be different, but the partition elements will be exactly the same. In fact, if $P_f = P_g$ then $g = \sigma \circ f$ for some permutation σ of the range. A lattice structure on (equivalence classes of) the functions mapping A to B can be defined by $f \ll g$ if $P_f \leq P_g$ (i.e., P_g refines P_f).

Bernoulli Generated Random Sets

By regarding different DFT frequencies as independent Bernoulli generated random variables, one can take advantage of the group structure on $\mathbb{Z}_N$. Thus, to each element $\omega \in \mathbb{Z}_N$ we assign a Bernoulli random variable $I(\omega)$—one taking values in $\{0,1\}$, each with the same fixed expected value $\mathbf{E}(I(\omega)) = \mathrm{prob}\,(I(\omega) = 1) = \tau \in (0,1)$. The set $\Omega = \{\omega \in \mathbb{Z}_N : I(\omega) = 1\}$ will be called a *Bernoulli generated random subset of* $\mathbb{Z}_N$. Its expected size is $\mathbf{E}(|\Omega|) = N\tau$. When referring to membership in a subset $\Omega \subset \mathbb{Z}_N$, we will say that "$\omega_j$ belongs to Ω with probability τ."

However, when membership in Ω is conditioned on some preassigned attribute of Ω such as its size, this probability can differ from τ. This trite observation will play an important role in what follows.

4.3.2 Traces of Powers of $H_{S\Omega}$

Recall that $A_{S\Omega} = D_S \mathscr{F}_N^{-1} D_\Omega \mathscr{F}_N$ is the finite analogue of a band- and time-limiting operator $Q_S P_\Sigma$. One has $A_{S\Omega} A_{S\Omega}^* = D_S \mathscr{F}_N^{-1} D_\Omega \mathscr{F}_N D_S$. Defining D_S as the matrix of the identity restricted to S, we can express the matrix $H = H_{S\Omega} = N A_{S\Omega} A_{S\Omega}^* - |\Omega| D_S$ in (4.7) by setting, for s and t in the fixed set S,

$$H(s,t) = \begin{cases} 0, & s = t \\ c(s-t) & s \neq t,\ c(u) = \sum_{\omega \in \Omega} e^{\frac{2\pi i}{N}\omega u} \end{cases}.$$

Given a multi-index $\mathbf{t} = (t_1, \ldots, t_{2n}) \in \mathbb{Z}_N^{2n}$, we use the convention $t_{2n+1} \equiv t_1$ and let $\Delta t_j = t_j - t_{j+1}$. In sums over t-values we will use the shorthand "$\Delta t_j \neq 0$" to mean $\Delta t_j \neq 0$ for *each index* j appearing in the sum. A diagonal element of the $2n$th power of H can be written

$$H^{2n}(t_1, t_1) = \sum_{t_2, \ldots, t_{2n} : \Delta t_j \neq 0} \prod_{j=1}^{2n} c(\Delta t_j) .$$

The expected trace of H^{2n} corresponding to the Bernoulli generated random set Ω is the sum over these diagonal elements, namely

$$\begin{aligned} \mathbf{E}(\mathrm{tr}(H^{2n})) &= \sum_{\mathbf{t}:\Delta t_j \neq 0} \mathbf{E}\Big(\sum_{\omega \in \Omega^{2n}} e^{\frac{2\pi i}{N} \sum_{j=1}^{2n} \omega_j \Delta t_j} \Big) \\ &= \sum_{\mathbf{t}:\Delta t_j \neq 0} \sum_{\omega \in \mathbb{Z}_N^{2n}} e^{\frac{2\pi i}{N} \sum_{j=1}^{2n} \omega_j \Delta t_j} \mathbf{E}\Big(\prod_{j=1}^{2n} I(\omega_j) \Big) \end{aligned}$$

where one has applied the linearity of expectation together with the definition $I(\omega) = 1$ if $\omega \in \Omega$, where $I(\omega)$ is regarded as a $\{0,1\}$-Bernoulli random variable with expected value $\tau \in (0,1)$.

Expectation Conditioned on a Partition

Any vector $\omega = (\omega_1, \ldots, \omega_{2n}) \in \mathbb{Z}_N^{2n}$ can be regarded as a function from $\{1, 2, \ldots, 2n\}$ to $\mathbb{Z}_N$. As such, ω induces a partition P_ω of $\{1, \ldots, 2n\}$ whose partition elements are the pre-image indices of the frequency elements $\omega_j \in \mathbb{Z}_N$. Once $\omega = (\omega_1, \ldots, \omega_{2n})$ is fixed, $I(\omega_j) = I(\omega_k)$ if $\omega_j = \omega_k$, so one can think of ω and its partition $P = P_\omega$ as inducing a *conditional expectation* on $\{1, 2, \ldots, 2n\}$ by regarding $I(\omega)$ as a function of the partition element. More precisely, if $U \in P$, let ω_U denote the common frequency element of $\mathbb{Z}_N$ assigned by ω to each member of U. Then, once ω is fixed,

$$\mathbf{E}\Big(\prod_{j=1}^{2n} I(\omega_j)\Big) = \mathbf{E}\Big(\prod_{U\in P} I(\omega_U)\Big) = \tau^{|P|}$$

whenever $P_\omega = P$. Coarser partitions yield higher expected values.

Denote by $W(P)$ the *pullback* $W(P) = \{\omega \in \mathbb{Z}_N^{2n} : P_\omega = P\}$. Then

$$\mathbf{E}(\mathrm{tr}(H^{2n})) = \sum_{\mathbf{t}:\Delta t_j \neq 0} \sum_{P\in\mathscr{P}} \tau^{|P|} \sum_{\omega\in W(P)} \mathrm{e}^{\frac{2\pi i}{N}\sum_{j=1}^{2n}\omega_j \Delta t_j}\,.$$

The inner sum can be analyzed further by taking advantage of symmetry.

4.3.3 Combinatorics for Exponential Sums

Candès, Romberg, and Tao introduced at this stage certain *inclusion–exclusion formulas* that allow simplification of the exponential sums inside the sum over the partitions. In what follows, for $U \subset A$ and $P \in \mathscr{P}(A)$, let $P(U) = \{U \cap V : V \in P\}$ denote the partition of U induced by P. The following is a special case of Candès et al.'s formula in the case $A = \{1,\dots,2n\}$.

Lemma 4.3.1 (Inclusion–exclusion). *For $P \in \mathscr{P}(\{1,\dots,2n\})$ and $f : \mathbb{Z}_N^{2n} \to \mathbb{C}$,*

$$\sum_{\omega\in W(P)} f(\omega) = \sum_{Q\leq P} (-1)^{|P|-|Q|} \prod_{U\in Q} (|P(U)|-1)! \sum_{R\leq Q} \sum_{\omega\in W(R)} f(\omega)\,.$$

The proof of this lemma will be given in Sect. 4.4.2. By splitting the set $A = \{1,\dots,2n\}$ into its partition elements with respect to a *fixed* Q, one has

$$\sum_{P\geq Q} \tau^{|P|}(-1)^{|P|-|Q|} \prod_{U\in Q} (|P(U)|-1)! = \prod_{U\in Q} \sum_{k=1}^{|U|} \mathrm{St}\,(|U|,k)\, \tau^k\, (-1)^{|U|-k}(k-1)!$$

where, as before, $\mathrm{St}\,(m,k)$ is the Stirling number. The inner sum depends only on $|U|$ and τ so, upon defining

$$F(m,\tau) = \sum_{k=1}^{m} \mathrm{St}\,(m,k)\, \tau^k\, (-1)^{m-k}(k-1)!\,, \tag{4.10}$$

the identity becomes

$$\sum_{P:Q\leq P} \tau^{|P|}(-1)^{|P|-Q|}\Big(\prod_{U\in Q} (|P(U)|-1)!\Big) = \prod_{U\in Q} F(|U|,\tau)\,.$$

By Lemma 4.3.1, one has the following.

Lemma 4.3.2.

$$\sum_{P\in\mathscr{P}} \tau^{|P|} \sum_{\omega\in W(P)} f(\omega) = \sum_{Q\in\mathscr{P}} \left(\sum_{R\leq Q} \sum_{\omega\in W(R)} f(\omega) \right) \prod_{U\in Q} F(|U|,\tau)\,.$$

Specializing to $f(\omega) = \mathrm{e}^{\frac{2\pi i}{N}\sum_{1\leq j\leq 2n}\omega_j \Delta t_j}$ yields the following formula for the expected trace of H^{2n}:

$$\mathbf{E}(\mathrm{tr}(H^{2n})) = \sum_{P\in\mathscr{P}} \sum_{\mathbf{t}:\Delta t_j\neq 0} \sum_{R\leq P} \sum_{\omega\in W(R)} \mathrm{e}^{\frac{2\pi i}{N}\sum_{j=1}^{2n}\omega_j \Delta t_j} \prod_{U\in P} F(|U|,\tau)\,.$$

Fix P and $U \in P$ and set $\Delta t_U = \sum_{j\in U} \Delta t_j$. For $\omega \in W(R)$ ($R \leq P$) and index j, the value of ω_j depends only on the partition set containing j so we write $\omega_j = \omega_U$; however, as R ranges over $R \leq P$ and ω ranges over $W(R)$, the set of $\{\omega_U = (\omega_{U_1},\ldots,\omega_{U_{|P|}})\}$ ranges precisely over $\mathbb{Z}_N^{|P|}$. In light of this, we have

$$\sum_{R\leq P} \sum_{W(R)} \mathrm{e}^{\frac{2\pi i}{N}\sum_{U\in P}\omega_U \Delta t_U} = \sum_{\omega_U\in\mathbb{Z}_N^{|P|}} \mathrm{e}^{\frac{2\pi i}{N}\sum\omega_U \Delta t_U} = \begin{cases} N^{|P|}, & \Delta t_U = 0 \text{ all } U \\ 0, & \text{otherwise.} \end{cases}$$

Substituting this back into the expected trace formula above leads to the following dramatically simplified estimate for the expected trace of H^{2n}.

Lemma 4.3.3.

$$\mathbf{E}(\mathrm{tr}(H^{2n})) = \sum_{P\in\mathscr{P}} \sum_{\mathbf{t}:\Delta t_j\neq 0 \text{ and } \forall U\in P,\, \Delta t_U=0} N^{|P|} \prod_{U\in P} F(|U|,\tau)\,.$$

For the estimates that follow, it will be important to observe that if P contains a singleton $U = \{j\}$ then, since $\Delta t_j \neq 0$, one cannot have $\Delta t_U = 0$. Consequently, the *partition sum in Lemma 4.3.3 is naturally restricted to*

$$\widetilde{\mathscr{P}} = \{P \in \mathscr{P} : \forall U \in P, |U| > 1\}. \tag{4.11}$$

The number of $\mathbf{t}$ for which Δt_U vanishes for all $U \in P$ is estimated as follows.

Lemma 4.3.4. *For any $P \in \mathscr{P}$,*

$$\#\{\mathbf{t} \in S^{2n} : \Delta t_U = 0 \text{ for all } U \in P\} \leq |S|^{2n-|P|+1}\,.$$

Proof. The sum Δt_U is a linear combination of the coordinates of the quantities $\Delta t_j \in S - S$ which are linearly independent of one another except for the *circular sum* constraint that $\sum_{j=1}^{2n} \Delta t_j = 0$. The added constraints $\Delta t_U = 0$ for all $U \in P$ amount to $|P|$ independent linear constraints. Each constraint diminishes the solution set by a factor $|S|$. The lemma follows. □

The estimate is sharp for the trivial partition—in this case the solution set is all of S^{2n}—and for the complete partition—in this case solutions must be *constants* $\mathrm{t} = (t,t,\dots,t)$, $t \in S$.

Any further estimates on $\mathbf{E}(\mathrm{tr}(H^{2n}))$ reduce to estimating $F(n,\tau)$ in (4.10) and to counting the number of partitions of $\{1,\dots,2n\}$ whose partition set sizes have a given distribution. The starting point is the following formula for $F(n,\tau)$, which follows from the recursion relation for the Stirling numbers.

Lemma 4.3.5. *If $n \geq 1$ and $0 \leq \tau < 1/2$ then, with $\tau^* = \tau/(1-\tau)$,*

$$F(n,\tau) = \sum_{k=1}^{\infty} (-1)^{n-k} k^{n-1} (\tau^*)^k . \tag{4.12}$$

Proof. The identity is proved by induction. For $n = 1$, $F(1,\tau) = \tau$ whereas

$$\sum_{k=1}^{\infty} (-1)^{1-k} (\tau^*)^k = 1 - \sum_{k=0}^{\infty} \left(\frac{\tau}{\tau - 1}\right)^k = \tau$$

so (4.12) holds for $n = 1$. Assuming (4.12) holds for n, applying $(\tau^2 - \tau)\frac{\mathrm{d}}{\mathrm{d}\tau}$ to both sides, reindexing and noting that $\mathrm{St}\,(n,0) = 0$, gives

$$\sum_{k=0}^{n+1} [\mathrm{St}\,(n,k-1) + k\,\mathrm{St}\,(n,k)](-1)^{n+1-k}(k-1)!\,\tau^k = \sum_{k=0}^{\infty} (-1)^{n+1-k} k^n (\tau^*)^k .$$

By (4.9), this is (4.12) with n replaced by $n+1$. □

Once τ is fixed, (4.12) can be used to bound the magnitude of $F(n,\tau)$.

Lemma 4.3.6. *Let $n \geq 1$ and $0 \leq \tau < 1/2$. Define*

$$G(n+1,\tau) \equiv \begin{cases} \tau^* & \text{if } \tau^* \leq \mathrm{e}^{-n} \\ \mathrm{e}^{n(\log n - \log\log\frac{1}{\tau^*} - 1)} & \text{if } \tau^* > \mathrm{e}^{-n} . \end{cases}$$

Then $|F(n,\tau)| \leq G(n,\tau)$.

Proof. Set $g(x) = (\tau^*)^x x^{n-1}$, which has a unique maximum value at $x_* = (1-n)/\log \tau^*$. If $\tau^* \leq \mathrm{e}^{1-n}$ then $x_* \leq 1$ and

$$F(n,\tau) = \sum_{k=1}^{\infty} (-1)^{n+k} g(k)$$

is a convergent alternating series with sum at most $g(1) = \tau^*$. Otherwise the series has magnitude at most $g(x_*) = G(n,\tau)$. □

Lemma 4.3.7. *If $\tau \leq 1/(1+\mathrm{e})$ then G satisfies*

$$G(n+1,\tau) \leq n\,G(n,\tau) .$$

Proof. Since G is log convex in n, for $G(n) = G(n,\tau)$ one has

$$\log G(n+1) \leq \log G(n) + \frac{\mathrm{d}}{\mathrm{d}n} \log G(n+1)\,.$$

The condition on τ implies that $\log\log(1/\tau^*) \geq 0$ which, in turn, implies that

$$\frac{\mathrm{d}}{\mathrm{d}n} \log G(n+1) \leq \log n \,\square$$

Applying Lemmas 4.3.3 and 4.3.4, abbreviating $G(n) = G(n,\tau)$, and defining $\widetilde{\mathscr{P}}$ as in (4.11), one concludes that

$$\mathbf{E}(\mathrm{tr}(H^{2n})) \leq \sum_{k=1}^{n} N^k |S|^{2n-k+1} \sum_{P\in\widetilde{\mathscr{P}}:\,|P|=k} \prod_{U\in P} G(|U|). \tag{4.13}$$

To get a more tangible bound on $\mathbf{E}(\mathrm{tr}(H^{2n}))$ we work on the term

$$\Phi(n,k) = \sum_{P\in\widetilde{\mathscr{P}}(\mathbb{N}_n):\,|P|=k} \prod_{U\in P} G(|U|)\,.$$

(The corresponding term in (4.13) is $\Phi(2n,k)$.)

Lemma 4.3.8. *For $n \geq 2$ and $k \geq 1$,*

$$\Phi(n+1,k) \leq n\big(\Phi(n,k) + G(2)\,\Phi(n-1,k-1)\big) \tag{4.14}$$

while, for $n > 2k$,

$$\Phi(n+1,k) \leq 2n\,\Phi(n,k)\,. \tag{4.15}$$

Proof. First consider (4.14). Recall that if $P \in \widetilde{\mathscr{P}}$ then any $U \in P$ has at least two elements. The estimate will come from considering the partition element containing the index "$n+1$."

There are $\mathrm{St}(n-1,k-1)$ k-partitions of $\{1,\ldots,n\}$ in which any fixed $j \in \{1,\ldots,n\}$ is a singleton partition element. Let P_j' denote one of these partitions and let P_j denote the corresponding partition of $\{1,\ldots,n+1\}$ obtained by grouping "$n+$ 1" with the singleton $\{j\}$. Since P' contains a singleton, the product $\prod_{V\in P_j'} G(|V|)$ is equal to $G(1)$ times a product arising in the sum defining $\Phi(n-1,k-1)$. But $\prod_{V\in P_j'} G(|V|)$ is multiplied by $G(2)/G(1)$ to produce $\prod_{U\in P_j} G(|U|)$. Summing over all such partitions P_j' when j is fixed, then summing over j, thus gives rise to the term $n\,G(2)\,\Phi(n-1,k-1)$ on the right-hand side of (4.14).

Any remaining k-partition P' of $\{1,\ldots,n\}$ has the property that its elements $\{V_1,\ldots V_k\}$ each satisfy $|V_j| \geq 2$. Appending "$n+1$" to one of the V_i results in a partition P_{V_i} of $\{1,\ldots,n+1\}$ such that $\prod_{U\in P_{V_i}} G(|U|) \leq |V_i| \prod_{V\in P'} G(|V|)$ since $G(m+1) \leq m\,G(m)$ by Lemma 4.3.7. Summing over the k choices of extension P_{V_i} of P, observing that $\sum |V_i| = n$, then summing over all such P' results in a bound of $n\,\Phi(n,k)$. This takes care of all the terms in the sum defining Φ, thus proving (4.14).

Next consider (4.15). By (4.14), it is enough to show that $G(2)\Phi(n-1,k-1) \le \Phi(n,k)$ when $n > 2k$. Let $P' \in \widetilde{\mathscr{P}}(\mathbb{N}_{n-1})$ be a $(k-1)$-partition. Since $n > 2k$, some element $V \in P'$ satisfies $|V| > 2$. Such an element gives rise to $|V|$ distinct partitions in $\widetilde{\mathscr{P}}(\mathbb{N}_n)$ by removing a single element $j \in V$ then appending $\{j,n\}$ to get $P \in \widetilde{\mathscr{P}}(\mathbb{N}_n)$. Different P' give rise to different elements of $\widetilde{\mathscr{P}}(\mathbb{N}_n)$ in this way. The partition product $\prod_{U'\in P'} G(|U'|)$ is replaced by

$$\prod_{U\in P} G(|U|) = G(2)\frac{G(|V|-1)}{G(|V|)} \prod_{U'\in P'} G(|U'|) \ge \frac{G(2)}{|V|-1} \prod_{U'\in P'} G(|U'|)\,.$$

Since $|V|$ k-partitions in $\widetilde{\mathscr{P}}_n$ arise from each $(k-1)$-partition in $\widetilde{\mathscr{P}}(\mathbb{N}_{n-1})$ in this manner, this proves that $G(2)\Phi(n-1,\,k-1) \le \Phi(n,k)$ when $n > 2k$. This completes the proof of the lemma. □

By applying (4.15) repeatedly, one has

$$\Phi(n,k) \le (n-1)(n-2)\cdots 2k\, 2^{n-k}\, \Phi(2k,k).$$

However, since $\widetilde{\mathscr{P}}_n$ has no k-partitions if $n < 2k$, it follows from repeated application of (4.14) that

$$\Phi(2k,k) \le (2k-1)\, G(2)\Phi(2k-2,k-1) \le \cdots \le \frac{(2k-1)!G^k(2)}{2^{k-1}(k-1)!}\,.$$

Putting these observations together one has, for $n > 2k$,

$$\Phi(n,k) \le \frac{(n-1)!}{(k-1)!} 2^{n-2k+1}\, G^k(2)\,. \tag{4.16}$$

Setting

$$B(n,k) = \frac{(n-1)!}{(k-1)!} N^k |S|^{n-k+1} 2^{n-2k+1}\, G^k(2),$$

it follows from (4.16), (4.13), and the definition of Φ that

$$\mathbf{E}(\mathrm{tr}(H^{2n})) \le \sum_{k=1}^{n} B(2n,k)\,.$$

Supposing that $n \le NG(2)/(4|S|)$ and noting that

$$\frac{B(2n,k)}{B(2n,k-1)} = \frac{NG(2)}{4|S|(k-1)},$$

it follows that $B(2n,k)$ is nondecreasing with $k \le n$, so

$$\mathbf{E}(\mathrm{tr}(H^{2n})) \le n\,B(2n,n) = n\frac{2n!}{n!} G^n(2)|S|^{n+1}N^n\,.$$

The classical Stirling approximation allows one to estimate

$$\frac{(2n)!}{n!} \leq 2^{2n+1}\left(\frac{n}{\mathrm{e}}\right)^n .$$

Recalling that $G(2) = \tau^*$, altogether this gives

$$\mathbf{E}(\mathrm{tr}(H^{2n})) \leq n\,2^{2n+1}\left(\frac{n}{\mathrm{e}}\right)^n (\tau^*)^n |S|^{n+1} N^n .$$

This completes the proof of Proposition 4.2.3, except for the proof of Lemma 4.3.1, which can be found in Sect. 4.4.2.

4.4 Notes and Auxiliary Results

4.4.1 Miscellaneous Multiband Issues

Multiband Usage

In this chapter, the term multiband has referred primarily to signals whose Fourier transforms are supported in finite unions of pairwise disjoint intervals. The term "multiband" arises frequently in the applied sciences and has different meanings in different contexts. For example, in electroencephalography, different frequency bands present in EEG scalp readings; i.e., Delta (0 – 4 Hz), Theta (4 – 7 Hz), Alpha (8 – 12 Hz), Beta (approx. 12 – 30 Hz) and Gamma (approx. 30 – 100 Hz) are attributed to different mental processes. In automated speech recognition (see, e.g., [218]) multiband methods refer to uniform or relatively uniform divisions of the full band into subbands for subsequent multirate filtering. In image processing, particularly in remote sensing applications, multispectral or multi-band imaging typically refers to the collection of image data at different wavelengths across the electromagnetic spectrum by means of instruments or filters sensitive to particular ranges of the electromagnetic spectrum; see, e.g., [242]. In audio encoding, multi-band excitation refers to a representational method developed first by Osamu Fujimura in 1967 [104]. In communications, multiband radio refers specifically to communications devices operating under FCC license rules that enable communication simultaneously across several radio bands.

Optimal Sets for Time and Band Limiting

The "$o(\log a)$" term in Theorem 4.1.1 can be substantial in the *non-asymptotic* regime for two reasons. First, Szegő's theorem is only an approximation. A concrete spectrum estimate is required when time and frequency supports are fixed and one is not taking an asymptotic limit. Second, the linear distribution of the time and frequency supports can make the "$O(1)$" term dominant in the eigenvalue estimates. It was proved by Donoho and Stark [86] that among all compact time and frequency supports S and Σ of a given area $|S||\Sigma|$, if $|S||\Sigma| < 0.8$ then the $P_\Sigma Q_S$ operator having the largest norm occurs when S and Σ are intervals. This was proved by means of an inequality for rearrangements in [86], stating that if $2\Omega T < 0.8$ and if f is supported on a set of Lebesgue measure T then its symmetric, decreasing rearrangement $|f|^\star$ satisfies

$$\int_{-\Omega}^{\Omega} |\widehat{f}(\xi)|^2 \, d\xi \le \int_{-\Omega}^{\Omega} |(|f|^\star)^\wedge(\xi)|^2 \, d\xi \, .$$

Since $|f|^\star$ is concentrated in $[-|S|/2, |S|/2]$ when f is supported in a set of measure $|S|$, this indicates that the frequency concentration over Σ is optimized when Σ is an interval, at least when $|S||\Sigma|$ is not too big. The question of whether an interval optimizes joint concentration when $T|\Sigma| \gg 0.8$ remains open.

4.4.2 *Proof of the Inclusion–Exclusion Relation*

If a partition P of $\{1,\dots,2n\}$ is fixed, then $\omega \in W(P)$ means that $\omega_i = \omega_j$ precisely when i and j belong to the same partition element of P. Since the left-hand side of the formula in Lemma 4.3.1 depends only on the values of f on $W(P)$, and since one can regard $W(P)$ as a quotient space of $\mathbb{Z}_N^{2n}$ by choosing a representative of each partition set, one can think of this quotient of f instead as a function, also denoted f, on the subset $\mathring{\mathbb{Z}}_N^m$ of $\mathbb{Z}_N^m$, where $m = |P|$, consisting of all $\omega = (\omega_1, \dots, \omega_m)$ having no repeated coordinates, that is, such that $\omega_i \neq \omega_j$ when $i \neq j$. On the other hand, the right-hand side of the identity of Lemma 4.3.1 involves sums over aggregations or *anti-refinements* of P. Under this quotientization of P, the clusters $Q \le P$ become ordinary partitions of $\mathbb{N}_m = \mathbb{N}_{|P|}$. The identity of Lemma 4.3.1 is restated here as follows:

$$\sum_{\omega \in \mathring{\mathbb{Z}}_N^m} f(\omega) = \sum_{P \in \mathscr{P}(\mathbb{N}_m)} (-1)^{m-|P|} \prod_{U \in P} (|U|-1)! \sum_{Q \le P} \sum_{\omega \in W(Q)} f(\omega) \, . \tag{4.17}$$

Proof (of (4.17)). We proceed by induction on m. When $m = 1$, there is only one partition and for it, both sides equal $\sum_{\omega \in \mathbb{Z}_N} f(\omega)$. Suppose that (4.17) is true for $m-1$. Expressing $\omega = (\omega_1, \dots, \omega_m)$ as $\omega = (\omega', \omega_m)$ where $\omega' = (\omega_1, \dots, \omega_{m-1})$, the left-hand side of (4.17) can be written

$$\sum_{\omega_m \in \mathbb{Z}_N} \sum_{\omega' \in \mathring{\mathbb{Z}}_N^{m-1}} \Big(f(\omega', \omega_m) - \sum_{j=1}^{m-1} f(\omega', \omega_j) \Big)$$

while the induction hypothesis implies that this can also be rewritten as

$$\sum_{P' \in \mathscr{P}(\mathbb{N}_{m-1})} (-1)^{m-|P'|} \prod_{U \in P'} (|U|-1)! \sum_{Q \le P'} \sum_{\omega' \in W(Q)} \Big(\sum_{\omega_m \in \mathbb{Z}_N} f(\omega', \omega_m) - \sum_{j=1}^{m-1} f(\omega', \omega_j) \Big) \, . \tag{4.18}$$

We need to show that (4.18) equals the right-hand side of (4.17). Given a partition P' of $\mathbb{N}_{m-1}$, denote by $P'\{m\} \in \mathscr{P}(\mathbb{N}_m)$ the partition obtained from P' by adding the singleton $\{m\}$ as a new partition element, and by $P'\{m,j\} \in \mathscr{P}(\mathbb{N}_m)$ the partition obtained by adding the index m to $U(j)$ where $U(j) \in P'$ is the partition element containing the index j. It is clear that, with this notation, any scalar-valued function g with domain $\mathscr{P}(\mathbb{N}_m)$ will satisfy

$$\sum_{P \in \mathscr{P}(\mathbb{N}_m)} g(P) = \sum_{P' \in \mathscr{P}(\mathbb{N}_{m-1})} g(P'\{m\}) + \sum_{P' \in \mathscr{P}(\mathbb{N}_{m-1})} \sum_{j=1}^{m-1} \frac{1}{|U(j)|} g(P'\{m,j\})$$

where the factor $1/|U(j)|$ weights repeated values. Applying this rule to g defined by the right-hand side of (4.17) allows one to express the latter as the sum over $P' \in \mathscr{P}(\mathbb{N}_{m-1})$ of the terms

$$\begin{aligned} g(P'\{m\}) &= (-1)^{m-|P'\{m\}|} \prod_{U \in P'\{m\}} (|U|-1)! \sum_{Q \le P'\{m\}} \sum_{\omega \in W(Q)} f(\omega', \omega_m) \\ &= (-1)^{m-1-|P'|} \prod_{U' \in P'} (|U|-1)! \sum_{Q' \le P'} \sum_{\omega' \in W(Q')} \sum_{\omega_m \in \mathbb{Z}_N} f(\omega', \omega_m) \end{aligned}$$

and the terms

$$\sum_{j=1}^{m-1} \frac{1}{|U(j)|} g(P'\{m,j\}) = \sum_{j=1}^{m-1} \frac{1}{|U(j)|} (-1)^{m-|P'\{m,j\}|} \prod_{U \in P'\{m,j\}} (|U|-1)! \sum_{Q \leq P'\{m,j\}} \sum_{\omega \in W(Q)} f(\omega)$$

$$= -\sum_{j=1}^{m-1} (-1)^{m-1-|P'|} \prod_{U \in P'} (|U|-1)! \sum_{Q' \leq P'} \sum_{\omega' \in W(Q')} f(\omega', \omega_j) .$$

Here we have used the fact that $\prod_{U \in P'\{m,j\}} (|U|-1)! = |U(j)| \prod_{U \in P'} (|U|-1)!$ and that, if $P'\{m,j\}$ is a refinement of Q then removing m from $U(j)$ results in an anti-refinement Q' of P' such that any $\omega \in W(Q)$ has the form (ω', ω_j). Thus, the two types of terms from the right-hand side of (4.17) match up exactly with the corresponding terms in (4.18). This completes the induction step and thereby completes the proof of Lemma 4.3.1. □

Chapter 5
Sampling of Band-limited and Multiband Signals

5.1 Sampling of Band-limited Signals

We provide here an overview of sampling theory that emphasizes real-variable aspects and functional analytic methods rather than analytic function-theoretic ones. While this approach does not justify the most powerful mathematical results, it does provide the basis for practical sampling techniques for band-limited and multiband signals. It also serves to establish links between sampling and time–frequency localization, such as Theorem 5.1.12, whose proof is based largely on the methods that were used in Chap. 4 to count the number of eigenvalues of $P_\Sigma Q_T$ larger than $1/2$.

5.1.1 The Roles of Harmonic Analysis and Function Theory

Sampling theory, particularly encoding norm properties of Paley–Wiener (PW) spaces in discrete sets of samples, has become perhaps the most developed aspect of the theory of PW spaces. The first part of this chapter provides an overview of sampling theory for the subspace PW of $L^2(\mathbb{R})$ consisting of those f whose Fourier transforms vanish outside $[-1/2,1/2]$. More specifically, we consider conditions on an increasing sequence $\Lambda=\{\lambda_k\}_{k=-\infty}^{\infty}$ of real numbers such that any $f\in \mathrm{PW}$ is determined by, or can be recovered from, its samples $\{f(\lambda_k)\}$. Convergence of representations of functions $f\in \mathrm{PW}_\Sigma$ in terms of their samples $\{f(\lambda_k)\}$ is encoded via the Fourier transform as convergence of representations of functions $F\in L^2(\Sigma)$ in terms of the exponentials $\mathscr{E}(\Lambda)=\{\mathrm{e}^{2\pi i\lambda_k\omega}:\lambda_k\in\Lambda\}$. When Σ is an interval I, the question of when $\mathscr{E}(\Lambda)$ forms a Riesz basis for $L^2(I)$ was settled by Pavlov [157] in the late 1970s, see p. 174, and the question of when $\mathscr{E}(\Lambda)$ forms a frame for $L^2(I)$ was settled by Ortega-Cerdà and Seip [251] in 2002, see p. 175. In both cases, some machinery from complex function theory is needed just to state the characterizations and much more is needed to prove them. A comprehensive and rigorous development of the sampling theory of PW would require a treatise all its own. An excellent source for function-theoretic aspects is Seip's monograph [290]. The characterizations themselves are deep and difficult and remain open in the multiband case, partly because convexity or connectedness of Σ plays some role in the proofs.

While function theory will be de-emphasized here, it plays such a decisive role in sampling theory that it would be impossible to gain an understanding of the scope of sampling theory without at least having outlined the role of function theory at critical junctures. One of these junctures is right at the very beginning and it provides us the opportunity to explain why PW is called a *Paley–Wiener space*. The following result is called *the Paley–Wiener* theorem.

Theorem 5.1.1. *If $f \in L^2(\mathbb{R})$ then $f \in \mathrm{PW}_{[-A,A]}$ if and only if f extends to an entire function on $\mathbb{C}$ of exponential type $2\pi A$, that is, there is a constant C such that $|f(z)| \leq C e^{2\pi A|z|}$ for all $z \in \mathbb{C}$.*

The extension of f is defined in terms of the Fourier inversion formula

$$f(z) = \int_{-A}^{A} \widehat{f}(\xi)\, e^{2\pi i z\xi}\, d\xi\,.$$

One applies the Lebesgue dominated convergence theorem to justify differentiation under the integral sign, which implies that f defines an entire function. Since $|e^{2\pi i z\xi}| \leq e^{2\pi A|z|}$, one can take $C = \int |\widehat{f}|$. That such a growth condition implies that $f \in \mathrm{PW}_{[-A,A]}$ was proved by Paley and Wiener in [252]; cf. also [316]. Convexity of the Fourier support is required for this converse implication.

Analyticity guarantees that f is equal to its average value in the neighborhood of a point. A consequence of this principle is the following special case of the Plancherel–Pólya inequality; cf. [25, 369].

Proposition 5.1.2. *Suppose that $f \in \mathrm{PW}_I$. Suppose that $\Lambda = \{\lambda_k\}$ is a real sequence such that* $\inf\{|\lambda_k - \lambda_\ell| : k \neq \ell\} = \delta > 0$. *Then there is a constant C_I such that*

$$\sum_{k=-\infty}^{\infty} |f(\lambda_k)|^2 \leq C_I \|f\|^2\,.$$

A function-theoretic proof yields $C_I = 4(e^{\pi|I|\delta/2} - 1)/(2\pi^2|I|\delta^2)$. The following real-variable argument also yields the Plancherel–Pólya inequality.

Proof (using real-variable techniques). We follow the method used in Gröchenig and Razafinjatovo [121]. Let φ be a Schwartz function whose Fourier transform $\widehat{\varphi}$ satisfies $\widehat{\varphi}(\xi) = 1$ on I and vanishes outside of $2I$. If $f \in \mathrm{PW}_I$ then $f = f * \varphi$ and an elementary estimate shows that $f^{\#}(x) \leq |f| * \varphi^{\#}(x)$, where $f^{\#}(x) = \sup\{|f(y)| : |x-y| \leq 1\}$. Since φ is rapidly decreasing, $\varphi^{\#} \in L^1(\mathbb{R})$ so that $\|f^{\#}\| \leq \||f| * \varphi^{\#}\| \leq \|f\|\|\varphi^{\#}\|_{L^1}$. Without loss of generality, we may assume that $\delta \leq 1$. Then

$$|f(\lambda_k)|^2 \leq \frac{1}{\delta} \int_{\lambda_k - \delta/2}^{\lambda_k + \delta/2} (f^{\#}(t))^2\, dt$$

and, using the fact that Λ is δ*-separated*, it follows that

$$\sum_k |f(\lambda_k)|^2 \le \frac{1}{\delta}\int_{-\infty}^{\infty} |f^{\#}(t)|^2\,\mathrm{d}t \le \frac{\|\varphi^{\#}\|_{L^1}^2}{\delta}\|f\|^2 .$$

This proves the proposition. □

The dependence of C_I on I in this version of the Plancherel–Pólya inequality is reflected in $\|\varphi^{\#}\|_{L^1}$.

5.1.2 The Shannon Sampling Theorem

The exponentials $\mathrm{e}^{2\pi ik\xi}$, ($k \in \mathbb{Z}$) form an orthonormal basis for $L^2[-1/2,1/2]$. If $f \in \mathrm{PW}$ then $\widehat{f} \in L^2[-1/2,1/2]$ so one has $\widehat{f}(\xi) = \sum_{k=-\infty}^{\infty} c_k \mathrm{e}^{2\pi ik\xi}$ converging in $L^2[-1/2,1/2]$. Here, $c_k = \int_{-1/2}^{1/2} \widehat{f}(\eta)\mathrm{e}^{-2\pi ik\eta}\,\mathrm{d}\eta = f(-k)$ by the Fourier inversion formula. Therefore, $\widehat{f}(\xi) = \sum_{k=-\infty}^{\infty} f(k)\,\mathrm{e}^{-2\pi ik\xi}$. Applying Fourier inversion once more,

$$\begin{aligned} f(t) &= \int_{-1/2}^{1/2} \widehat{f}(\xi)\,\mathrm{e}^{2\pi it\xi}\,\mathrm{d}\xi = \sum_{k=-\infty}^{\infty} f(k)\int_{-1/2}^{1/2} \mathrm{e}^{2\pi i(t-k)\xi}\mathrm{d}\xi \\ &= \sum_{k=-\infty}^{\infty} f(k)\frac{1}{\pi(t-k)}\frac{1}{2i}\mathrm{e}^{2\pi i(t-k)\xi}\Big|_{-1/2}^{1/2} = \sum_{k=-\infty}^{\infty} f(k)\frac{\sin\pi(t-k)}{\pi(t-k)} . \end{aligned}$$

This observation is usually called the *Shannon sampling theorem* [293] because Shannon was able to explain its importance in communications theory. However, Kotel′nikov [185] made much the same observations as Shannon regarding this sampling identity, whose history goes back to J.M. Whittaker [353], E.T. Whittaker, [352] and well beyond; cf. [28, 110, 147, 149, 150]. Setting $\mathrm{sinc}\,(t) = \sin(\pi t)/(\pi t)$ the sampling theorem can be stated as follows.

Theorem 5.1.3 (Shannon sampling theorem). *If* $f \in \mathrm{PW}$ *then*

$$f(t) = \sum_{k=-\infty}^{\infty} f(k)\,\mathrm{sinc}\,(t-k)$$

where the partial sums of the sampling series converge to f *in* L^2*-norm.*

The rest of this chapter addresses two types of generalizations of the Shannon sampling theorem. In the first case, one seeks general conditions on a discrete sequence of sample points that allow one to recover any $f \in \mathrm{PW}$ from its samples at those points in a stable manner and, preferably, with interpolation formulas. In the second case, one seeks specific conditions on a discrete sequence of sample points that allow one to recover $f \in \mathrm{PW}_\Sigma$, where Σ is multiband, from its sample values.

5.1.3 Criteria for Reconstructing Band-limited Signals

Sets of Uniqueness, Sampling, and Interpolation

Let Σ be a finite union of intervals with finite measure $|\Sigma|$. One calls a set Λ a *set of uniqueness* for PW_Σ if (i) $f \in \mathrm{PW}_\Sigma$ and (ii) $f(\lambda) = 0$ for all $\lambda \in \Lambda$ together imply that f is identically equal to zero. Uniqueness, by itself, is not strong enough to quantify important characteristics such as signal strength. One calls a countable set $\Lambda = \{\lambda_k\}_{k\in\mathbb{Z}}$ a *set of stable sampling* for PW_Σ, or simply a set of sampling, for short, provided $\|f\|_2^2 \le C\sum|f(\lambda_k)|^2$ for some fixed constant C independent of $f \in \mathrm{PW}_\Sigma$. One might also ask whether, given any sequence $\{a_k\} \in \ell^2(\mathbb{Z})$, one can construct an $f \in \mathrm{PW}_\Sigma$ with $f(\lambda_k) = a_k$. If so, then Λ is called a *set of interpolation* for PW_Σ. The precise relationship between sampling and interpolation will be discussed in what follows. We will also refer to $\Lambda = \{\lambda_k\}$ as a Bessel sequence for PW_Σ provided that there is a constant $B > 0$ such that $\sum|f(\lambda_k)|^2 \le B\|f\|^2$ for all $f \in \mathrm{PW}_\Sigma$, that is, if the linear mapping $f \mapsto \{f(\lambda_k)\}_{k\in\mathbb{Z}}$ is continuous from PW_Σ to $\ell^2(\mathbb{Z})$. Proposition 5.1.2 states that any separated sequence is a Bessel sequence for PW_I, for any interval I.

Sequences of Exponentials

The uniqueness, sampling, and interpolation properties that a set Λ might possess with respect to a PW space can be encoded in terms of the set of exponentials $\mathscr{E}(\Lambda) = \{\mathrm{e}_\lambda : \lambda \in \Lambda\}$ where $\mathrm{e}_\lambda(\xi) = \mathrm{e}^{-2\pi i\lambda\xi}$. If Σ is compact then the Fourier inversion formula says that

$$f(\lambda) = \int_\Sigma \widehat{f}(\xi)\,\mathrm{e}^{2\pi i\lambda\xi}\,\mathrm{d}\xi = \langle \widehat{f}, \mathrm{e}_\lambda\rangle$$

where $\langle\cdot,\cdot\rangle$ denotes the Hermitian inner product on $L^2(\Sigma)$. Thus, the statement that Λ is a set of uniqueness for PW_Σ is equivalent to the statement that $\mathscr{E}(\Lambda)$ is complete in $L^2(\Sigma)$, while the statement that $\Lambda = \{\lambda_k\}$ is a Bessel sequence for PW_Σ is equivalent to the statement that $\mathscr{E}(\Lambda)$ is a Bessel family for $L^2(\Sigma)$, that is, $\sum_k |\langle g, \mathrm{e}_{\lambda_k}\rangle|^2 \le B\|g\|^2$ for all $g \in L^2(\Sigma)$.

Frames and Riesz Bases

While a basis for a separable Hilbert space $\mathscr{H}$ expresses the ability to synthesize an element of $\mathscr{H}$ uniquely as a superposition of a countable set of generators, the notion of a *Riesz basis* goes a step further, relating the norm of an element of $\mathscr{H}$ with the ℓ^2-norm of its coefficients.

Definition 5.1.4. A sequence $\{\mathbf{x}_k\}_{k\in\mathbb{Z}}$ in a separable Hilbert space $\mathscr{H}$ is said to be a Riesz basis for $\mathscr{H}$ provided that there exist constants $0 < A \le B < \infty$ such that for all $\{c_k\} \in \ell^2(\mathbb{Z})$ one has

$$A\sum|c_k|^2 \leq \left\|\sum c_k \mathbf{x}_k\right\|_{\mathscr{H}}^2 \leq B\sum|c_k|^2.$$

More generally, one defines a *Riesz sequence* to be a set that is not necessarily complete in $\mathscr{H}$, but forms a Riesz basis for its closed linear span. The question of which sequences of exponentials form Riesz bases for $L^2[-1/2,1/2]$ was a long-standing open problem. Sufficiency criteria were obtained by Paley and Wiener [252] using perturbation arguments and the sharp estimate under the perturbation condition $\lambda_k \approx k$ is known as the *Kadec quarter theorem* [167] (cf. [369]). It states that if $|\lambda_k - k| < 1/4$ then $\mathscr{E}(\Lambda)$ forms a Riesz basis for $L^2[-1/2,1/2]$. The problem of characterizing *all* Riesz basic sequences Λ—not only those that are uniform perturbations of the integers—was solved by Pavlov [157] in 1978. Pavlov's characterization is not entirely intuitive, but it does provide a criterion that is computable, in principle.

In a sense, there are more sets of sampling than sets of interpolation, so it should be relatively simple to establish criteria sufficient for Λ to be a set of sampling for a PW space. Partly as an attempt to do so, Duffin and Schaeffer [88] introduced the concept of a *frame* for a Hilbert space in 1952, although the *frame inequalities* for systems of exponentials actually appeared earlier in Paley and Wiener's AMS Colloquium Publication [252, p. 115, inequalities (30.56)].

Definition 5.1.5. A sequence $\{\mathbf{x}_k\}_{k\in\mathbb{Z}}$ in a separable Hilbert space $\mathscr{H}$ is said to be a frame for $\mathscr{H}$ provided that there exist constants $0 < A \leq B < \infty$ such that for all $\mathbf{x} \in \mathscr{H}$ one has

$$A\|\mathbf{x}\|_{\mathscr{H}}^2 \leq \sum|\langle \mathbf{x}, \mathbf{x}_k\rangle|^2 \leq B\|\mathbf{x}\|_{\mathscr{H}}^2.$$

Any Riesz basis is a frame, but frames can be redundant. For example, adding a unit vector to a Riesz basis results in a frame. The frame concept turned out to provide the right context in which to phrase several important issues in applied harmonic analysis; some historical aspects are discussed in Daubechies [77] and Benedetto [27] and some basic expositions include Christensen [73], Han et al. [132], and Casazza [63]. Beyond Duffin and Schaeffer's foundational introduction of frames in the context of Fourier representations, the work of Daubechies and others in the late 1980s, e.g., [76,78], brought frames to bear on problems in broader contexts, and the potential of frames is still being realized; see, e.g., [29,31]. Even though Duffin and Schaeffer defined frames in the context of an abstract Hilbert space, they were particularly interested in the problem of quantifying those sequences Λ whose exponentials $\mathscr{E}(\Lambda) = \{\mathrm{e}_\lambda\}_{\lambda\in\Lambda}$ form a frame for $L^2(I)$, where I is an interval. Duffin and Schaeffer referred to such sequences as *Fourier frames*. Equivalently, they wished to identify those Λ that form both a Bessel sequence and a set of sampling for PW_I.

The Density of a Sequence

The Shannon sampling theorem takes advantage of the fact that the exponentials $\mathrm{e}^{-2\pi i n t}$ form a complete orthogonal system for $L^2[-1/2,1/2]$ in a transparent way. A number of applications, ranging from the mathematical theory of partial differential equations to the concrete engineering problem of reconstructing an analog signal from digital samples, call for an identification of those sequences of exponentials $\mathscr{E}(\Lambda)$, where Λ is a countable, discrete set, that provide a complete system, or stable system, or basis for $L^2(\Sigma)$, where $\Sigma \subset \mathbb{R}$ is compact. The e_λ will not, in general, be orthogonal, so completeness or stability has to be expressed in terms of norm inequalities relating a discrete subset of values to all values. Such conditions will depend on the linear distribution of Λ. When $\Lambda = \mathbb{Z}/|I|$, the exponentials $\mathscr{E}(\Lambda)$ form an orthogonal family in $L^2(I)$. For a more general discrete set Λ, one can try to formulate stability properties of $\mathscr{E}(\Lambda)$ in terms of the average number of points in Λ per unit length, or *density* of Λ.

By a *discrete* subset $\Lambda \subset \mathbb{R}$ we mean a *locally finite* set, that is, one whose intersection with any compact set is finite. Any such set is countable and so can be expressed as a *sequence* $\Lambda = \{\lambda_k\}_{k=-\infty}^{\infty}$. In fact, any discrete set can be *linearly ordered* such that $\lambda_k < \lambda_{k+1}$ for all $k \in \mathbb{Z}$. This is done by associating to Λ a *counting function* $n_\Lambda(t) = \#\Lambda \cap (0,t]$ if $t > 0$ and $n_\Lambda(t) = -\#\Lambda \cap (t,0]$ if $t < 0$. Then $n_\Lambda(b) - n_\Lambda(a)$ denotes the number of elements of Λ in the half-open interval $(a,b]$ and Λ can be expressed as an ordered sequence by setting $\lambda_k = \inf\{t : n_\Lambda(t) = k\}$. We will always assume that any discrete set is ordered in this way. It should be clear now that if Λ is discrete then $n_\Lambda(S) = \#\{\Lambda \cap S\}$ is defined for any compact set S. A discrete set Λ is said to be uniformly separated or, simply, *separated* if $\inf_k\{\lambda_k - \lambda_{k-1}\} = \delta > 0$. In this case, δ is called the separation constant and one says that Λ is δ-separated. In some cases, it is useful to consider sequences that can be expressed as finite unions of separated sequences. Such sequences are said to be *relatively separated* where *relatively* refers to the ability to partition a sequence like $\mathbb{Z} \cup \{n + 1/n : n = 1,2,\dots\}$ into a finite set of subsequences that are each uniformly separated.

There is not a canonical way to assign a density to an ordered sequence. In fact, different densities are useful for characterizing different properties of sequences of exponentials. The *natural density* of a sequence is defined (cf. [24, 369]) as

$$D_{\mathrm{nat}}(\Lambda) = \lim_{R\to\infty} \frac{n_\Lambda[-R,R]}{2R}.$$

A less obvious density is the *Pólya maximum density*

$$D_{\text{Pólya}}(\Lambda) = \lim_{\alpha\to 1^-} \limsup_{R\to\infty} \frac{n_\Lambda([-R,R]) - n_\Lambda([-\alpha R,\alpha R])}{2R(1-\alpha)}.$$

The Pólya density provides a bound on the length of an interval over which $\mathscr{E}(\Lambda)$ is complete, as was proved by Levinson [199], cf. [369].

Theorem 5.1.6. *The sequence of exponentials $\mathscr{E}(\Lambda)$ is complete over any interval I of length less than $D_{\text{Pólya}}(\Lambda)$.*

To capture the type of stability required in order for Λ to be a set of sampling or interpolation, some *translational invariance* has to be reflected in the density.

The *Beurling lower density* $D_-(\Lambda)$ and *Beurling upper density* $D_+(\Lambda)$ are defined, respectively, as

$$D_-(\Lambda) = \lim_{R\to\infty} \frac{\inf_x(n_\Lambda(x+R) - n_\Lambda(x))}{R} \quad \text{and}$$
$$D_+(\Lambda) = \lim_{R\to\infty} \frac{\sup_x(n_\Lambda(x+R) - n_\Lambda(x))}{R}. \tag{5.1}$$

It is simple to construct sequences Λ such that $D_+(\Lambda) > D_-(\Lambda)$. If $D_\pm = D_+(\Lambda) = D_-(\Lambda)$ then we call $D(\Lambda) = D_+(\Lambda)$ the *Beurling density* of Λ.

Landau [188, 189] provided necessary conditions for sampling and interpolation in terms of the Beurling densities, as follows.

Theorem 5.1.7. *If Λ is a set of sampling for* PW_Σ *then $D_-(\Lambda) \geq |\Sigma|$. If Λ is a set of interpolation for* PW_Σ *then $D_+(\Lambda) \leq |\Sigma|$.*

In order to obtain sufficient conditions for sampling, the local behavior of Λ has to be quantified more explicitly. One says that Λ has uniform density $\Delta > 0$ if there is a fixed $L > 0$ such that, for all $k \in \mathbb{Z}$,

$$\left(\lambda_k - \frac{k}{\Delta}\right) \leq L. \tag{5.2}$$

The simplest example corresponds to a shifted lattice $\lambda_k = h + (k/\Delta)$ where h is fixed. Λ is uniformly dense if it is the image of the integers under a map of the form $\psi(t) = h + (t/\Delta) + B(t)$ where $B(t)$ is uniformly bounded. That is, Λ is a uniformly bounded perturbation of a lattice. Duffin and Schaeffer proved the following in [88] using complex function-theoretic techniques.

Theorem 5.1.8. *If Λ is a separated sequence with uniform density $\Delta > 0$ then $\mathscr{E}(\Lambda)$ forms a frame for $L^2(I)$ whenever $|I| < \Delta$.*

Time–Frequency Localization of Multiband Signals

In Sect. 1.2.5, a ballpark estimate of the number of eigenvalues of PQ_T between α and $1-\alpha$, $\alpha \in (0,1)$, was obtained by comparing the trace norm and Hilbert–Schmidt norm of PQ_T. In [189], Landau used the same method to estimate the decay of eigenvalues of the time–frequency localization operator $P_\Sigma Q_S$ for general time and frequency supports sets S and Σ. The nth eigenvalue–eigenvector pair $(\lambda_n(S,\Sigma), \varphi_n)$ of $P_\Sigma Q_S$ satisfies

$$\lambda_n(S,\Sigma)\varphi_n(t) = \int_S \varphi_n(u)\left(\mathbb{1}_\Sigma\right)^\vee(t-u)\,du. \tag{5.3}$$

Here, $K(t-s) = \left(\mathbb{1}_\Sigma\right)^\vee(t-s)$ is the reproducing kernel for PW_Σ. Thus,

$$\sum_{n=0}^{\infty} \lambda_n(S,\Sigma) = \int_S K(t,t)\,\mathrm{d}t = |\Sigma| \int_S 1\,\mathrm{d}s = |S||\Sigma|.$$

When $S = [-T,T]$ one can apply Q_T to both sides of (5.3) and then take Fourier transforms to obtain

$$\lambda_n(2T,\Sigma)\widehat{\varphi}_n(\xi) = \int_\Sigma \widehat{\varphi}_n(\eta) \frac{\sin 2\pi T(\xi-\eta)}{\pi(\xi-\eta)}\,\mathrm{d}\eta\,. \tag{5.4}$$

Thus, $\widehat{\varphi}_n$ is an eigenfunction of $P_{2T}Q_\Sigma$ with kernel $K(\xi,\eta) = \mathbb{1}_\Sigma(\eta)\frac{\sin 2\pi T(\xi-\eta)}{\pi(\xi-\eta)}$. Since $\widehat{\varphi}_n$ has its support in Σ, applying the Hilbert–Schmidt criterion to (5.4) tells us that

$$\sum_n \lambda_n^2(2T,\Sigma) = \int_\Sigma \int_\Sigma \left|\frac{\sin 2\pi T(\xi-\eta)}{\pi(\xi-\eta)}\right|^2 \mathrm{d}\xi\,\mathrm{d}\eta\,.$$

When $\Sigma = \cup_{\nu=1}^M I_\nu$ with pairwise disjoint intervals I_ν of length $|I_\nu| = \ell_\nu$,

$$\begin{aligned}
\sum_n \lambda_n^2(2T,\Sigma) &= \sum_{\mu,\nu} \int_{I_\mu} \int_{I_\nu} \left|\frac{\sin 2\pi T(\xi-\eta)}{\pi(\xi-\eta)}\right|^2 \mathrm{d}\xi\,\mathrm{d}\eta \\
&\geq \sum_{\nu=1}^M \int_0^{\ell_\nu} \int_0^{\ell_\nu} \left|\frac{\sin 2\pi T(\xi-\eta)}{\pi(\xi-\eta)}\right|^2 \mathrm{d}\xi\,\mathrm{d}\eta \\
&\geq \sum_{\nu=1}^M \left(2T\ell_\nu - \frac{1}{\pi^2}\log^+ 2T\ell_\nu - 1\right) \\
&= 2T|\Sigma| - A\log^+(2T) - M,
\end{aligned}$$

see p. 21, where $\log^+(x) = \max(\log(x),0)$. Here, A depends only on the linear distribution of Σ. Combining the estimates for $\sum \lambda_n$ and $\sum \lambda_n^2$ one obtains

$$\sum_n \lambda_n(1-\lambda_n) \leq A\log^+(2T) - M\,. \tag{5.5}$$

In particular, for Σ fixed, the number of eigenvalues of $P_\Sigma Q_T$ lying between two values $\alpha > 0$ and $\beta < 1$ can grow at most logarithmically with T. Specifically, if $\alpha(1-\alpha) \leq \beta(1-\beta)$ then the number of eigenvalues between α and β is at most $(A\log^+(2T)+M)/(\alpha(1-\alpha))$.

Sampling and Interpolation: Landau's Criteria

In [189], Landau proved that one cannot improve upon the Nyquist sampling rate. That is, one cannot obtain an arbitrary $f \in \mathrm{PW}$ from an irregular sampling expansion with samples taken, on average, at a rate lower than one sample per unit time. In fact, a consequence of Landau's methods is that any stable sampling sequence for PW_Σ has to have a lower Beurling density $D^- \geq 1/|\Sigma|$. The following approach, based on Landau [189], uses the method presented in Sect. 4.1 to bound the number of *large* eigenvalues of $P_\Sigma Q_S$ in terms of $|S||\Sigma|$. We will need to return temporarily to using the symbol λ_n to denote the nth largest eigenvalue of $P_\Sigma Q_S$. We will use "$\{t_k\}$" instead of "$\{\lambda_k\}$" to denote a real sampling sequence.

Lemma 5.1.9. *Suppose that Σ is bounded and $\{t_k\}$ is a δ-separated sampling set for* PW_Σ. *For a compact set $S \subset \mathbb{R}$, let $N_+(S,\delta) = \#\{k : \mathrm{dist}(t_k, S) \leq \delta/2\}$. Then there is a $\beta < 1$ depending only on Σ and $\{t_k\}$ such that $\lambda_{N_+(S,\delta)}(S,\Sigma) \leq \beta < 1$.*

A dual estimate applies to sets of interpolation.

Lemma 5.1.10. *Suppose that Σ is bounded and $\{t_k\}$ is a δ-separated set of interpolation for* PW_Σ. *Let $N_-(S,\delta) = \#\{k : \mathrm{dist}(t_k, \mathbb{R}\setminus S) > \delta/2\}$. Then $\lambda_{N_-(S,\delta)-1}(S,\Sigma) \geq \alpha$ for some $\alpha > 0$ depending only on Σ and $\{t_k\}$.*

Proof (of Lemma 5.1.9). By the Weyl–Courant lemma,

$$\lambda_n(S,\Sigma) \leq \sup_{f\in \mathrm{PW}_\Sigma,\, f\perp \mathscr{S}_n} \left(\|Q_S f\|^2/\|f\|^2\right)$$

whenever $\mathscr{S}_n$ is a subspace of PW_Σ of dimension n. To prove Lemma 5.1.9, then, it suffices to prove that if f is orthogonal to $N_+(S)$ suitably chosen functions in PW_Σ then a certain fraction of the energy of f has to lie outside of S. To accomplish this, fix $h \in L^2(\mathbb{R})$ supported in $[-\delta/2, \delta/2]$ whose Fourier transform $\widehat{h}$ has modulus at least one on Σ. A dilate of $\sqrt{2}h$ with h as in Lemma 4.1.4 will do.

Convolution with h preserves PW_Σ and, in fact, defines a continuous mapping on PW_Σ. Given $f \in \mathrm{PW}_\Sigma$, $f*h(t) = \int_{|t-u|<\delta/2} f(u)h(t-u)\,\mathrm{d}u$. By the Cauchy–Schwarz inequality,

$$|f*h(t)|^2 \leq \|h\|^2 \int_{|t-u|<\delta/2} |f(u)|^2\,\mathrm{d}u\,.$$

The hypothesis that $\{t_k\}$ is a set of sampling for PW_Σ implies that

$$\int_{-\infty}^{\infty} |f*h(t)|^2\,\mathrm{d}t \leq K\sum_k |f*h(t_k)|^2,$$

while our requirement that $|\widehat{h}| \geq 1$ on Σ implies that

$$\|f\|^2 = \|\widehat{f}\|^2 = \int_\Sigma |\widehat{f}(\xi)|^2\,\mathrm{d}\xi \leq \int_\Sigma |\widehat{f}(\xi)\widehat{h}(\xi)|^2\,\mathrm{d}\xi = \|f*h\|^2\,.$$

Imposing the $N_+(S)$ orthogonality conditions $f*h(t_k) = \int f(u)h(t_k-u)\,\mathrm{d}u = 0$ whenever $t_k \in S_+ = \{t : \mathrm{dist}(t,S) \leq \delta/2\}$, we obtain

$$\begin{aligned}\int |f|^2 \leq \int |f*h|^2 &\leq K\sum_{t_k\notin S_+} |f*h(t_k)|^2 \\ &\leq K\|h\|^2 \sum_{t_k\notin S_+} \int_{|t_k-u|<\delta/2} |f(u)|^2\,\mathrm{d}u \leq K\|h\|^2 \int_{\mathbb{R}\setminus S} |f(u)|^2\,\mathrm{d}u\,.\end{aligned}$$

That is,

$$\left(\|Q_S f\|^2/\|f\|^2\right) \leq 1 - \frac{1}{K\|h\|^2} = \beta < 1$$

whenever f is orthogonal to $h(t_k - \cdot)$ for all $N_+(S)$ values of $t_k \in S_+$. The choice of h, and thus its norm, depended on δ and thus on $\{t_k\}$, and on Σ, but did not depend on S. The constant K was determined solely by the hypothesis that $\{t_k\}$ is a sampling set for PW_Σ. Therefore, β does not depend on S. Since the bound is obtained by taking f to be orthogonal to one particular subspace of PW_Σ of the given dimension, the lemma follows. □

Before proving Lemma 5.1.10 we need first some more precise information regarding the nature of interpolation.

Proposition 5.1.11. *Let Σ be bounded and let $\{t_k\}$ be a set of interpolation for* PW_Σ*. Then $\{t_k\}$ is uniformly separated and the interpolation can be performed in a stable manner in the sense that there is a subspace of* PW_Σ *and a $K > 0$ such that, for any square-summable sequence $\{\alpha_k\}$, there is an f in the subspace satisfying $f(t_k) = \alpha_k$ and $\|f\|^2 \leq K\sum|f(t_k)|^2$.*

Proof. Applying Fourier inversion and the Cauchy–Schwarz inequality gives

$$|f(t)|^2 \leq |\Sigma|\|f\|^2 \qquad \text{and}$$
$$|f'(t)|^2 \leq 2\pi\Big(\int_\Sigma |\xi|^2\Big)\|f\|^2.$$

First, to get a contradiction, suppose that $\{t_k\}$ is not uniformly separated. Then one can find subsequences $\{p_k\}$ and $\{q_k\}$ such that $|p_k - q_k| \leq 1/k^2$ and, since $\{t_k\}$ is an interpolation set for PW_Σ, a function $f \in \mathrm{PW}_\Sigma$ such that $f(p_k) = 1/k = -f(q_k)$ and such that $f(t_k) = 0$ for all other points t_k. However, this implies that $2/k = f(p_k) - f(q_k) \leq |p_k - q_k||f'(s_k)|$ for some s_k between p_k and q_k. That is, $|f'(s_k)| \geq 2k$, which contradicts the derivative bound once k is sufficiently large. This proves that $\{t_k\}$ is separated. The separation constant can be estimated through the derivative bound.

Proving that interpolation can be carried out in a stable manner amounts to: (i) identifying a closed subspace of PW_Σ such that there is a constant K for which $\|f\|^2 \leq K\sum|f(t_k)|^2$ whenever f belongs to this subspace and (ii) verifying that any sequence $\{a_k\}$ is the set of values $\{f(t_k)\}$ for some f in this space. Let $\mathrm{PW}_{\Sigma,0}$ denote those $f \in \mathrm{PW}_\Sigma$ that vanish identically on $\{t_k\}$. This subspace is closed since the bound $|f(t)|^2 \leq |\Sigma|\|f\|^2$ implies that convergence in L^2 in turn implies uniform convergence in PW_Σ. Now, given a sequence of values $\{a_k\}$ and $f \in \mathrm{PW}_\Sigma$ such that $f(t_k) = a_k$ for all k, one can form the affine space $\{f + \phi : \phi \in \mathrm{PW}_{\Sigma,0}\}$. Any element of this space also interpolates $\{a_k\}$ and one can minimize $\{\|f + \phi\| : \phi \in \mathrm{PW}_{\Sigma,0}\}$ by computing the orthogonal projection of f onto the orthogonal complement $\mathrm{PW}_{\Sigma,0}^\perp$ of $\mathrm{PW}_{\Sigma,0}$ inside PW_Σ. Interpolating inside $\mathrm{PW}_{\Sigma,0}^\perp$ then provides a unique, L^2-minimizing interpolant.

Consider then the mapping that sends a sequence $\{a_k\} \in \ell^2$ to its interpolant inside of $\mathrm{PW}_{\Sigma,0}^\perp$. Since $|f(t)|^2 \leq |\Sigma|\|f\|^2$, this mapping has a closed graph, so it is bounded. That is, there exists a K independent of $\{a_k\}$ such that the solution of $f(t_k) = a_k$ in $\mathrm{PW}_{\Sigma,0}^{\perp}$ satisfies $\|f\|^2 \leq K\sum_k |a_k|^2$, which was to be shown. □

Proof (of Lemma 5.1.10). Now one uses the Weyl–Courant lemma in the form

$$\lambda_{n-1}(S,T) \geq \inf_{f\in \mathrm{PW}_\Sigma, f\in \mathscr{S}_n, f\neq 0} (\|Q_S f\|^2/\|f\|^2),$$

whenever $\mathscr{S}_n$ is an n-dimensional subspace of PW_Σ.

According to Proposition 5.1.11, one can interpolate inside $\mathrm{PW}_\Sigma(\{t_k\},K) = \{f \in \mathrm{PW}_\Sigma : \|f\|^2 \leq K\sum |f(t_k)|^2\}$. For each k, let $\psi_k \in \mathrm{PW}_\Sigma(\{t_k\},K)$ be a function with value one at t_k and vanishing at each t_ℓ with $\ell \neq k$. These ψ_k are linearly independent.

As in the proof of Lemma 5.1.9, we can choose h supported in $[-\delta/2,\delta/2]$ such that $|\widehat{h}| \geq 1$ on Σ. As before, δ is the separation constant for $\{t_k\}$. Thus, convolution with h is a bounded and continuously invertible mapping on PW_Σ. Therefore, we can write $\psi_k = \phi_k * h$ such that the ϕ_k are also linearly independent. The subspace of PW_Σ spanned by $\{\phi_k : k \in S_-\}$ then has dimension $N_-(S)$.

For any f in this subspace, $f*h$ is a linear combination of the ψ_k's and $\|f\|^2 \leq \|f*h\|^2$ since $|\widehat{h}| \geq 1$ on Σ. Since $f*h$ is constructed solely from those ψ_k's with $t_k \in S_-$, $f*h(t_k)$ vanishes at each $t_k \notin S_-$. That is,

$$\|f*h\|^2 \leq K \sum_{t_k\in S_-} |f*h(t_k)|^2 .$$

In addition, the convolution integrals defining $f*h(t_k)$ are taken over pairwise disjoint sets, each contained in S, so that

$$\sum_{t_k\in S_-} |f*h(t_k)|^2 \leq \|h\|^2 \int_S |f|^2 .$$

Altogether,

$$\|Q_S f\|^2 \geq \frac{1}{\|h\|^2} \sum_{t_k\in S_-} |f*h(t_k)|^2 \geq \frac{1}{K\|h\|^2} \int |f*h|^2 \geq \frac{1}{K\|h\|^2} \int |f|^2 .$$

Taking $\alpha = 1/(K\|h\|^2)$ gives the desired lower bound for $\lambda_{N_-(S)-1}$. Since neither K nor h depends on S, neither does α. This proves the lemma. □

Landau applied the eigenvalue estimates of Lemmas 5.1.9 and 5.1.10 to obtain strong uniform density estimates for sets of sampling and interpolation. In order to apply the lemmas, one wishes to take advantage of the existence of a set of sampling and interpolation—namely, the integers—for the case in which $\Sigma = [-1/2, 1/2]$. When Σ is a single interval and S is a finite union of M intervals, one can reduce to the case in which $\Sigma = [-1/2,1/2]$ by applying appropriate modulations and dilations that do not change the time–frequency area. Lemma 5.1.9 applied to $\{t_k\} = \mathbb{Z}$, with the roles of the time and frequency supports reversed, yields

$$\lambda_{[|S||\Sigma|]+2M} \leq \beta_0 < 1 \tag{5.6}$$

for fixed β_0 while, applying Lemma 5.1.10 to this case yields

$$\lambda_{[|S||\Sigma|]-2M-1} \geq \alpha_0 > 0 \tag{5.7}$$

for a fixed α_0, since each interval in the finite union increases or omits at most two integer points when counting only those integers that are distance less than $1/2$ from S or, respectively, $\mathbb{R} \setminus S$.

Theorem 5.1.12. *Let $\{t_k\}$ be a separated sampling set for* PW_Σ*, where Σ is a pairwise disjoint union of M intervals. Then there are constants $A > 0$ and $B > 0$ such that every interval of length T must contain at least $T|\Sigma| - A\log^+ T - B$ elements of $\{t_k\}$.*

Proof. Let $\delta < 1/2$ be the separation constant for $\{t_k\}$. For an interval I of length T, let $N(I) = \#\{k : t_k \in I\}$ so $N_+(I,\delta) \leq N(I)+2$. By Lemma 5.1.9, there is a fixed $\beta < 1$ such that $\lambda_{N(I)+2}(T,\Sigma) \leq \beta < 1$. But (5.7) says that

$$\lambda_{[T|\Sigma|]-2M-1}(T,\Sigma) \geq \alpha_0 > 0\,.$$

Since, by (5.5), the number of eigenvalues $\lambda_k(T,\Sigma)$ such that $\alpha_0 < \lambda_k(T,\Sigma) \leq \beta$ increases at most logarithmically with T, it follows that for suitable constants A' and B',

$$(T|\Sigma| - 2M - 1) - (N(I)+2) \leq A' \log^+(T) + B' \qquad \text{or}$$

$$N(I) \geq T|\Sigma| - A\log^+ T - B$$

when $A = A'$ and $B = B' - 2M - 3$. This proves the theorem. □

Theorem 5.1.13. *If $\{t_k\}$ is a set of interpolation for* PW_Σ *then no interval of length T can contain more than $T|\Sigma| + C\log^+ T + D$ elements of $\{t_k\}$, for appropriate constants C and D independent of T.*

The proof of Theorem 5.1.13 is similar to that of Theorem 5.1.12 with Lemma 5.1.9 replaced by Lemma 5.1.10.

5.1.4 Sampling and Interpolation: Riesz Bases

In this section our focus will turn to families of exponentials defined in terms of discrete sets. Since we will no longer need to refer to eigenvalues of $P_\Sigma Q_S$, we will return to the convention of indexing a discrete sample set by the symbols $\Lambda = \{\lambda_k\}_{k\in\mathbb{Z}}$.

Theorems 5.1.12 and 5.1.13 imply that if Λ is a set of sampling for PW_Σ then its lower density satisfies $D_-(\Lambda) \geq |\Sigma|$, while if Λ is a set of interpolation for PW_Σ then $D_+(\Lambda) \leq |\Sigma|$. Landau's estimates actually provide local information that is lost in these density bounds. In view of the proof of Proposition 5.1.11, a set of interpolation for PW_Σ need not be a set of uniqueness for PW_Σ. In the event that an interpolation set Λ is also a set of uniqueness, we will call Λ a *complete interpolation set*. Such sets correspond to Riesz bases of exponentials, as the following proposition asserts.

Proposition 5.1.14. *Λ is a set of sampling and complete interpolation for* PW_Σ *if and only if $\mathscr{E}(\Lambda)$ forms a Riesz basis for $L^2(\Sigma)$.*

Proof. Write $\Lambda = \{\lambda_k\}_{k\in\mathbb{Z}}$. First, if $\mathscr{E}(\Lambda)$ forms a Riesz basis for $L^2(\Sigma)$ then it forms a basis, so any $g \in L^2(\Sigma)$ has an expansion $g(\xi) = \sum c_k \mathrm{e}_{\lambda_k}$ in which the c_k are uniquely determined. In addition, the Gram matrix defined by $G_{k\ell} = \langle \mathrm{e}_{\lambda_k}, \mathrm{e}_{\lambda_\ell}\rangle_{L^2(\Sigma)} = \int_\Sigma \mathrm{e}^{2\pi\mathrm{i}(\lambda_\ell-\lambda_k)\xi}\,\mathrm{d}\xi$ is bounded and continuously invertible on $\ell^2(\mathbb{Z})$. The sequence $\{\langle g, \mathrm{e}_{\lambda_k}\rangle\}$ is the image of $\{c_k\}$ under G^T. That is, the mapping that sends $\{c_k\}$ to $\{g^\vee(\lambda_k)\}$ is bounded and continuously invertible, and we have $A'\sum|g^\vee(\lambda_k)|^2 \le \|g\|^2 \le B'\sum|g^\vee(\lambda_k)|^2$, where A' and B' depend on the Riesz bounds A, B and the norms of G and G^{-1}. This shows that Λ is a set of sampling and complete interpolation for PW_Σ.

Conversely, suppose that Λ is a set of sampling and complete interpolation for PW_Σ. Then, for each sequence $\{a_k\} \in \ell^2(\mathbb{Z})$, there exists a unique $f \in \mathrm{PW}_\Sigma$ such that $f(\lambda_k) = a_k$. In addition, by Proposition 5.1.11, one has $\|f\|^2 \le K\sum|a_k|^2$ for a fixed constant K. On the other hand, the statement that $\sum_k |a_k|^2 \le C\|f\|^2$ is just the statement that the exponentials form a Bessel sequence for PW_Σ, since $a_k = f(\lambda_k)$. This was shown in Proposition 5.1.2. Strictly speaking, that argument applies to the case in which Σ is an interval. However, a Bessel sequence for $L^2(I)$ is also a Bessel sequence for $L^2(\Sigma)$ whenever $\Sigma \subset I$. Since Σ is assumed to be compact, we conclude that to each $\{a_k\}$ there is a unique $f \in \mathrm{PW}_\Sigma$ such that $\sum|a_k|^2 \le C\|f\| \le KC\sum|a_k|^2$. However, the statement that $\mathscr{E}(\Lambda)$ forms a Riesz basis is that $A\sum|c_k|^2 \le \|\sum c_k \mathrm{e}_k\|^2_{L^2(\Sigma)} \le B\sum|c_k|^2$. But the mapping that takes $\widehat{f} = \sum c_k \mathrm{e}_{\lambda_k}$ to the interpolant of $\langle \widehat{f}, \mathrm{e}_{\lambda_k}\rangle$ is defined in terms of the Gram matrix of the $\{\mathrm{e}_{\lambda_k}\}$ which, again, is bounded and continuously invertible, this time because of the sample and interpolation bounds. This proves the proposition. □

5.2 Fourier Frames

According to Landau's criteria, if Λ is a set of sampling and complete interpolation for PW_Σ, then Λ must have Beurling density $|\Sigma|$. However, not every set of Beurling density Σ will be a set of sampling and complete interpolation since it need not satisfy the more stringent criteria of Theorems 5.1.12 and 5.1.13.

In order that Λ forms a set of sampling and complete interpolation for PW, it must be some sort of perturbation of a lattice. The precise conditions on this perturbation were deduced by Pavlov as stated in Theorem 5.2.18. The proof of that theorem requires techniques from harmonic analysis and function theory extending beyond the scope of this book. In this section we will focus, instead, on the relationship between Fourier frames and Riesz bases of exponentials. It is tempting to guess that if $\mathscr{E}(\Lambda)$ is a frame for $L^2[-1/2,1/2]$ then Λ has the form $\Lambda = \Lambda_1 \cup \Lambda_2$ where $\mathscr{E}(\Lambda_1)$ is a Riesz basis for $L^2[-1/2,1/2]$. This is not quite true, but it is *nearly* true in a sense that was quantified by Ortega-Cerdà and Seip and will be discussed below. What is true is that if Λ is a Fourier frame sequence then it must contain a subsequence of suitable uniform density, as was shown by Jaffard [160]. In what follows we will outline Jaffard's characterization of the *frame diameter* $\mathrm{diam}(\Lambda)$ of Λ, that is, the supremum of those lengths I such that $\mathscr{E}(\Lambda)$ is a frame for $L^2(I)$. See also [30] for further related insights and techniques.

5.2.1 Frame Diameter: Jaffard's Criterion

Since shifting an exponential is the same as multiplying it by a unimodular constant, if $\mathscr{E}(\Lambda)$ forms a frame for $L^2(I)$, where I is an interval, then $\mathscr{E}(\Lambda)$ forms a frame for $L^2(I+\alpha)$ for any $\alpha \in \mathbb{R}$, with the same frame bounds. Also, when $L^2(I)$ is regarded as a subspace of $L^2(\mathbb{R})$ consisting of those $f \in L^2(\mathbb{R})$ that vanish outside of I, $L^2(J) \subset L^2(I)$ when $J \subset I$. Consequently, a frame for $L^2(I)$ will automatically induce a frame for $L^2(J)$. In the case of a family of exponentials, then, it makes sense to define the *frame diameter* $\mathrm{diam}(\Lambda)$ as the supremum of those $\delta > 0$ such that $\mathscr{E}(\Lambda)$ forms a frame for $L^2[0,\delta)$. As stated in Theorem 5.1.8, for a uniformly dense sequence, the frame diameter is no smaller than the uniform density. It is not obvious, however, that a Fourier frame sequence possesses a uniformly dense subsequence. Jaffard's primary contribution was to show that a Fourier frame sequence contains subsequences of uniform density arbitrarily close to the frame diameter. The first step is a *relative separation criterion*. While Proposition 5.1.2 implies that any separated sequence is a Bessel sequence, not every frame sequence needs to be uniformly separated. However, a Fourier frame sequence needs to be relatively separated, that is, a finite union of separated sequences.

Lemma 5.2.1. *If $\mathscr{E}(\Lambda)$ is a Bessel sequence for $L^2(I)$ for some interval I then there exists M such that, for any $k \in \mathbb{Z}$, $n_\Lambda([k,k+1)) \le M$.*

Proof. It suffices to consider intervals of the form $[0,r]$. Fix $r > 0$ and suppose that $\{\mathrm{e}_\lambda\}_{\lambda\in\Lambda}$ is a Bessel sequence for $L^2[0,r]$. Choose $\varepsilon > 0$ small enough that, whenever $|\eta| < \varepsilon$ one has $|\int_0^r \mathrm{e}^{2\pi i\eta t}\,\mathrm{d}t|^2 \ge r/2$. Suppose, to get a contradiction, that $n_\Lambda([k,k+1))$ is unbounded. That is, for each M there exists $k \in \mathbb{Z}$ such that $n_\Lambda([k,k+1)) \ge M$. Then there exists a sequence $\{\alpha_\nu\} \subset \mathbb{R}$, $\nu = 1,2,\dots$ such that $n_\Lambda((\alpha_\nu,\alpha_\nu+\varepsilon)) > \nu$. Let $f_\nu(t) = \mathrm{e}^{-2\pi i\alpha_\nu t}$. Then

$$\sum_{\lambda\in\Lambda} \left|\langle f_\nu, \mathrm{e}_\lambda\rangle_{L^2[0,r]}\right|^2 \ge$$

$$\sum_{\lambda\in(\alpha_\nu,\alpha_\nu+\varepsilon)} \left|\langle f_\nu, \mathrm{e}_\lambda\rangle_{L^2[0,r]}\right|^2 = \sum_{\lambda\in(\alpha_\nu,\alpha_\nu+\varepsilon)} \left|\int_0^r \mathrm{e}^{2\pi i(\lambda-\alpha_\nu)t}\,\mathrm{d}t\right|^2 \ge \nu r/2,$$

whereas $\|f_V\|^2_{L^2(0,r)} = r$. This contradicts the hypothesis that

$$\sum_{\lambda\in\Lambda} \left|\langle f, \mathrm{e}_\lambda\rangle_{L^2[0,r]}\right|^2 \leq B\|f\|_{L^2[0,r]}\,.$$

Thus, if $\{\mathrm{e}_\lambda\}_{\lambda\in\Lambda}$ is a Bessel sequence for $L^2[0,r]$ then $n_\Lambda([k,k+1))$ is uniformly bounded in k. □

The lemma says that Λ is a union of at most $2M$ 1-separated sequences. Let $\mathrm{UD}(\Lambda)$ denote the collection of all uniformly dense subsequences of Λ and let $\mathrm{RS}(\Lambda)$ denote the collection of all relatively separated subsequences of Λ.

Lemma 5.2.2. *With D_- and Δ defined as in (5.1) and (5.2), if Λ is relatively separated then*

$$\sup_{\Theta\in\mathrm{RS}(\Lambda)} D_-(\Theta) = \sup_{\Theta\in\mathrm{UD}(\Lambda)} \Delta(\Theta)\,.$$

The proof is a straightforward exercise in real variables and is left to the reader. The following lemma is also left as a slightly more challenging exercise.

Lemma 5.2.3. *If Λ is relatively separated then $D_-(\Lambda)$ exists.*

Theorem 5.1.8 and the preceding lemmas imply the following corollary.

Corollary 5.2.4. *Let Λ be a relatively separated sequence. Then Λ generates a Fourier frame for $L^2(I)$ whenever $|I| < D_-(\Lambda)$.*

If Λ is uniformly dense, then its Beurling density is equal to its uniform density, $D = \Delta$. Removal of a finite subset of frame elements leaves either a frame or an incomplete set. Such a removal does not change D_- or Δ, though it affects the choice of L in the definition of Δ. This suggests the possibility of removing an infinite subsequence Λ' in such a way that $D_-(\Lambda\setminus\Lambda') = D_-(\Lambda)$ but $\Lambda\setminus\Lambda'$ is no longer uniformly dense. Conversely, one might add a sparse sequence without affecting D_- but resulting in a sequence that is not uniformly dense. Viewing Λ as a uniformly dense sequence plus a remainder allowed Jaffard to obtain the *right* notion of frame density for characterizing the frame diameter in Theorem 5.2.6.

Definition 5.2.5. Let $\Lambda = \{\lambda_k\}$ be a sequence that contains a subsequence having uniform density and containing at most M elements per unit length. The frame density $D_{\mathrm{frame}}(\Lambda)$ of Λ is defined as the supremum of the uniform densities taken over all subsequences of Λ having uniform density.

Theorem 5.2.6. *Let $\Lambda = \{\lambda_k\}$ be a relatively separated sequence that contains a subsequence having uniform density. Then the frame diameter of Λ is equal to the frame density of Λ, that is,* $\mathrm{diam}(\Lambda) = D_{\mathrm{frame}}(\Lambda)$.

Lemma 5.2.7. *The following are equivalent:*

(a) There is an interval I such that $\mathscr{E}(\Lambda)$ is a frame for $L^2(I)$,

(b) Λ is a disjoint union of a sequence Λ_0 of uniform density and a finite number of separated sequences. In this case, $\mathrm{diam}(\Lambda) \geq D_{\mathrm{frame}}(\Lambda_0)$.

Proof. To show that $(b) \Rightarrow (a)$, suppose that $\Lambda = \Lambda_0 \cup \Lambda_1 \cup \cdots \cup \Lambda_M$ where Λ_0 is uniformly dense and Λ_j is separated for $j = 1, \ldots, M$. By Duffin and Schaeffer's Theorem 5.1.8, $\mathscr{E}(\Lambda_0)$ is a frame for any interval of length at most $\Delta(\Lambda_0)$. The lower frame bound for $\mathscr{E}(\Lambda_0)$ provides a corresponding lower frame bound for $\mathscr{E}(\Lambda)$. On the other hand, since any separated sequence gives rise to a Bessel sequence for $L^2(I)$, for any interval I, and since Bessel sequences are closed under finite unions, it follows that $\mathscr{E}(\Lambda)$ is also a frame for $L^2(I)$.

To show that $(a) \Rightarrow (b)$, suppose that $\mathscr{E}(\Lambda)$ is a frame for $L^2(I)$. Then $\mathscr{E}(\Lambda)$ is a Bessel sequence for $L^2(I)$ and Lemma 5.2.1 implies that Λ is relatively separated. If it can be shown that there is a $\rho > 0$ such that $n_\Lambda([\rho k, \rho(k+1))) \geq 1$ then a uniformly dense subsequence Λ_0 can be obtained choosing one element of Λ from $[\rho k, \rho(k+1))$ for each $k \in \mathbb{Z}$. Since $\Lambda' = \Lambda \setminus \Lambda_0$ will be relatively uniformly separated, the claim will follow.

To get a contradiction, assume that for each $\rho > 0$ we can find a $k = k(\rho)$ such that $n_\Lambda([\rho k, \rho(k+1))) = 0$. Let ξ_ρ be the midpoint of this interval and let $g_\rho(t) = \mathrm{e}^{2\pi i \xi_\rho t}$. Then, in $L^2(I)$,

$$|\langle g_\rho, \mathrm{e}_{\lambda_k}\rangle|^2 = \left|\frac{\sin(\pi(\lambda_k - \xi_\rho)|I|)}{\pi(\lambda_k - \xi_\rho)|I|}\right|^2 \leq \frac{1}{\pi^2 |I|^2} \frac{1}{|\lambda_k - \xi_\rho|^2} .$$

Setting $M = \max_\ell n_\Lambda([\ell, \ell+1))$ and noting that $n_\Lambda([\ell, \ell+1)) = 0$ if $|\ell - \xi_\rho| < \rho/4$, where $\rho > 4$, one obtains

$$\begin{aligned}
\sum_k |\langle g_\rho, \mathrm{e}_{\lambda_k}\rangle|^2 &= \sum_{\ell \in \mathbb{Z}} \sum_{\lambda_k \in [\ell, \ell+1)} |\langle g_\rho, \mathrm{e}_{\lambda_k}\rangle|^2 \\
&\leq M \sum_{\ell : |\ell - \xi_\rho| > \rho/4} \frac{1}{\pi^2 |I|^2} \frac{1}{(|\ell - \xi_\rho| - 1)^2} \\
&\leq C \sum_{|\ell| > \rho/4} \frac{1}{\pi^2 |I|^2} \frac{1}{(|\ell| - 2)^2} \leq C'/\rho .
\end{aligned}$$

Here we have normalized the inner product in $L^2(I)$ so that any pure exponential has norm one. In particular, $\|g_\rho\| = 1$. The last inequality then contradicts the lower frame bound once ρ is large enough. Therefore it follows that (a) implies (b). □

Lemma 5.2.7 shows that $\mathrm{diam}(\Lambda) \geq D_{\text{frame}}(\Lambda)$. In what follows we will provide an outline of the main steps needed to show that $D_{\text{frame}}(\lambda) \geq \mathrm{diam}(\Lambda)$ as well. This reduces to showing that adding a sparse, separated sequence does not increase the frame diameter. The following lemma was proved by Duffin and Schaeffer [88] using analytic function theory.

Lemma 5.2.8. *If $\Lambda = \{\lambda_k\}$ generates a Fourier frame for $L^2(I)$ then there is a $\delta > 0$ such that, if $|\mu_k - \lambda_k| \leq \delta$ for all k, then $\{\mu_k\}$ also generates a Fourier frame for $L^2(I)$.*

Lemma 5.2.9. *If Λ is a Fourier frame sequence and if Λ', obtained from Λ by removal of a finite number of elements, remains a frame sequence, then* $\mathrm{diam}(\Lambda') = \mathrm{diam}(\Lambda)$.

Proof (of Lemma 5.2.9). It is clear that $\mathrm{diam}(\Lambda') \leq \mathrm{diam}(\Lambda)$. The simple argument that, for any $\varepsilon > 0$, $\mathrm{diam}(\Lambda') > \mathrm{diam}(\Lambda) - \varepsilon$, is taken from Casazza et al. [64]. Let I be an interval of length smaller than $\mathrm{diam}(\Lambda)$. Then by Theorem 5.1.12, $D_{-}(\Lambda) > |I|$ and so, by Corollary 5.2.4, Λ generates a Fourier frame for $L^2(I)$. Since a finite set does not affect density, $D_{-}(\Lambda') = D_{-}(\Lambda)$ so Λ' also generates a Fourier frame for $L^2(I)$. Since the only condition on I was that $|I| < \mathrm{diam}(\Lambda)$, it follows that $\mathrm{diam}(\Lambda') > \mathrm{diam}(\Lambda) - \varepsilon$ whenever $\varepsilon > 0$. That is, $\mathrm{diam}(\Lambda') \geq \mathrm{diam}(\Lambda)$. □

If one *duplicates* some elements of a frame then one obtains a new frame, but with different frame bounds: the only change in the frame inequalities is to multiply some of the terms $\sum |\langle \mathbf{x}, \mathbf{x}_k \rangle|^2$ by a factor two. For a Fourier frame, the frame diameter is unchanged. The next *perturbation* lemma states that the Fourier frame diameter of the union of a frame sequence and a small perturbation of itself is no larger than that of the original sequence.

Lemma 5.2.10. *Suppose that $\Lambda^{(1)}$ and $\Lambda^{(2)}$ satisfy $|\lambda_k^1 - \lambda_k^2| < 1/(1+k^2)$. Then* $\mathrm{diam}(\Lambda^1) = \mathrm{diam}(\Lambda^1 \cup \Lambda^2)$.

Proof. If $g \in L^2(I)$ then

$$|\langle g, \mathrm{e}_{\lambda_k^1} \rangle - \langle g, \mathrm{e}_{\lambda_k^2} \rangle| \leq \|g\| \|\mathrm{e}_{\lambda_k^1} - \mathrm{e}_{\lambda_k^2}\| \leq C \|g\| |\lambda_k^1 - \lambda_k^2| \,.$$

Here e_λ is normalized to have norm one in $L^2(I)$ and the inequality follows from direct integration and the estimate $\sin u/u \approx 1 - u^2/2$. Let C' be such that

$$\big| |\langle g, \mathrm{e}_{\lambda_k^1} \rangle|^2 - |\langle g, \mathrm{e}_{\lambda_k^2} \rangle|^2 \big| \leq C' \|g\|^2 |\lambda_k^1 - \lambda_k^2| \,.$$

Suppose now that $\Lambda^1 \cup \Lambda^2$ is a frame sequence for $L^2(I)$ with frame bounds A and B. Choose K so that $C' \sum_{|k|>K} 1/(1+k^2) \leq A/2$. Then

$$\sum_{|k|>K} |\langle g, \mathrm{e}_{\lambda_k^1} \rangle|^2 + |\langle g, \mathrm{e}_{\lambda_k^2} \rangle|^2 \leq 2 \sum_{|k|>K} |\langle g, \mathrm{e}_{\lambda_k^1} \rangle|^2 + \frac{A}{2} \|g\|^2$$

so that, by the frame hypothesis on $\Lambda^1 \cup \Lambda^2$,

$$\begin{aligned} A\|g\|^2 &\leq \Big\{ \sum_{|k|\leq K} + \sum_{|k|>K} \Big\} |\langle g, \mathrm{e}_{\lambda_k^1} \rangle|^2 + |\langle g, \mathrm{e}_{\lambda_k^2} \rangle|^2 \\ &\leq \Big(\sum_{|k|\leq K} |\langle g, \mathrm{e}_{\lambda_k^1} \rangle|^2 + |\langle g, \mathrm{e}_{\lambda_k^2} \rangle|^2 \Big) + \Big(2 \sum_{|k|>K} |\langle g, \mathrm{e}_{\lambda_k^1} \rangle|^2 + \frac{A}{2} \|g\|^2 \Big) \\ &\leq \frac{A}{2} \|g\|^2 + 2B \|g\|^2 . \end{aligned}$$

Subtracting $A\|g\|^2/2$ from both sides and dividing by two shows that $\mathscr{E}(\Lambda^1 \cup \{\lambda_k^2 : |k| \le K\})$ is a frame with lower bound $A/4$ and upper bound B. Since the frame diameter is unaffected by removal of the finite set $\{\lambda_k^2 : |k| \le K\}$, the lemma follows. □

Combining Lemmas 5.2.10 and 5.2.8 one has the following.

Lemma 5.2.11. *If* $\lim_{k\to\infty} |\lambda_k^1 - \lambda_k^2| = 0$ *then* $\mathrm{diam}(\Lambda^1) = \mathrm{diam}(\Lambda^2)$.

Proof. Let $\delta > 0$ be as in Lemma 5.2.8. Suppose that $\Lambda^1 \cup \Lambda^2$ is a frame sequence for I. Substitute $\widetilde{\lambda}_k = \lambda_k^1 + 1/(1+k^2)$ for λ_k^2 whenever $1/(1+k^2) \le \delta/2$ and $|\lambda_k^1 - \lambda_k^2| \le \delta/2$. Otherwise let $\widetilde{\lambda}_k = \lambda_k^2$. Then $\Lambda^1 \cup \widetilde{\Lambda}$ remains a frame sequence for $L^2(I)$, where $\widetilde{\Lambda} = \{\widetilde{\lambda}_k\}$. Since $|\widetilde{\lambda}_k - \lambda_k^1| \le 1/(1+k^2)$ for large enough k, it follows from Lemma 5.2.10 that Λ^1 is a frame for $L^2(I')$ whenever $|I'| < |I|$. The same argument applies to Λ^2. This proves the lemma. □

Lemma 5.2.12. *If* Λ *is relatively separated then* $\mathrm{diam}(\Lambda) \le D_-(\Lambda)$.

Proof. Let Λ be relatively separated and let M be such that $[k, k+1)$ contains at most M elements of Λ. Given any $\delta \in (0, 1/M)$ we can choose a subsequence Λ' from a grid of size δ in such a way that $|\lambda_k - \lambda_k'| < M\delta$ for any k. Here we assume, as before, that the elements λ_k of Λ are linearly ordered. If I is a fixed interval and δ is small enough then, by Lemma 5.2.8, if $\mathscr{E}(\Lambda)$ is a frame for $L^2(I)$ then so is $\mathscr{E}(\Lambda')$. Additionally, $D_-(\Lambda) = D_-(\Lambda')$ since $n_{\Lambda'}(J) - 2M \le n_\Lambda(J) \le n_{\Lambda'}(J) + 2M$ for any interval J. By Theorem 5.1.12, $D_-(\Lambda') \ge \mathrm{diam}(\Lambda')$. Since I was arbitrary, the lemma follows. □

Completion of the Proof of Theorem 5.2.6

To prove the theorem, it remains to prove that $\mathrm{diam}(\Lambda) \le D_{\mathrm{frame}}(\Lambda)$. To begin, suppose that Λ can be written as a union of M uniformly separated sequences. It suffices to show that for any $\varepsilon > 0$, one can decompose Λ into the union of (i) a sequence Λ_0 of uniform density at least $D_{\mathrm{frame}}(\Lambda) - \varepsilon$, (ii) a finite collection of separated sequences, each of which tends to a subsequence of Λ_0 at infinity—and hence does not extend $\mathrm{diam}(\Lambda)$—and (iii) a remainder sequence Θ that has density at most $2M\varepsilon$. Since the behavior of the separated subsequences has to be controlled at infinity, it is convenient to proceed by subdividing $\Lambda \cap [0, \infty)$, then applying a similar subdivision to $\Lambda \cap (-\infty, 0]$.

Fix a separated subsequence Λ_0 having uniform density at least $D_{\mathrm{frame}}(\Lambda) - \varepsilon$. Let $E_1 = \cup_{k=1}^{\infty} [\lambda_k^0 - 1, \lambda_k^0 + 1]$ and let Θ^1 consist of those $\lambda \in \Lambda, \lambda > 0$ that are not in E_1. Then Θ^1 can be expressed as the union of M separated sequences, none of which has density larger than 2ε since, if one did, then its union with Λ_0 would have density at least $D_{\mathrm{frame}}(\Lambda) + \varepsilon$. Therefore, $D_-(\Theta^1) \le 2M\varepsilon$. By definition of D_-, we can find an interval I_1 large enough that $n_{\Theta^1}(I_1) \le 3M\varepsilon|I_1|$. If $I_1 \cap E_1 = \emptyset$, let $A_1 = \sup(I_1)$. Otherwise, let $A_1 = \lambda_k^0 + 1$ where k is the largest integer such that

I_1 intersects $[\lambda_k^0 - 1, \lambda_k^0 + 1]$. Begin building Θ by including those $\theta_k^1 \in \Theta^1$ with $\theta_k^1 \leq A_1$.

One proceeds with the construction by induction. Suppose that we have identified the elements of Θ that are no larger than A_{m-1}. Define $E_m = \cup_{\lambda_k^0 \geq A_{m-1}} [\lambda_k^0 - 1/m, \lambda_k^0 + 1/m]$ and let Θ^m consist of those $\lambda_k \geq A_{m-1}$ that are not in E_m. By the same density argument as above, we can choose a compact interval $I_m \subset [A_{m-1}, \infty)$ of length at least m that contains at most $3M\varepsilon|I_m|$ elements of Θ^m. Let $A_m = \sup I_m$ if $I_m \cap E_m = \emptyset$ and, otherwise, let $A_m = \lambda_k^0 + 1/m$ with k the largest integer such that I_m intersects $[\lambda_k^0 - 1/m, \lambda_k^0 + 1/m]$.

The sequence Θ thus constructed has density at most $3M\varepsilon$. The $\lambda_k > 0$ that are contained in $\Lambda \setminus (\Lambda_0 \cup \Theta)$ will lie in $[A_m, A_{m+1}]$ for some m, in which case they must lie within $1/m$ of an element of Λ^0. Therefore, $\text{diam}(\Lambda \setminus \Theta) = \text{diam}(\Lambda^0)$ by Lemma 5.2.11.

Once the corresponding subdivision is performed for $\Lambda \cap (-\infty, 0]$, it follows from Lemma 5.2.12 that $\text{diam}(\Lambda) \leq D_{\text{frame}} + 3M\varepsilon$. Since $\varepsilon > 0$ was arbitrary, the proof of Theorem 5.2.6 is complete. □

5.2.2 Characterizing Fourier Frames

Discussion: Riesz Bases of Exponentials for $L^2[0,1]$, Reduction and Extension

In a survey of irregular sampling and frames published in the early 1990s [25], Benedetto observed that, even after 40 years, Duffin and Schaeffer's Theorem 5.1.8 "remains difficult to prove." That still holds true well into the third millennium. However, our understanding of Fourier frames and Riesz bases for PW has improved dramatically due to the work of Pavlov (cf. [157]), Jaffard, Ortega-Cerdà and Seip [251], and others. A frame that has the property that the removal of any of its elements leaves an incomplete set is called an *exact frame*. An exact frame is, in fact, a Riesz basis. By Proposition 5.1.14 the frame sequence Λ of an exact Fourier frame $\mathscr{E}(\Lambda)$ is then a set of sampling and complete interpolation, while $\mathscr{E}(\Lambda)$ is a Riesz basis for $L^2(I)$, where $|I| = \text{diam}(\Lambda)$.

Suppose that I is an interval of length $|I| = \text{diam}\,\Lambda$. Then Λ might be a frame for $L^2(I)$ but, then again, it might not. Seip [289] quantified stability at this critical length in terms of the possibility of adding points to Λ or taking points away from Λ in such a way that the resulting sequence generates a Riesz basis for $L^2(I)$. We state his results as follows.

Theorem 5.2.13. *If Λ is a uniformly discrete set satisfying $D_-(\Lambda) > |I|$ then $\mathscr{E}(\Lambda_-)$ is a Riesz basis for $L^2(I)$ for some $\Lambda_- \subset \Lambda$.*

Theorem 5.2.14. *If Λ is a uniformly discrete set satisfying $D_+(\Lambda) < |I|$ then $\mathscr{E}(\Lambda_+)$ is a Riesz basis for $L^2(I)$ for some $\Lambda_+ \supset \Lambda$.*

As we will see, Theorems 5.2.13 and 5.2.14 follow from Avdonin's *quarter in the mean* extension of Kadec's quarter theorem [12]. Avdonin's theorem can be stated as follows.

Theorem 5.2.15. *If* $\lambda_k = k + \delta_k$ *is uniformly discrete and if there exists an* N *and* $d < 1/4$ *such that*

$$\left|\sum_{k=N\ell+1}^{N(\ell+1)} \delta_k\right| \leq Nd$$

for all $\ell \in \mathbb{Z}$*, then* Λ *generates a Riesz basis* $\mathrm{e}_\lambda = \mathrm{e}^{-2\pi i \lambda t}$ *for* $L^2[0,1]$.

Proof (of Theorem 5.2.13, assuming Theorem 5.2.15). We will consider the case $|I| = 1$. Let $D_-(\Lambda) > 1$ and choose L large enough so that $n_\Lambda(J) \geq L+1$ whenever J is an interval of length L. Set $N = 2(L+1)L$. Consider the problem of choosing N points

$$\lambda_{k,-} = k + \delta_k, \qquad k = mN+1, \ldots, (m+1)N$$

from the set

$$\Lambda(m) = \Lambda \cap [mN + 1/2, (m+1)N + 1/2).$$

Let $\underline{S}(m)$ and $\bar{S}(m)$ denote the minimum and maximum possible values of

$$S(m) = \sum_{k=mN+1}^{(m+1)N} \delta_k.$$

Divide $[mN+1/2, (m+1)N+1/2) = 2m(L+1)L + 1/2 + [0, 2(L+1)L)$ into $2L+2$ intervals of length L. The smallest possible choice of λ_k in each of these intervals is its left endpoint. Thus,

$$\underline{S}(m) \leq (L+1)\sum_{j=1}^{2L}\left(jL + \frac{1}{2}\right) - \sum_{k=1}^{2L(L+1)} k = -L(L^2+1)$$

whereas, since λ_k is at most a right endpoint,

$$\bar{S}(m) \geq (L+1)\sum_{j=1}^{2L}\left(jL + L + \frac{1}{2}\right) - \sum_{k=1}^{2L(L+1)} k = L(L^2+1).$$

Since $|\lambda_{k+1} - \lambda_k| \leq 2L$, it follows that $\lambda_{k,-}$ can be chosen in such a way that

$$\left|\sum_{k=N\ell+1}^{N(\ell+1)} \delta_k\right| \leq L = \frac{N}{2(L+1)} = Nd \tag{5.8}$$

with $d = 1/(2(L+1)) < 1/4$. Therefore, Λ_- satisfies the hypotheses of Theorem 5.2.15 so $\mathscr{E}(\Lambda_-)$ forms a Riesz basis for $L^2[0,1]$. This proves Theorem 5.2.13. □

The proof of Theorem 5.2.14 is much the same. In this case one chooses L such that $n_\Lambda(J) \leq L-1$ whenever $|J| = L$ and sets $N = L(L+1)$. Then for each m one adjoins to Λ those points of the set

$$R(m) = \{mN + 1/2 + k/(2L)\}_{k=0}^{2NL-1}$$

that are *admissible* in the sense of having distance at least $1/(4L)$ from any point in Λ. The same bound (5.8) is shown to be satisfied, and thus the extended family Λ_+ is also a Riesz basis for $L^2[0,1]$.

The proof of Theorem 5.2.13 (and 5.2.14) relied on the strict density inequality in order to choose an appropriate reduction (or extension) of Λ. In order that $\mathscr{E}(\Lambda)$ forms a Riesz basis for $L^2(I)$ it is, of course, necessary that $D_+(\Lambda) = D_-(\Lambda)$. However, equality of these densities does not guarantee that Λ has an extension or reduction whose exponentials form a Riesz basis. Still, it is tempting to conjecture that, in this event, if Λ is incomplete then there is an extension Λ_+ such that $\mathscr{E}(\Lambda_+)$ is a Riesz basis for $L^2(I)$, and similarly for a reduction of Λ in the overcomplete case.

Seip [289] proved that such processes are not possible, in general.

Theorem 5.2.16. *If* $\Lambda_1 = \{k(1 - 1/|k|^{1/2})\}_{|k|>0}$, *then* $\mathscr{E}(\Lambda_1)$ *is a frame for* $L^2[0,1]$ *but* $\mathscr{E}(\Lambda_-)$ *is not a Riesz basis for any subsequence* Λ_- *of* Λ_1.
If $\Lambda_2 = \{k(1+1/|k|^{1/2})\}_{|k|>0}$, *then* $\mathscr{E}(\Lambda_2)$ *is a Riesz sequence for* $L^2[0,1]$ *but* $\mathscr{E}(\Lambda_+)$ *is not a Riesz basis for any supersequence* Λ_+ *of* Λ_2.

The proof of Theorem 5.2.16 requires too much machinery to reproduce here. A curious consequence of the proof, however, is that removal of any finite subset of Λ_1 leaves a set of sampling for PW but reduction by any infinite subsequence does not. Similarly, addition of a finite complementary set to Λ_2 results in an (incomplete) interpolating set for PW while the interpolation property fails upon extension by any infinite set.

The situation for frames is more complicated than for Riesz bases. For example, the following theorem, due to Casazza, Christensen, Li and Lindner [64], shows that if the upper and lower Beurling densities of Λ are unequal then there might be a gap in the lengths of intervals over which $\mathscr{E}(\Lambda)$ forms a frame for $L^2(I)$.

Theorem 5.2.17. *Given* $0 < d < D$ *there is a sequence* Λ *such that* $D_-(\Lambda) = d$, $D_+(\Lambda) = D$, *and* $\mathscr{E}(\Lambda)$ *is a subsequence of an orthonormal basis of* $L^2[0,D]$ *and spans* $L^2[0,A]$ *for any* $A \in (0,D)$. *As a consequence,* $\mathscr{E}(\Lambda)$ *is a frame for* $L^2[0,A]$ *whenever* $A \in (0,d]$ *and is a Riesz sequence in* $L^2[0,A]$ *whenever* $A \geq D$. *However,* $\mathscr{E}(\Lambda)$ *is not a frame sequence for* $L^2[0,A]$ *whenever* $A \in (d,D)$.

Discussion: Pavlov's Characterization of Exponential Riesz Bases and Seip and Ortega-Cerdà's Characterization of Fourier Frames for PW

The Paley–Wiener (PW) spaces form a scale that increases with containment of the frequency support sets. However, this scale is not sufficiently fine to quantify precisely what happens as the interval length increases to the frame diameter, or when a Riesz sequence criterion breaks down for decreasing length. This is precisely the issue with Theorem 5.2.16: any reduction subsequence $\Lambda' \subset \Lambda_1$ in that theorem that remains complete is too big for $\mathscr{E}(\Lambda')$ to be a Riesz basis, while any extension $\Lambda' \supset \Lambda_2$ to a complete sequence cannot remain a Riesz sequence. In a sense, Λ_1 grows *locally uniformly too slowly* while Λ_2 grows *locally uniformly too rapidly*, and Beurling densities are not sensitive to this local structure.

Rather than reducing or enlarging the sequence Λ at the critical length, one can consider instead the possibility of reducing or extending a scale of spaces containing the PW spaces. This observation was made precise in the work of Ortega-Cerdà and Seip [251], who observed that within a certain class of entire functions it is possible to capture the sampling property of a sequence Λ in the zeros of a specific entire function. There is some parallel here with the manner in which Pavlov was able to quantify the interpolation property.

In order to state Pavlov's characterization of complete interpolating sequences for PW [157], one needs to define the *John–Nirenberg space* BMO of functions $f: \mathbb{R} \to \mathbb{C}$ having bounded mean oscillation, that is, such that $\frac{1}{|I|}\int_I |f - f_I|^2$ is uniformly bounded independent of the interval I. Here, f_I denotes the mean value of f over the interval I. Let $P(x,t) = \frac{t}{\pi}\int_{\mathbb{R}} \frac{f(u)}{(x-u)^2+t^2}\,du$ denote the harmonic Poisson extension of $f \in L^2(\mathbb{R})$ to the upper half-plane (UHP), $\{(x,t) : t > 0\}$. Pavlov's theorem [157] can be formulated as follows.

Theorem 5.2.18. *A separated sequence Λ of real numbers is a complete interpolating sequence for* PW *if and only if $(n_\Lambda(t) - t)$ belongs to* BMO*, and its Poisson extension $P(x,s)$ for $s = 1$ can be expressed as $P(x,1) = \mathrm{conj}\,(u) + v + C$ where C is constant and $u, v \in L^\infty(\mathbb{R})$ with $\|v\|_\infty < 1/4$.*

Here, $\mathrm{conj}\,(u)$ denotes the harmonic conjugate of the harmonic function u. Functions in BMO can be unbounded. The condition that $(n_\Lambda(t) - t)$ belongs to BMO implies that its growth at infinity is at most logarithmic: $\ln t$ is the standard example of an unbounded function in BMO. Pavlov's theorem relies on deep connections among complex analysis, the theory of Muckenhoupt A_p weights, and BMO. These connections are outlined in the work of Seip and Lyubarskii [214].

Fourier frame sequences for PW were characterized by Ortega-Cerdà and Seip [251] in terms of sequences of zeros of analytic functions in a certain *de Branges class*. A Hilbert space $\mathscr{H}$ of entire functions is called a de Branges space [79] if: (i) the $\mathscr{H}$-norm of $f(z) \times (z - \bar{\zeta})/(z - \zeta)$ equals the $\mathscr{H}$-norm of f whenever $f(\zeta) = 0$; (ii) $f \mapsto L_\zeta(f) = f(\zeta)$ defines a continuous linear functional whenever $\mathrm{Re}\,\zeta \neq 0$; and (iii) the $\mathscr{H}$-norm is preserved under $f^*(z) = \overline{f(\bar{z})}$. The space PW is a prime example of a de Branges space, but a broader class of relevance here can be constructed via the *Hermite–Biehler class* HB consisting of all entire functions that are zero free in the UHP and satisfy $|E(z)| \geq |E(\bar{z})|$ whenever $\mathrm{Im}\,z > 0$. To each $E \in$ HB, one can associate a Hilbert space $H(E)$ consisting of all entire functions f such that $f(z)/E(z)$ and $f^*(z)/E(z)$ belong to the Hardy class H^2(UHP), defined as those functions satisfying $\sup_{s>0}\int |f(x+\mathrm{i}s)|^2\,dx < \infty$. The norm on $H(E)$ is defined as

$$\|f\|_{\mathrm{E}}^2 = \int_{-\infty}^{\infty} \frac{|f(t)|^2}{|E(t)|^2}\,dt.$$

In the case of PW_Ω, $E(z) = \mathrm{e}^{\pi \mathrm{i} \Omega z}$. A fundamental theorem of de Branges states that all de Branges spaces arise in this way. Ortega-Cerdà and Seip's characterization of sampling sequences for PW can be stated as follows.

Theorem 5.2.19. *A separated sequence $\Lambda \subset \mathbb{R}$ is a sampling sequence for PW if and only if there exist two entire functions E and F in HB such that*
(i) $H(E) = \mathrm{PW}$ and
(ii) Λ constitutes the zero sequence of $EF + E^*F^*$.
If Λ is a complete interpolating sequence for PW then one can take $F = 1$.

The last statement suggests that the function F accounts for the *redundant* part of the sampling sequence. Since, as [289] suggests, a sampling sequence will not always contain a complete interpolating subsequence, one can instead think of *enlarging the space*, and this is the role of F: the sampling set Λ turns out to solve the complete interpolation problem for the larger space $H(EF)$. Although a sampling sequence may not have a complete interpolating subsequence, as Corollary 5.2.21 will show, any sampling sequence nevertheless has a subsequence that *intertwines* a complete interpolating sequence. This fact can be proved by viewing Λ as the set of points at which $\cos(\varphi_{EF}) = 0$, where $EF = |EF|\mathrm{e}^{\mathrm{i}\varphi_{EF}}$ is the polar decomposition of EF. This is precisely the statement that Λ is the zero set of $EF + E^*F^*$.

Returning to the question of the sense in which sampling and interpolating sequences can be viewed as perturbations of a lattice, one seeks a more concrete description of the growth of the zero set of an element of the class HB such that $\mathrm{PW} = H(E)$. Such a growth criterion was also obtained by Ortega-Cerdà and Seip, based on ideas of Lyubarskii and Malinnikova [213]. *The purpose is to describe those sampling sequences that do not have a complete interpolating subsequence.* Suppose that ψ is a nondecreasing function such that $\psi(t) = 0$ for $t \leq 0$ and $\psi \to \infty$ as $t \to \infty$, but $\psi'(t) \to 0$ as $t \to \infty$. An example is $\psi(t) = \log^+(1+t)$. Seip showed in [289] that if λ_k is defined as the solution of $\lambda_k + \psi(\lambda_k) = k$ ($k \in \mathbb{Z}$) then $\Lambda = \{\lambda_k\}$ is a set of uniqueness for PW, none of whose subsequences is a complete interpolation set. To state the criterion that such a Λ also forms a sampling set, first one says that ψ *induces a regular logarithmic partition* if (i) the sequence t_k such that $\psi(t_k) = k$ ($k = 0, 1, \ldots$) is separated, (ii) $t_{k+1} - t_k$ is uniformly bounded, and (iii)

$$\sup_{x>0} \sum_{x/2<t_k<2x} \frac{(t_{k+1}-t_k)^2}{(x-t_k)^2 + (t_{k+1}-t_k)^2} < \infty.$$

Theorem 5.2.20. *Suppose that $\psi(t) = 0$ for $t \leq 0$ and that ψ is increasing on $(0, \infty)$. Let $\Lambda(\psi)$ be the sequence λ_k defined by $\lambda_k + \psi(\lambda_k) = k$.*
(i) If $\psi'(t) = O(1/t)$ as $t \to \infty$, and ψ induces a logarithmically regular partition, then $\Lambda(\psi)$ is a sampling set for PW.
(ii) If $\psi'(t) = o(1/t)$ as $t \to \infty$, then $\Lambda(\psi)$ is not a sampling set for PW.

If ψ vanishes on $(-\infty, 0]$ then $\lambda_k = k$ for $k < 0$, but the result is easily generalized so that the behavior of ψ on $(-\infty, 0]$ reflects the behavior on $[0, \infty)$. The theorem says that ψ has to be growing fast enough. In particular, ψ has to tend to infinity at least as fast as $\log^+$. This means that a sampling sequence that does not have an interpolating subsequence cannot be close to being a lattice. Nevertheless, any sampling sequence will have an *intertwined* interpolating sequence in the sense of the following corollary of Theorem 5.2.19.

Corollary 5.2.21. *If* Λ *is a separated sampling sequence for* PW *then there exists a complete interpolating sequence* $\Gamma = \{\gamma_k\}$ *such that for every* $k \in \mathbb{Z}$ *there is at least one* $\lambda \in \Lambda$ *such that* $\gamma_k \leq \lambda < \gamma_{k+1}$.

5.3 Sampling Schema for Multiband Signals

In the *single band* case, we considered the problem of finding necessary and sufficient conditions in order that a sequence Λ is a set of sampling or interpolation for PW. Much of the machinery for solving this problem relied on connectedness or convexity of the frequency support, though not always in an apparent way. In the multiband case, such machinery is not available, and the emphasis shifts entirely onto the problem of finding sufficient conditions for *some* sequences to be sampling and interpolation sets for PW_Σ. The sufficiency criteria depend on the structure of Σ. When Σ is a finite union of intervals, the criteria almost always boil down to inversion, or pseudoinversion, of some irregular discrete Fourier matrix, or simultaneous inversion of some system of such matrices. The interpolating functions have Fourier transforms that are dual to the exponentials $\mathscr{E}(\Lambda)$ in $L^2(\Sigma)$ in an appropriate sense. The irregular Fourier matrices arise from subdividing or *slicing* the spectral support set Σ in such a way that slices of one of the intervals comprising Σ alias onto other intervals in a tractable way.

This section consists mainly of *case studies*. The first case addresses frequency supports that are unions of M intervals each having the same length L, but spaced arbitrarily. The corresponding sampling sets will be unions of M phase-shifted copies of a regular lattice of density L. We present in detail work of Lyubarskii and Seip [215] that addresses this case. In the second case, Σ can be divided into M pieces whose lengths depend on the linear distribution of Σ, such that each piece has at most a fixed number P of aliases when divided appropriately. In this case, one can build a uniform sampling lattice corresponding to the Nyquist rate for the convex hull of Σ, then remove *sublattices* corresponding to pieces that have no aliases in this hull. We present in detail work of Bresler and Venkataramani [332] that addresses this situation. In Sect. 5.4 , variations of these two basic cases will be outlined, along with further methods that are fundamentally different from these two cases. Particular further methods are found in the work of Behmard and Faridani [21], cf. Behmard, Faridani, and Walnut [22], that addresses iterative reconstructions based on resampling over a union of *incommensurate* lattices, and the work of Avdonin and Moran [10, 11], cf. [13], based on ideas from control theory, among other approaches.

Brown's Aliasing Criteria

A subset $F \subset \mathbb{R}$ will called an *Ω-packing set* if, for all $k \in \mathbb{Z}$, $|(F + k\Omega) \cap F| = 0$. That is, any shift of F by a multiple of Ω intersects F in a set of measure zero. If, in addition, (almost) any $\xi \in \mathbb{R}$ is contained in $F + k\Omega$ for some $k \in \mathbb{Z}$ then F will be called an *Ω-tiling set*. An Ω-tiling set itself cannot have measure larger than Ω. If one thinks of $F + k\Omega$ as an *alias* of F, the packing criterion says that F has measurably trivial intersection with its aliases. It has long been accepted that if $f \in L^2(\mathbb{R})$ has its Fourier transform supported in an Ω-tiling set, then f can be recovered from its samples $f(k/\Omega)$. Normalizing so that $\Omega = 1$, one can write $f = \sum f(k)(\mathbb{1}_F)^\vee(x-k)$ with convergence in $L^2(\mathbb{R})$. In the engineering literature, however, this observation was not recognized as a rigorous fact before J.L. Brown provided a proof in 1985 [53]. We will reproduce parts of Brown's reasoning here because the important observation from his perspective is that a Shannon-type sampling formula applies to multiband Σ provided that none of the intervals in Σ *alias* one another. The sampling approaches outlined in the subsequent sections amount to figuring out what to do when frequency bands *do* alias one another. However, it must be noted first that Kluvánek [180] produced Brown's result in the much more general context of locally compact abelian groups in 1965. In Kluvánek's language, a tiling set is called a *measurable transversal*; cf. Beaty et al. [20] who proved a *converse* to Kluvánek's theorem characterizing transversals as those sets for which an analogue of the Shannon sampling formula applies. For the case of the real line, let $\Sigma \subset \mathbb{R}$ be a finite union of closed and bounded intervals, and fix a parameter $\sigma > 0$ such that Σ is a σ-packing set.

Theorem 5.3.1. *Set $T = 1/\sigma$. The following are equivalent:*
(i) Any $f \in \mathrm{PW}_\Sigma$ is uniquely determined by its samples $f(kT)$, $k \in \mathbb{Z}$,
(ii) $\mathrm{e}^{2\pi i k T \xi}$ is complete in $L^2(\Sigma)$,
(iii) Σ is a σ-packing set,
(iv) for each $f \in \mathrm{PW}_\Sigma$, $\|f(\cdot) - T\sum_{k=-N}^{N} f(kT)(\mathbb{1}_\Sigma)^\vee(\cdot - kT)\| \to 0$ as $N \to \infty$, and
(v) for each pair $f, g \in \mathrm{PW}_\Sigma$, one has the generalized Parseval relation

$$\int_{-\infty}^{\infty} f\bar{g} = \sum f(kT)\overline{g(kT)}.$$

Proof. We will only verify the equivalence of (i) and (ii), and the implication (iii) implies (iv). By (i) we mean that if $f(kT) = 0$ for all $n \in \mathbb{Z}$ then f is identically equal to zero. Since $|\Sigma| < \infty$, $L^2(\Sigma) \subset L^1(\Sigma)$ so, by the Fourier inversion formula, if $\widehat{f} \in L^2(\Sigma)$ then $f(kT) = \int_\Sigma \widehat{f}(\xi)\,\mathrm{e}^{2\pi i k T \xi}\,\mathrm{d}\xi$. Thus, if the exponentials $\mathrm{e}^{2\pi i k T \xi}$ are complete then the vanishing of $f(kT)$ for all T implies that $\widehat{f} = 0$. Then $f = 0$ by Fourier uniqueness. Conversely, if the exponentials $\mathrm{e}^{2\pi i k T \xi}$ are not complete, then there is some nontrivial $\varphi \in L^2(\Sigma)$ that is orthogonal to $\mathrm{e}^{2\pi i k T \xi}$ for all $k \in \mathbb{Z}$. That is, $(\varphi)^\vee(kT) = 0$ for all $k \in \mathbb{Z}$. So $(\varphi)^\vee$ is an element of PW_Σ that is not determined by its samples. Thus (i) and (ii) are equivalent.

To show that (iii) implies (iv), let $f \in PW_\Sigma$. The tiling hypothesis (iii) implies that $\sum_k \widehat{f}(\xi + k\sigma) = \widehat{f}(\xi)$ for $\xi \in \Sigma$ so that, by the Poisson summation formula, $\widehat{f}(\xi) = T\sum f(kT)\mathrm{e}^{-2\pi ik\xi T}$ with convergence in $L^2(\Sigma)$. Taking inverse Fourier transforms of both sides, which preserves L^2-convergence, one then has that

$$f(t) = T\sum f(kT)(\mathbb{1}_\Sigma)^\vee(t-kT)\,\square$$

5.3.1 When Σ is a Union of Commensurate Intervals

Kohlenberg [182] first considered the possibility of determining conditions under which a union of shifted lattices could form a sampling and interpolation set for PW_Σ when Σ is a union of two intervals of equal length. His ideas were extended by Lyubarskii and Seip [215]. They determined a sufficient condition under which the union of M shifted lattices could form a sampling and interpolation set for PW_Σ when Σ is a pairwise disjoint union of M intervals of equal length $\rho = |\Sigma|/M$—a situation commonly encountered in applications—and provided formulas for interpolating functions (Corollary 5.3.6). We will review Lyubarskii and Seip's construction now.

To fix ideas, suppose that Σ can be written $\Sigma = \cup_{j=1}^M [\delta_j, \delta_j + \rho)$ where $\delta_{j+1} - \delta_j \geq \rho$. It will be convenient to regard Σ as a subset of $[-1/2, 1/2)$ such that $\delta_1 = -1/2$ and $\delta_M = 1/2 - \rho$. The goal is to identify sampling sequences of the form

$$\Lambda = \left\{\lambda_{k\ell} = \frac{k}{\rho} + \beta_\ell : k \in \mathbb{Z}, \ell = 1, \dots, M\right\} \tag{5.9}$$

such that the interpolation problem $f(\lambda_{k\ell}) = a_{k\ell}$ has a unique solution $f \in PW_\Sigma$, with $\|f\|^2 \leq K\sum_{\ell=1}^M \sum_{k=-\infty}^\infty |a_{k\ell}|^2$, with K independent of $\{a_{k\ell}\}$.

For the moment, consider the slightly more general sample series

$$f \mapsto g(t) = \sum_{\ell=1}^M g_\ell(t); \qquad g_\ell(t) = \sum_{k\in\mathbb{Z}} f(\alpha_\ell k + \beta_\ell) s_\ell(t - \alpha_\ell k - \beta_\ell). \tag{5.10}$$

Here, α_ℓ represents a *step size*, β_ℓ a *phase*, and s_ℓ is an interpolating function. The following is an immediate consequence of the Poisson summation formula.

Proposition 5.3.2. *The function g_ℓ in (5.10) has Fourier transform*

$$\widehat{g_\ell}(\xi) = \widehat{s_\ell}(\xi)\left(\frac{1}{\alpha_\ell}\sum_{k\in\mathbb{Z}} \widehat{f}\left(\xi + \frac{k}{\alpha_\ell}\right)\mathrm{e}^{2\pi ik\beta_\ell/\alpha_\ell}\right).$$

Proof. By the Poisson summation formula, the Fourier transform of g_ℓ is

$$\widehat{g_\ell}(\xi) = \sum_{k\in\mathbb{Z}} f(\alpha_\ell k + \beta_\ell)\widehat{s_\ell}(\xi)\mathrm{e}^{-2\pi i(\alpha_\ell k + \beta_\ell)\xi} = \frac{\widehat{s_\ell}(\xi)}{\alpha_\ell}\sum_{k\in\mathbb{Z}} \widehat{f}\left(\frac{k}{\alpha_\ell} + \xi\right)\mathrm{e}^{2\pi ik\beta_\ell/\alpha_\ell}. \,\square$$

Returning to the special case in which $\alpha_\ell = 1/\rho$ for each ℓ, we obtain

$$\widehat{g}(\xi) = \rho \sum_{\ell=1}^{M} \widehat{s_\ell}(\xi) \sum_{k\in\mathbb{Z}} \widehat{f}(\xi + k\rho) \mathrm{e}^{2\pi \mathrm{i} k\rho\beta_\ell} .$$

In order that such a sum should reproduce $f \in \mathrm{PW}_\Sigma$, the functions $S_\ell = \widehat{s_\ell}$ should satisfy

$$h(\xi) = \rho \sum_{k\in\mathbb{Z}} h(\xi + k\rho) \sum_{\ell=1}^{M} S_\ell(\xi) \mathrm{e}^{2\pi \mathrm{i} k\rho\beta_\ell} \tag{5.11}$$

whenever $h \in L^2(\Sigma)$. Thus, the inner terms should sum to one when $k = 0$ and to zero for all other k. The functions S_ℓ will be taken as piecewise constant and supported in Σ. The pieces will be obtained by subdividing the intervals $I_j = [\delta_j, \delta_j + \rho)$ according to their linear distribution, see Fig. 5.1

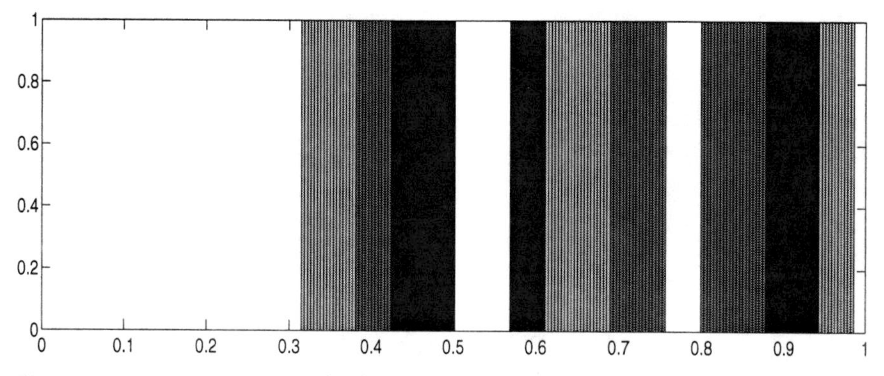

Fig. 5.1 Multiband spectrum $\Sigma \subset [0,1]$. Intervals have equal length 0.1875. Shaded subintervals correspond to $I_{j,\nu}$

To keep track of the separation between I_j and I_ℓ, we define $m_{j\ell}$ to be the number of times δ_ℓ must be shifted by ρ to end up in I_j; that is, so that

$$\delta_{j\ell} = \delta_\ell - m_{j\ell}\rho \in I_j.$$

Then $m_{jj} = 0$ and $\delta_{jj} = \delta_j$. Since $I_\ell = [\delta_\ell, \delta_\ell + \rho)$, for $\xi \in I_j$,

$$\xi + (m_{j\ell} + 1)\rho \in I_\ell \quad \text{if } \xi < \delta_{j\ell} \quad \text{and} \quad \xi + m_{j\ell}\rho \in I_\ell \quad \text{if } \xi \geq \delta_{j\ell}. \tag{5.12}$$

To keep notation simple, we will *assume* that the $\delta_{j\ell}$ are all distinct. Having fixed j, we index $\{1, \ldots, M\}$ by $\ell(\nu)$ in such a way that $\delta_{j,\ell(\nu)}$ is increasing in ν. This indexing, of course, depends on j. Express $I_j = \cup I_{j,\nu}$ where $I_{j,\nu} = [\delta_{j,\ell(\nu)}, \delta_{j,\ell(\nu+1)})$. Thus, if $\xi \in I_{j,\ell(\nu)}$ then $\xi \geq \delta_{j,\ell(\mu)}$ when $\mu \leq \nu$ so, by (5.12), $\xi + m_{j,\ell(\mu)}\rho \in I_{\ell(\mu)}$, whereas $\xi + (m_{j,\ell(\mu)} + 1)\rho \in I_{\ell(\mu)}$ for $\mu > \nu$, and

$$\xi + m\rho \notin \Sigma \qquad \text{if} \qquad m \notin \{m_{j,\ell(\mu)} : \mu \leq \nu\} \cup \{m_{j,\ell(\mu)} + 1 : \mu > \nu\}.$$

When $h = 0$ outside of Σ, the expression in (5.11) becomes

$$h(\xi) = \sum_{1\le\mu\le\nu} h(\xi + m_{j,\ell(\mu)}\rho) \sum_{\ell=1}^{M} S_\ell(\xi)\,\mathrm{e}^{2\pi i\rho\beta_\ell m_{j,\ell(\mu)}} + \sum_{\nu<\mu\le M} h(\xi + (m_{j,\ell(\mu)}+1)\rho) \sum_{\ell=1}^{M} S_\ell(\xi)\,\mathrm{e}^{2\pi i\rho\beta_\ell (m_{j,\ell(\mu)}+1)}\,.$$

Since $m_{jj} = 0$, our indexing $\ell(\mu)$ yields $m_{j,\ell(1)} = 0$ for each j. Therefore, the expansion of h can be expressed as

$$h(\xi) = h(\xi)\sum_{\ell=1}^{M} S_\ell(\xi) + \sum_{2\le\mu\le\nu}\cdots + \sum_{\nu<\mu\le M}\cdots$$

where the second and third sums contain terms $h(\xi + m\rho)$, $m \neq 0$. Therefore, in order for (5.13) to reproduce every $h \in L^2(\Sigma)$ it is sufficient and, in fact, necessary, that the system

$$\begin{cases} \sum_{\ell=1}^{M} S_\ell(\xi) = 1 & \\ \sum_{\ell=1}^{M} S_\ell(\xi)\,\mathrm{e}^{2\pi i\rho\beta_\ell m_{j,\ell(\mu)}} = 0, & \mu = 2,\dots,\nu \\ \sum_{\ell=1}^{M} S_\ell(\xi)\,\mathrm{e}^{2\pi i\rho\beta_\ell (m_{j,\ell(\mu)}+1)} = 0, & \mu = \nu+1,\dots,M \end{cases} \tag{5.13}$$

is satisfied. Set $\boldsymbol{\beta} = (\beta_1,\dots,\beta_M)$ and for j,ν fixed set

$$a_{\mu\ell}^{(j,\nu)}(\boldsymbol{\beta}) = \begin{cases} 1, \ \mu = 1 \\ \mathrm{e}^{2\pi i\rho\beta_\ell m_{j,\ell(\mu)}}, \ \mu = 2,\dots,\nu \\ \mathrm{e}^{2\pi i\rho\beta_\ell (m_{j,\ell(\mu)}+1)}, \ \mu = \nu+1,\dots,M\,. \end{cases} \tag{5.14}$$

Let $A^{(j,\nu)}(\boldsymbol{\beta})$ be the $M\times M$ matrix with entries $a_{\mu\ell}^{(j,\nu)}(\boldsymbol{\beta})$, let $D^{(j,\nu)}(\boldsymbol{\beta})$ be the determinant of $A^{(j,\nu)}(\boldsymbol{\beta})$, and let $D_\ell^{(j,\nu)}(\boldsymbol{\beta})$ be the cofactor of $a_{1\ell}^{(j,\nu)}(\boldsymbol{\beta})$. Our goal is to identify a choice of $\boldsymbol{\beta}$ such that $D^{(j,\nu)}(\boldsymbol{\beta}) \neq 0$ for each choice of j and ν. In this case, the functions

$$S_\ell(\xi) = \sum_{j,\nu} \frac{D_\ell^{(j,\nu)}(\boldsymbol{\beta})}{D^{(j,\nu)}(\boldsymbol{\beta})}\mathbb{1}_{I_{j,\nu}}(\xi), \qquad \ell = 1,\dots,M \tag{5.15}$$

provide interpolant solutions of (5.13). Since the intervals are pairwise disjoint, these separate solutions for each j,ν are sufficient to verify (5.15).

It remains to verify that the *phase vector* $\boldsymbol{\beta}$ can be chosen such that $D^{(j,\nu)}(\boldsymbol{\beta}) \neq 0$ for each choice of j and ν. Unfortunately, the method for doing so requires some facts from several complex variables which, although fundamental, require too much machinery to reproduce here. For the record, we summarize the relevant facts from [215] as follows.

Lemma 5.3.3. *(i) The mapping $\boldsymbol{\beta} \mapsto D^{(j,\nu)}(\boldsymbol{\beta})$ extends to a holomorphic function of M complex variables. (ii) The restriction of a nontrivial holomorphic function on $\mathbb{C}^M$ to $\mathbb{R}^M$ does not vanish identically.*

Since the product of the $D^{(j,\nu)}(\boldsymbol{\beta})$ taken over all $1 \leq j, \nu \leq M$ also extends holomorphically, there exist values of the phase $\boldsymbol{\beta}$ such that $D^{(j,\nu)}(\boldsymbol{\beta}) \neq 0$ for each pair (j,ν). Actually, such $\boldsymbol{\beta}$ are *typical*, since $D^{(j,\nu)}(\boldsymbol{\beta})$ can vanish locally only on a lower dimensional subset. Thus, for almost all phase vectors one can obtain a unique solution of the system (5.13), and the interpolating functions are the inverse Fourier transforms of the functions S_ℓ in (5.15).

Given a phase vector $\boldsymbol{\beta}$ solving (5.13), now we will see that the set Λ in (5.9) is a set of sampling and complete interpolation for $L^2(\Sigma)$. Since it is a union of M shifts of the regular lattice $\mathbb{Z}/\rho$, Λ is separated and, hence, the exponentials $\mathscr{E}(\Lambda)$ form a Bessel sequence for $L^2(I)$ for any I.

First, we show that an upper Riesz bound is satisfied.

Lemma 5.3.4. *With $s_\ell = (S_\ell)^\vee$ in (5.15), the mapping*

$$\mathbf{c} \mapsto \sum_{\ell=1}^{M} \sum_{k=-\infty}^{\infty} c_{k\ell}\, s_\ell\Big(t - \frac{k}{\rho} - \beta_\ell\Big)$$

is bounded from $\ell^2(\mathbb{Z} \times \{1,\dots,M\})$ to PW_Σ.

Proof. Since the sum over ℓ is finite, it suffices to prove that

$$\Big\| \sum_{k=-\infty}^{\infty} c_{k\ell}\, s_\ell\Big(t - \frac{k}{\rho} - \beta_\ell\Big) \Big\|^2 \leq C \sum_k |c_{k\ell}|^2$$

for each $\ell = 1,\dots,M$. By Plancherel's theorem it suffices to prove that

$$\Big\| \sum_{k=-\infty}^{\infty} c_{k\ell}\, \mathrm{e}^{-2\pi \mathrm{i}\xi\left(\frac{k}{\rho}+\beta_\ell\right)} \sum_{j,\nu} \frac{D_\ell^{(j,\nu)}(\boldsymbol{\beta})}{D^{(j,\nu)}(\boldsymbol{\beta})} \mathbb{1}_{I_{j,\nu}}(\xi) \Big\|^2 \leq C \sum_k |c_{k\ell}|^2 .$$

Since the sum over (j,ν) is finite and since $D_\ell^{(j,\nu)}(\boldsymbol{\beta})/D^{(j,\nu)}(\boldsymbol{\beta})$ is bounded for each ℓ, j, ν, it suffices simply to show that

$$\Big\| \sum_{k=-\infty}^{\infty} c_{k\ell}\, \mathrm{e}^{-2\pi \mathrm{i}\xi\left(\frac{k}{\rho}+\beta_\ell\right)} \mathbb{1}_{I_{j,\nu}}(\xi) \Big\|^2 \leq C \sum_n |c_{k\ell}|^2$$

for each choice of j, ν, ℓ. This estimate follows from the fact that, for ℓ fixed, the functions $\exp(2\pi \mathrm{i}\xi(k/\rho + \beta_\ell))$ form a complete orthogonal family for any interval of length ρ and that the L^2-norm on the left-hand side represents an integral over a subinterval of such an interval. This proves the lemma. □

The preceding estimates are imprecise, but good enough for the Riesz bound. We claim now that any $f \in \mathrm{PW}_\Sigma$ is determined uniquely by its samples $\{f(n/\rho + \beta_\ell)\}$.

Lemma 5.3.5. *Given a phase vector $\boldsymbol{\beta}$ solving (5.13), any $f \in \mathrm{PW}_\Sigma$ is determined uniquely by its samples along Λ in (5.9).*

Proof. We need to show that if

$$g(t) \equiv \sum_{\ell=1}^{M} \sum_{k=-\infty}^{\infty} f\Big(\frac{k}{\rho}+\beta_\ell\Big) s_\ell\Big(t-\frac{k}{\rho}-\beta_\ell\Big)$$

vanishes identically, then $f(k/\rho+\beta_\ell)=0$ for each k and ℓ. For shorthand, let $f(k/\rho+\beta_\ell)=c_{k\ell}$. Taking the Fourier transform of g then gives

$$\begin{aligned}\widehat{g}(\xi) &= \sum_{\ell=1}^{M}\sum_{k=-\infty}^{\infty} c_{k\ell}\, \mathrm{e}^{-2\pi\mathrm{i}\xi\left(\frac{k}{\rho}+\beta_\ell\right)} \sum_{j,\nu} \frac{D_\ell^{(j,\nu)}(\boldsymbol{\beta})}{D^{(j,\nu)}(\boldsymbol{\beta})} \mathbb{1}_{I_{j,\nu}}(\xi)\\
&= \sum_{j,\nu} \mathbb{1}_{I_{j,\nu}}(\xi) \sum_{\ell=1}^{M} \frac{D_\ell^{(j,\nu)}(\boldsymbol{\beta})}{D^{(j,\nu)}(\boldsymbol{\beta})}\, \mathrm{e}^{-2\pi\mathrm{i}\xi\beta_\ell} \sum_k c_{k\ell}\, \mathrm{e}^{-2\pi\mathrm{i}\xi\frac{k}{\rho}}\\
&= \sum_{j,\nu} \mathbb{1}_{I_{j,\nu}}(\xi) \sum_{\ell=1}^{M} \frac{D_\ell^{(j,\nu)}(\boldsymbol{\beta})}{D^{(j,\nu)}(\boldsymbol{\beta})}\, \mathrm{e}^{-2\pi\mathrm{i}\xi\beta_\ell} \widehat{f_\ell}(\xi)\end{aligned}$$

where $\widehat{f_\ell}=\sum_k c_{k\ell}\exp(-2\pi\mathrm{i}\xi k/\rho)$. By the Fourier uniqueness criterion, it suffices to prove that if g vanishes identically then $f_\ell=0$ for each ℓ. Since each $\widehat{f_\ell}$ is ρ-periodic, it suffices to prove that $\widehat{f_\ell}$ vanishes on an interval of length ρ, say on I_1. The restriction of $\widehat{g}$ to the interval $I_{1,1}$ is

$$\sum_{\ell=1}^{M} \frac{D_\ell^{(1,1)}(\boldsymbol{\beta})}{D^{(1,1)}(\boldsymbol{\beta})}\, \mathrm{e}^{-2\pi\mathrm{i}\xi\beta_\ell} \widehat{f_\ell}\,.$$

The shift of $I_{1,1}$ by $-m_{j1}\rho$ is I_{j,ν_j}—this is how we decomposed I_j. Since each $\widehat{f_\ell}$ is ρ-periodic, the vanishing of $\widehat{g}(\xi)$ for $\xi \in I_{j,\nu(j)}$ shifts back to $I_{1,1}$ to give the relation

$$\sum_{\ell=1}^{M} \frac{D_\ell^{(j,\nu_j)}(\boldsymbol{\beta})}{D^{(j,\nu_j)}(\boldsymbol{\beta})}\, \mathrm{e}^{-2\pi\mathrm{i}\xi\beta_\ell} \widehat{f_\ell}(\xi)=0 \quad \text{on} \quad I_{1,1}, \quad j=1,\dots,M. \tag{5.16}$$

This provides a linear system for $\mathrm{e}^{-2\pi\mathrm{i}\xi\beta_\ell}\widehat{f_\ell}(\xi)$ on $I_{1,1}$. As indicated by Lyubarskii and Seip, a straightforward calculation based on the fact that the intervals $I_{1,\nu}$, $\nu=1,\dots,M$ undergo a cyclic rearrangement when translated onto I_j then shows that the coefficients of this system coincide with the elements of the inverse of the matrix $A^{(1,1)}(\boldsymbol{\beta})$. This shows that the system (5.16) has no nontrivial solutions. That is, $\widehat{f_\ell}=0$ on $I_{1,1}$. The other intervals $I_{1,\nu}$ are addressed in the same manner. □

The uniqueness property leads to the interpolating property of the functions $s_m=(S_m)^\vee$ in (5.15). Since $s_m \in \mathrm{PW}_\Sigma$,

$$s_m(t-\beta_m) = \sum_{\ell=1}^{M}\sum_n s_m\Big(\frac{k}{\rho}+\beta_\ell-\beta_m\Big)s_\ell\Big(t-\frac{k}{\rho}-\beta_\ell\Big).$$

Since the left-hand side corresponds to the term $k=0$ and $\ell=m$ on the right-hand side, with coefficient $s_m(0)$, the uniqueness result Lemma 5.3.5 implies that $s_m(0)=1$ but also that $s_m(k/\rho+\beta_\ell-\beta_m)=0$ whenever $k\neq 0$ or $\ell\neq m$.

Corollary 5.3.6. *The functions s_ℓ satisfy*

$$s_m\Big(\frac{k}{\rho}+\beta_\ell-\beta_m\Big)=\delta_{0k}\delta_{\ell m}.$$

As a second corollary of Lemma 5.3.5 one has the following.

Corollary 5.3.7. *The function*

$$f(t)=\sum_{\ell=1}^{M}\sum_k c_{k\ell}\, s_\ell\Big(t-\frac{k}{\rho}-\beta_\ell\Big)$$

solves the interpolation problem $f(k/\rho+\beta_\ell)=c_{k\ell}$ in PW_Σ.

5.3.2 Venkataramani and Bresler's Multicoset Method

Venkataramani and Bresler [332] identified interpolating functions for *periodic nonuniform sampling* or *multicoset sampling* of general multiband signals. Let $\Sigma=\cup_{k=1}^{N}[a_k,b_k]$ where $b_k<a_{k+1}$ and, for convenience, let $a_1=0$ and $b_N=1/T$. Then $\mathrm{PW}_\Sigma\subset\mathrm{PW}_{[0,1/T]}$ and any $f\in\mathrm{PW}_\Sigma$ can be recovered by sampling once every T units of time. Fix a large integer L. For $f\in\mathrm{PW}_\Sigma$ denote $\mathbf{x}=\mathbf{x}(f)$ the sequence with nth coordinate $x_n=\tilde f(nT)=f(-nT)$. For $k=\{0,1,\dots,L-1\}$ we define the kth *sample coset* $\mathbf{x}_k$ of $\mathbf{x}$ to be the sequence with nth term $x_{kn}=x_n$ if $n=k+mL$ for some $m\in\mathbb{Z}$ and $x_{kn}=0$ otherwise. Thus, $\mathbf{x}_k=\sigma^k U_L D_L\sigma^{-k}\mathbf{x}$ where σ is the shift operator $(\sigma\mathbf{x})_n=x_{n-1}$ and D_L and U_L are the respective operators of *downsampling* and *upsampling* by the factor L. To each $\mathbf{x}_k$ one associates its $1/(LT)$-periodic reversed Fourier series

$$\begin{aligned} X_k(\xi) &= \sum_n x_{kn}\mathrm{e}^{2\pi i nT\xi}=\mathrm{e}^{2\pi i kT\xi}\sum_{m\in\mathbb{Z}}\tilde f((k+mL)T)\mathrm{e}^{2\pi i mLT\xi}=\cdots\\ &=\frac{1}{LT}\sum_{\ell=0}^{L-1}\widehat f\Big(\xi+\frac{\ell}{LT}\Big)\mathrm{e}^{-2\pi i k\ell/L},\ \xi\in\Big[0,\frac{1}{LT}\Big). \end{aligned} \tag{5.17}$$

The last equality is seen by multiplying both sides by $\mathrm{e}^{-2\pi i kT\xi}$: its left-hand side becomes the Fourier series of $\sum_{\ell=0}^{L-1}g(\xi+\ell/(LT))$ with $g(\xi)=\widehat f(\xi)\mathrm{e}^{-2\pi i kT\xi}/(LT)$.

One would like to be able to reconstruct f from P out of L of its sample cosets $\mathbf{x}_k$ where, ideally, $P/L\approx T|\Sigma|$. Then, on average, approximately $|\Sigma|$ samples per unit time are needed to reconstruct f. In what follows we will outline a special case of Venkataramani and Bresler's methods in which one reconstructs from the *first* P out of L cosets. The corresponding sampling method is often referred to as *bunched sampling*. The method for subdividing Σ in general is called *spectral slicing*.

Spectral Slicing

To subdivide the spectrum efficiently first denote the moduli of the endpoints of $\Sigma \subset [0,1/T]$ modulo $1/LT$ as $0=\gamma_0<\gamma_1<\cdots<\gamma_M$ where $M\le 2N$, since multiple endpoints can have the same modulus. For each $m=1,\dots,M$ set $\Gamma_m=[\gamma_{m-1},\gamma_m]$. For each $\ell\in\{0,\dots,L-1\}$, $\Gamma_m+\ell/(LT)$ is either contained in one of the intervals in Σ or is disjoint from Σ. Define index sets $\mathscr{I}_m=\{\ell:\Gamma_m+\ell/(LT)\subset\Sigma\}$ and $\mathscr{J}_m=\{\ell:\Gamma_m+\ell/(LT)\cap\Sigma=\emptyset\}$. Spectral slicing is illustrated in Figs. 5.2 and 5.3.

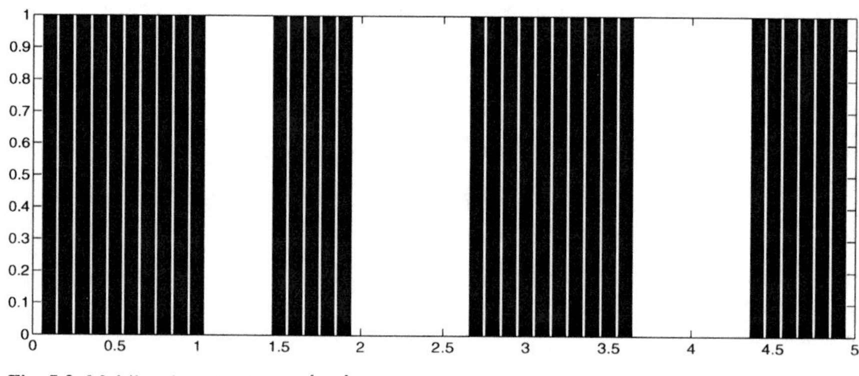

Fig. 5.2 Multiband spectrum $\Sigma\subset[0,5]$

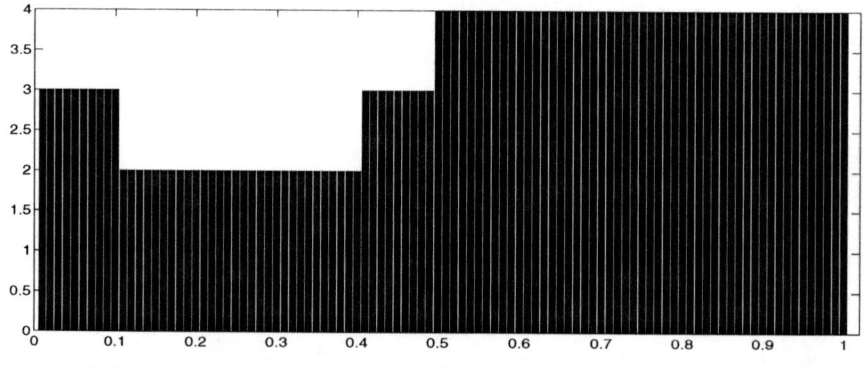

Fig. 5.3 Pullback of multispectrum, $L=5$, $T=1/5$, $1/(LT)=1$

Interpolating Functions

Let $P = \max_{m=1,\dots,M} |\mathscr{I}_m|$. For each $m = 1,\dots,M$ define a partial Fourier matrix F_m by retaining entries in the first P rows and the columns of the $L \times L$ DFT matrix corresponding to $\ell \in \mathscr{I}_m$, and setting all other entries equal to zero (here one labels the columns from 0 to $L-1$). For the purpose of constructing interpolating functions it is not strictly necessary to use the first P rows—any fixed choice of P rows such that each of the resulting F_m have full rank $|\mathscr{I}_m|$ is sufficient. Denote by F_m^{-1} a fixed partial left inverse of F_m (that is, $F_m^{-1}F_m\mathbf{y} = \mathbf{y}$ whenever $y_\ell = 0$, $\ell \notin \mathscr{I}_m$) and let G_m be any left annihilator of F_m, that is, $G_m F_m = 0$.

The first step in defining appropriate interpolating functions is to extract the spectral components. For each $\ell \in \{0,\dots,L-1\}$ set

$$\widehat{f_\ell}(\xi) = \widehat{f}\Big(\xi + \frac{\ell}{LT}\Big)\mathbb{1}_{\Gamma_m}, \qquad (\ell \in \mathscr{I}_m)$$

(one can take $\widehat{f_\ell} = 0$ if $\ell \notin \cup \mathscr{I}_m$). Since (5.17) says that $X_k(\xi)$ is, up to normalization, the $L \times L$ DFT of the sequence $\{\widehat{f}(\xi + \ell/LT)\}$, one can recover $\widehat{f_\ell}$ $(\ell \in \mathscr{I}_m)$ from the components X_k, $k = 0,\dots,P-1$ via

$$\widehat{f_\ell} = LT\sum_{k=0}^{P-1}(F_m)^{-1}_{\ell k}X_k \qquad \text{on } \Gamma_m\,.$$

Then expressing $f = \sum f_\ell$, one obtains for $\xi \in [0,1/T]$,

$$\widehat{f}(\xi) = \sum_{m=1}^{M}\sum_{\ell\in\mathscr{I}_m}\Big(LT\sum_{k=0}^{P-1}(F_m)^{-1}_{\ell k}X_k(\xi)\mathbb{1}_{\Gamma_m}\Big(\xi - \frac{\ell}{LT}\Big)\Big)\,.$$

Taking inverse Fourier transforms and applying (5.17) then yields

$$f(t) = \sum_{n=-\infty}^{\infty}\sum_{k=0}^{P-1} f((k+nL)T)\,s_k(t+nLT) \quad \text{where} \tag{5.18}$$

$$s_k(t) = LT\sum_{m=1}^{M}\sum_{\ell\in\mathscr{I}_m}(F_m)^{-1}_{\ell k}(\mathbb{1}_{\Gamma_m})^{\vee}(t+kT)\,\mathrm{e}^{2\pi i\ell(t+kT)/(LT)} \tag{5.19}$$

which expresses f as a sum over the first P of its L sample cosets. This choice corresponds to taking the left annihilator G_m above to be the zero matrix. One can define more general interpolating functions by taking

$$s_k(t-kT) = LT\sum_{m=1}^{N}\Big(\sum_{\ell\in\mathscr{I}_m}(F_m)^{-1}_{\ell k}\mathrm{e}^{2\pi i\ell t/(LT)} + \sum_{\ell\in\mathscr{J}_m}(G_m)_{\ell k}\mathrm{e}^{2\pi i\ell t/(LT)}\Big)\mathbb{1}^{\vee}_{\Gamma_m}(t)\,.$$

The matrices G_m might serve to construct interpolating functions with better localization properties. A detailed analysis of the aliasing errors that arise from these techniques is also provided in [332].

Discussion: Complexity of Spectral Slicing

The parameter $L \in \mathbb{N}$ is called the *slicing parameter*. Complexity of coset sampling depends on a judicious choice of L as the following examples illustrate.

Example. Consider $\Sigma = [0,1/p] \cup [1-1/q,1]$. Since $\Sigma \subset [0,1]$ the Nyquist rate is $T = 1$. The *Landau rate* on the other hand is $|\Sigma| = (p+q)/(pq)$ samples per unit time. This rate can be achieved with $L = pq$.

Example. Let $\Sigma = [0,1/2] \cup [2/3-\varepsilon, 1-\varepsilon]$. The Nyquist rate is $1-\varepsilon \approx 1$ for small ε and the Landau rate is $5/6$. Taking $L = 6$ will yield $P = 6$, so there is no improvement in the sampling rate. If $\varepsilon = p/q$ then $L = \mathrm{lcm}(6,q)$ will yield a sampling rate of $5/6$. In summary, the distribution of Σ plays a prominent role in the tradeoff between sampling efficiency and computational complexity. Other work closely related to Bresler and Venkataramani's work in [332] can be found in [47, 100, 333–335].

5.4 Notes and Auxiliary Results

5.4.1 *More on Sampling of Band-limited Functions*

Random Sampling

The following result, due to Gröchenig and Bass [17], is stated in its n-dimensional setting because its proof in n dimensions requires more sophisticated arguments than its one-dimensional case. We let $\mathrm{PW}(\mathbb{R}^n)$ denote those $f \in L^2(\mathbb{R}^n)$ whose Fourier transforms vanish outside $[-1/2,1/2]^n$, and let Q_T denote multiplication by $\mathbb{1}_{[-T,T]^n}$. Set

$$\mathrm{PW}(T,\delta) = \{f \in \mathrm{PW} : \|Q_T f\|^2 \geq (1-\delta)\|f\|^2\}.$$

Gröchenig and Bass proved the following random sampling theorem for such approximately time- and band-limited functions.

Theorem 5.4.1. *Suppose that $\{x_k : k = 1,2,\dots\}$ is a sequence of i.i.d random variables uniformly distributed over $[-T,T]^n$ and let $0 < \mu < 1-\delta$. Then there exist $A, B > 0$ such that for any $f \in \mathrm{PW}(T,\delta)$ one has*

$$\frac{r}{(2T)^n}(1-\delta-\mu)\|f\|^2 \leq \sum_{k=1}^{r} |f(x_j)|^2 \leq \frac{r}{(2T)^n}(1+\mu)\|f\|^2$$

with probability at least

$$1 - 2A\mathrm{e}^{-B\frac{r}{(2T)^n}\frac{\mu^2}{41+\mu}}.$$

One can take $B = \sqrt{2}/36$, and for large r one can take $A = \mathrm{e}^{cT^2}$, where c is a dimensional constant. The probability is taken with respect to the sampling sequence. The major difficulty is to establish the inequality uniformly over $\mathrm{PW}(T,\delta)$.

Necessary Conditions for Sampling and Interpolation, Revisited

Landau's necessary conditions for sampling and interpolation, Theorem 5.1.7, have a straightforward extension to $\mathbb{R}^n$ where the geometry of Σ can be complicated. Gröchenig and Razafinjatovo found a somewhat simpler proof of Landau's necessary conditions for sampling and interpolation of band-limited functions (in $\mathbb{R}^n$) under the mild constraint that the boundary of the spectrum has measure zero. The method is quite different from Landau's in that it uses Ramanathan and Steger's *homogeneous approximation* method [273] to prove the following general *comparison theorem*.

Theorem 5.4.2. *Let $g_1,\dots,g_p$ and $h_1,\dots,h_q$ be in $\mathrm{PW}_\Sigma(\mathbb{R}^n)$. Let $\Gamma_1,\dots,\Gamma_p$ and $\Lambda_1,\dots,\Lambda_q$ be discrete subsets of $\mathbb{R}^n$ such that $\cup_{k=1}^{p}\{g_k(\cdot-\gamma):\gamma\in\Gamma_k\}$ forms a Riesz basis for $\mathrm{PW}_\Sigma(\mathbb{R}^n)$ and $\cup_{\ell=1}^{q}\{h_\ell(\cdot-\lambda):\lambda\in\Lambda_\ell\}$ forms a frame for $\mathrm{PW}_\Sigma(\mathbb{R}^n)$. Then, for any $\varepsilon>0$, there is an $R>0$ such that, for any $r\geq 0$ and any $y\in\mathbb{R}^n$, one must have*

$$(1-\varepsilon)\sum_{k=1}^{p}\#\{\gamma\in\Gamma_k:|\gamma-y|<r\}\leq\sum_{\ell=1}^{q}\#\{\lambda\in\Lambda_\ell:|\lambda-y|<r+R\}.$$

In particular,

$$\sum_{\ell=1}^{q}D_+(\Lambda_\ell)\geq\sum_{k=1}^{p}D_-(\Gamma_k).$$

Theorem 5.1.7 follows readily from Theorem 5.4.2 under the added constraint that $|\partial\Sigma|=0$. Landau's density criterion is also a consequence of the John–Nirenberg theorem for BMO functions; see [251, Sect. 4].

5.4.2 Other Approaches to Sampling of Multiband Signals

Equiband Sampling (Bezuglaya and Katsnelson)

Bezuglaya and Katsnelson provided some of the first results for the case of sampling of multiband signals with an arbitrary number of intervals [37]; see also [172, 173], Beaty and Dodson [19], and Higgins [148]. The Fourier transform in [37] is normalized as $\int \mathrm{e}^{it\xi}f(t)\,\mathrm{d}t$. One assumes that Σ has the *equiband* form

$$\Sigma=\bigcup\Big[\mu_m-\frac{\Omega}{q},\mu_m+\frac{\Omega}{q}\Big],\quad \mu_m=-\Omega+(2\ell_m)\frac{\Omega}{q},\quad 0\leq\ell_1<\cdots<\ell_p\leq q.$$

In words, except for a dilation and shift, Σ is obtained by choosing a subset of intervals from q consecutive integer intervals $[\ell,\ell+1]$.

Theorem 5.4.3. *Let Σ be a union of equibands as above. Let γ_j, $j=1,\dots,p$ be a collection of real numbers such that $D=\det F(\gamma_1,\dots,\gamma_p)\neq 0$ where $F(\gamma_1,\dots,\gamma_p)$ is the $p\times p$ matrix with (m,j)th entry $\mathrm{e}^{\mathrm{i}\mu_m\gamma_j}$. Also, set*

$$t_{jk}=\frac{q\pi}{\Omega}k+\gamma_j,\quad j=1,\dots p,\quad k\in\mathbb{Z}$$

and

$$S_{j,k}(t) = (-1)^{k(q-1)} \frac{D_j(t)}{D} \pi sinc_{1/\pi}\Big(\frac{\Omega}{q}(t-t_{jk})\Big)$$

where $D_j(t)$ is the determinant of the matrix obtained from $F(\gamma_1,\dots,\gamma_p)$ by replacing the jth column with the column vector whose mth row is $e^{i\mu_m t}$.
Then each $f \in \mathrm{PW}_\Sigma$ can be written

$$f(t) = \sum_{j=1}^{p}\sum_{k=-\infty}^{\infty} f(t_{jk})S_{j,k}(t)$$

and, for constants m and M depending only on Ω,

$$m^2 \sum |f(t_{jk})|^2 \le \|f\|_{L^2}^2 \le \frac{M^2}{D^2}\sum |f(t_{jk})|^2 .$$

Lattice Layering (Behmard and Faridani)

Periodic nonuniform sampling is one means of reproducing multiband signals using a finite set of interpolating functions. An alternative approach is to identify a lattice that *fits* each frequency support interval and then peel off components of $f \in \mathrm{PW}_\Sigma$ by resampling remainder terms not accounted for by lattice samples from previous layers. As before, let $\Sigma = \cup [a_k, b_k]$. In algorithmic terms, first one samples on $\mathbb{Z}/(b_1 - a_1)$, then subtracts a suitable interpolant, and iterates on the remainder, sampling along $\mathbb{Z}/(b_2 - a_2)$ et cetera. Care must be taken when the lattices intersect—when the interval lengths are rational multiples of one another.

As an example, suppose that $\Sigma = I_1 \cup I_2$ with $|I_1| = 1/2$ and $|I_2| = 1/3$. To account for I_1, one should sample along $2\mathbb{Z}$, and for I_2, one should sample along $3\mathbb{Z}$. However, there is no way to disentangle information associated with I_1 and I_2 contained in the samples along the intersection $6\mathbb{Z}$. This suggests that, instead, one might sample along $2\mathbb{Z} \cup (3\mathbb{Z} + \alpha)$ for $\alpha \ne 0$.

This is the sort of approach that underlies the work of Behmard and Faridani [21]. Their results are phrased in the language of a locally compact abelian group G, which is a natural context for their approach. But the essence of their results is already captured in the case $G = \mathbb{R}$. The following is a corollary of Theorem 2 of [21].

Theorem 5.4.4. *Let $I_1 \subset I_2$ and let $\Sigma = I_2 \cup (\eta + I_1)$ express Σ as a disjoint union of two intervals. Fix t_1 and t_2 such that for all $k \in \mathbb{Z}$, $\eta(t_1 - t_2 + k/|I_1|) \notin \mathbb{Z}$. Set $\varphi_i = (\mathbb{1}_{I_i})^\vee$, $i = 1,2$, and let*

$$S_2 f(t) = \sum_{m\in\mathbb{Z}} f\Big(t_2 + \frac{m}{|I_2|}\Big)\varphi_2\Big(t - t_2 - \frac{m}{|I_2|}\Big).$$

Then one can write $f = S_2 + g$ where

$$g(t) = \big(1 - e^{2\pi i (t-t_2)\eta}\big)\sum_{k\in\mathbb{Z}} g\Big(t_1 + \frac{k}{|I_1|}\Big)\frac{\varphi_1(t - t_1 - k/|I_1|)}{1 - e^{2\pi i (t_1 - t_2 + k/|I_1|)\eta}} .$$

Example. Let $I_1 = [0,1/3] \subset [0,1/2] = I_2$ and set $\eta = 2/3 - \varepsilon$ so $\Sigma = [0,1/2] \cup [2/3-\varepsilon, 1-\varepsilon]$. Let $x_1 = 1/2$ and $x_2 = 0$. Then we can write $S_2 f(x) = \sum_{k\in\mathbb{Z}} f(2k)\varphi_{[0,1/2]}(x - 2k)$ where $\varphi_{[a,b]} = \mathbb{1}^\vee_{[a,b]}$, and then

$$g(x) = \big(1 - e^{2\pi i x(2/3-\varepsilon)}\big)\sum_{k\in\mathbb{Z}} \frac{g(3k+1/2)}{1 - c_\varepsilon e^{-6\pi i k\varepsilon}}\varphi_{[2/3-\varepsilon,1-\varepsilon]}\Big(x - \frac{1}{2} - 3k\Big)$$

where $c_\varepsilon = e^{\pi i (2/3-\varepsilon)}$. Expressing $f = g + S_2 f$ shows that any $f \in \mathrm{PW}_\Sigma$ can be recovered from its samples along $2\mathbb{Z} \cup (1/2 + 3\mathbb{Z})$, therefore at the optimal average rate of 5 out of every 6 samples. The sampling set here does not correspond to cosets as defined by Venkataramani and Bresler, though. Theorem 5.4.4 can be extended to unions of more than two lattices. Then one has to reiterate the process that defines g as a remainder.

Better Localization for Multiband Interpolation

The following theorem, due to Strohmer and Tanner [320], summarizes a multiple filter approach that leads to efficient approximate reconstructions.

Theorem 5.4.5. *Any $f \in \mathrm{PW}$ can be expressed in terms of its samples along $P > \lceil T \rceil$ uniform grids $\{kT + T_\ell\}_{k\in\mathbb{Z}}$, $\ell = 1,\ldots,P$. Here, f is decomposed into the translates of P atoms ψ_ℓ, each associated with a particular grid,*

$$f(t) = T\sum_{\ell=1}^{P}\sum_{k\in\mathbb{Z}} f(kT - T_\ell)\,\psi_\ell(t - (kT + T_\ell))\,.$$

The atoms ψ_k are defined in terms of their Fourier transforms

$$\Psi_k(\xi) = \sum_{j=1}^{K} c_{n_j,k}\Phi_{n_j}(\xi)$$

where Φ are partitioning functions in the sense that

$$[-1/2,1/2] \subset \bigcup_{j=1}^{K} I_{n_j};\quad \operatorname{supp}\Phi_{n_j} \subset I_{n_j},\quad \sum_{j=1}^{K}\Phi_{n_j} = 1 \text{ on } [-1/2,1/2].$$

The coefficients $\mathbf{c}_n = [c_{n1},\ldots,c_{nP}]^T$ are obtained by solving

$$AR_n\mathbf{c}_n = \mathrm{e}_{P-n+1}$$

where e_{P-n+1} is the standard basis vector, A is the $P\times P$ matrix $A_{k\ell} = \mathrm{e}^{2\pi \mathrm{i} T_\ell k/T}$ and R_n is the $P\times P$ diagonal matrix with ℓth diagonal entry $\mathrm{e}^{2\pi \mathrm{i} T_\ell (n-P-1)/T}$.

In summary, Strohmer and Tanner's method allows a smooth partition of the Fourier supports that leads, in turn, to interpolating functions that are much better localized than the sinc-type interpolating functions used in other approaches, e.g., [154]. This leads to explicit and satisfactory estimates in [320] for estimation of $f \in \mathrm{PW}$ from a finite number of periodic nonuniform sampling (PNS) samples when f is also time localized. A different approach, based on regular sampling, will be given in Chap. 6.

Discussion: Controllability, Observability and Multiband Riesz Bases—Toward a General Criterion

None of the multiband sampling techniques outlined so far leads to general criteria for nonuniform sampling. In order to generate a broader class of examples, it could be profitable to consider criteria under which $\mathscr{E}(\Lambda)$ is a basis or frame for $L^2(\Sigma)$, when Σ is a finite union of intervals, derived from other contexts. One such context is control theory. The control theory approach to finding general conditions on a finite union of intervals and a discrete set Λ such that $\mathscr{E}(\Lambda)$ forms a Riesz basis for $L^2(\Sigma)$ is the subject of work of Avdonin, Bulanova, and Moran, including [10, 11, 13].

The role of nonharmonic Fourier series in questions of observability and controllability of physical systems is the subject of the texts by Avdonin and Ivanov [14] and of Komornik and Loreti [183]. Operational formulations of the roles played by exponentials in issues of *controllability* and *reachability* are given in detail in Avdonin and Ivanov [14, p. 160]. The following discussion will be very general with the sole purpose of motivating and describing the relationship between the control problem for an appropriate, albeit nonphysical, *string* and the Riesz basis property in $L^2(\Sigma)$ for $\mathscr{E}(\Lambda)$ in which Λ is the set of eigenvalues for the string problem.

Controllability refers to the possibility of forcing a system into a specific state in a given amount of time by means of an appropriate control signal, while observability refers to the possibility of inferring the state of a system through certain output measurements. In the case of a first-order, linear, autonomous system of the form

$$\frac{\mathrm{d}Y}{\mathrm{d}t} = AY; \qquad Y(0) = Y_0, \tag{5.20}$$

controllability can be formulated in the following general terms. Let $\mathscr{H}$ be a Hilbert space on which A generates a strongly continuous semigroup e^{tA}, and let B be a linear operator mapping some subspace of $\mathscr{H}$, which contains the domain of A, into an auxiliary Hilbert space $\mathscr{K}$ such that $\|BY_0\|_{\mathscr{K}} \leq C\|AY_0\|_{\mathscr{H}}$ for all Y_0 in the domain of A. One calls B an *observability operator* provided there exists a time interval I and constant c_I such that the solution Y of (5.20) satisfies

$$\|BY\|_{L^2(I,\mathscr{K})} \leq c_I \|Y_0\|_{\mathscr{H}}. \tag{5.21}$$

Under the same hypotheses, one defines a problem dual to (5.20), namely

$$\frac{\mathrm{d}X}{\mathrm{d}t} = -A^*X + B^*Z; \qquad X(0) = X_0, \tag{5.22}$$

for all $X_0 \in \mathscr{H}'$ and $Z \in L^2_{\mathrm{loc}}(\mathbb{R}, \mathscr{K}')$. Here $\mathscr{H}', \mathscr{K}'$ denote the *Banach* duals of $\mathscr{H}, \mathscr{K}$. If Y solves (5.20) and (X,Z) solves (5.22) then, with $\langle \cdot, \cdot \rangle$ denoting the respective dual pairings on $(\mathscr{H}, \mathscr{H}')$ and $(\mathscr{K}, \mathscr{K}')$, one has, formally,

$$\langle X(T), Y(T)\rangle = \langle X_0, Y_0\rangle + \int_0^T \langle Z(t), BY(t)\rangle \, \mathrm{d}t. \tag{5.23}$$

Under the hypotheses, the problem (5.22) has a unique solution (see [183]). The adjoint B^* of the observability operator B is regarded as a mapping taking a control signal as input, and whose output can drive X to a desired state in a finite time, under suitable conditions. Thus, in a mathematical sense, observability and controllability are *dual properties*.

The role played by basis properties of trigonometric functions is, perhaps, simpler to describe in the case of observability. A physical example of observability discussed by Komornik and Loreti [183] is a string with free ends, satisfying

$$\begin{aligned} u_{tt} &= u_{xx}, \qquad t \in [0,\infty),\ x \in [0,L] \\ u_x(t,0) &= 0 = u_x(t,L), \qquad t > 0 \\ u(0,x) &= u_0;\ u_t(0,\cdot) = u_1, \qquad x \in (0,L) \end{aligned} \tag{5.24}$$

where the subscripts refer to differentiation in the given variables. One assumes an ability to observe the oscillations $u(t,0)$ of the left end of the string for some time interval $0 \leq t \leq T$. One would like to know whether these observations are sufficient to determine the initial data (u_0, u_1). This means: Is the mapping $(u_0, u_1) \to u(\cdot, 0)$ one to one between suitable spaces of real-valued functions?

Let $L = 1$. Set $L_0^2[0,1] = \{f \in L^2[0,1] : \int_0^1 f = 0\}$ and let $H_0^1[0,1] = \{f \in H^1[0,1] : \int_0^1 f = 0\}$ where H^1 denotes those functions with derivatives in $L^2[0,1]$. We take the norm of f in L_0^2 to be its L^2-norm and in H_0^1 to be the L^2-norm of f'. The functions $\cos \pi kx$, $k = 1, 2, \ldots$ form an orthogonal basis for L_0^2 and for H_0^1 so, using separation of variables, one can write

$$u(t,x) = \sum_{k=1}^{\infty} (a_k \cos 2\pi kt + b_k \sin 2\pi kt) \cos \pi kx.$$

Then

$$\|u_1\|_{L^2}^2 = \int_0^1 |u_t(0,x)|^2\,\mathrm{d}x = 4\pi^2 \sum_{k=1}^{\infty} k^2 b_k^2, \text{ and}$$
$$\|u_0\|_{H^1}^2 = \int_0^1 |u_x(0,x)|^2\,\mathrm{d}x = \pi^2 \sum_{k=1}^{\infty} k^2 a_k^2.$$

On the other hand, for any positive integer M, the functions $\{\cos 2\pi kt, \sin 2\pi kt\}_{k=1}^{\infty}$ also form an orthogonal system in $L^2[0,M]$, so that

$$\int_0^M |u_t(t,0)|^2\,\mathrm{d}t = 4\pi^2 M \sum_{k=1}^{\infty} (a_k^2 + b_k^2).$$

Therefore, if $E_0 = \|u_1\|_{L^2}^2 + 4\|u_0\|_{H^1}^2$ then

$$ME_0 \le \int_0^T |u_t(t,0)|^2\,\mathrm{d}t \le (M+1)E_0$$

whenever $M \le T \le M+1$. The observability property—that the mapping $(u_0,u_1) \mapsto u(t,0)$ is one to one—holds whenever $T \ge 1$. Observations of $u(t,0)$ over an interval of length smaller than one are insufficient to determine (u_0,u_1). Thus *observability can be expressed in terms of a basis property of sines and cosines.*

The general second-order scalar-valued case can be formulated as

$$\frac{\mathrm{d}^2 y}{\mathrm{d}t^2} + Ay(t) = f(t); \quad y(0) = y_0; \quad \left.\frac{\mathrm{d}y}{\mathrm{d}t}\right|_{t=0} = y_1. \tag{5.25}$$

The discussion of *controllability* of this system in [14, p. 160] can be summarized as follows. The control signal f in (5.25) is assumed to have the form $f = Bu$ where u lies in an auxiliary Hilbert space U, mapped continuously by B into a certain space of functions W whose expansions in terms of the eigenfunctions of A have coefficients in an ℓ^2-space weighted by appropriate powers of the eigenvalues. The *reachability set* $\mathscr{R}(T,y_0,y_1)$ consists of those vectors $(y,\dot{y}))\big|_{t=T}$ corresponding to $f = Bu$ for some $u \in U$. W-controllability of the system (5.25) is then defined in terms of the condition that the y-coordinates of the reachability set comprise a dense subspace of W if and only if the exponentials $\mathrm{e}^{\mathrm{i}\lambda_n t}$ form a Riesz basis for their closed linear span inside $L^2(0,T;U)$.

Control in the Multiband Setting (Avdonin, Moran, and Bulanova)

Avdonin and Moran [11] considered a special case of (5.25) in the form of the string equation

$$\begin{aligned} \rho^2(x)\,y_{tt}(x,t) &= y_{xx}(x,t), \\ y(0,t) &= u(t), \\ y_x(L,t) &= 0, \end{aligned} \tag{5.26}$$

with ρ regarded as a positive function on $[0,L]$ to be determined. In this case, $\mathscr{H} = L^2([0,L];\rho^2\,\mathrm{d}x)$. Ordinarily, one regards u as a function in $L^2[0,T]$ for some $T > 0$, but Avdonin et al. instead took u to be a locally square-integrable function supported in $\Sigma \subset [0,\infty)$. Let $\sigma_0 = \inf(\Sigma)$ and $\sigma_1 = \sup(\Sigma)$, and consider the initial conditions

$$y(x,\sigma_0) = y_0(x); \qquad y_t(x,\sigma_0) = y_1(x).$$

The eigenvalues of this system can be determined from the boundary value problem

$$\varphi''(x) + \lambda^2\rho^2(x)\varphi(x) = 0; \qquad 0 < x < L; \qquad \varphi(0) = \varphi'(L) = 0. \tag{5.27}$$

Since ρ has possible jumps at x_j, $j = 1, \ldots, N-1$, one imposes that the values of φ and φ' match up at the partition endpoints. The system (5.26) is said to be (exactly) controllable if, for any initial conditions $(y_0, y_1) \in L^2(0,L) \times H^{-1}(0,L)$, there is a unique control $u \in \mathrm{PW}_\Sigma$ that brings the system to the origin at time σ_1, that is, $y(\cdot, \sigma_1) = (\mathrm{d}y/\mathrm{d}t)(\cdot, \sigma_1) = 0$. In [11] the following was proved.

Theorem 5.4.6. *The system (5.26) with eigenvalues λ_n as in (5.27) is exactly controllable if and only if the family $\{\mathrm{e}^{\pm \mathrm{i}\lambda_n t}\}$ forms a Riesz basis for $L^2(\Sigma)$.*

Since $\lambda_n^2 > 0$, one can assume that $\lambda_n > 0$. In light of Theorem 5.4.6, the problem of constructing a Riesz basis of exponentials for $L^2(\Sigma)$ amounts to choosing a suitable weight function ρ such that (5.26) is exactly controllable. In [13] one takes ρ to be piecewise constant with $\rho(x) = \rho_j$ on (x_{j-1}, x_j) such that $\rho_j(x_j - x_{j-1}) = (b_j - a_j)/2$ where $\Sigma = \bigcup [a_j, b_j]$. Thus, one can take $\rho(0) = 0$ and $\rho(L) = |\Sigma|/2$, which is consistent with the condition for exact controllability in the single band case. The authors of [13] conjectured that these conditions on ρ are sufficient to provide exact controllability, and thus furnish $L^2(\Sigma)$ with a Riesz basis of exponentials corresponding to the eigenvalue problem (5.27). The conjecture was confirmed in [10], except for the ρ_j defining a *nowhere dense* set of ratios ρ_j/ρ_{j-1}.

The method outlined in [10] for solving the control problem (5.26) with data $y(x, a_1) = y_0(x)$ and $y_t(x, a_1) = y_1(x)$ with $\rho(x)$ piecewise constant as above can be written in an explicit, but complicated, form. The solution is worked out in detail in [13] for the case in which Σ is a union of two intervals. In this case, there is no loss in generality in taking $\Sigma = [0, \beta] \cup [\beta + \gamma, \beta + \gamma + \delta]$ with $\gamma > 0$. Thus, β is the length of the first interval, γ is the gap, and δ is the length of the second interval.

Consider the operator $\mathscr{W} : L^2[0, \beta] \to L^2[\gamma, \beta + \gamma]$ defined by

$$\mathscr{W}(f)(t) = \mathbb{1}_{[\gamma, \beta+\gamma]}(t) \sum_{r=0}^{\infty} \sum_{k=0}^{\infty} \sum_{q=0}^{k} (-1)^{r+k} \mu^k \frac{(r+k)!}{r!\,q!\,(k-q)!} f(t - \tau(r,k,q))$$

with $\tau(r,k,q) = \beta(r+k-q) + \gamma(r+q)$. Here, $\mu = (\rho_2 - \rho_1)/(\rho_2 + \rho_1)$ where ρ is the piecewise constant density function associated with Σ. Avdonin et al. prove that $\mathscr{W}$ is invertible if μ is sufficiently small. In turn, it is proved that the system (5.26) is exactly controllable if $\mathscr{W}$ is continuously invertible. One proceeds by reducing $\mathscr{W}$ to the sum of two operators, each containing only four terms of the series defining $\mathscr{W}$, and proves the invertibility of these operators for μ small. Even after all of these reductions, the proof is still combinatorially complicated. However, given the result, one can readily produce Riesz basic sequences Λ as follows.

Algorithm (Avdonin–Bulanova–Moran).
Input: two intervals $[0, \beta]$, $[\beta + \gamma, \beta + \gamma + \delta]$. Their union is Σ.

Generate two values $\rho_1 > 0$, $\rho_2 > 0$, $\rho_2 \neq \rho_1$
Set $\mu = (\rho_2 - \rho_1)/(\rho_2 + \rho_1)$
Set $x_1 = \beta/(2\rho_1)$
Set $\ell = x_1 + \delta/(2\rho_2)$
Define

$$\rho(x) = \begin{cases} \rho_1, & 0 < x < x_1 \\ \rho_2, & x_1 < x < \ell \end{cases}$$

Find the eigenvalues λ_n^2 of

$$\varphi''(x) + \lambda^2 \rho^2(x)\, \varphi(x) = 0, \quad 0 < x < \ell, \quad \varphi(0) = \varphi'(\ell) = 0.$$

Output: $\Lambda = \{\lambda_n\}$, such that $\mathscr{E}(\Lambda)$ forms a Riesz basis for $L^2(\Sigma)$.

Spectrum Blind Multiband Sampling (Eldar, Mishali, and Others)

In certain applications one might wish to acquire, through sampled data, a multiband signal for which the total bandwidth is known, but for which precise location of the active bands is not known. In this case it is desirable to sample along a set such that reconstruction is possible, no matter what spectral bands are active. The possibility of doing so was partly addressed in the work of Bresler and Venkataramani, who introduced the idea of a *universal* pattern for coset sampling. A choice of P rows of the DFT matrix is said to be *universal* if the subsequent choice of any P columns results in an invertible matrix. This can also be phrased in terms of the *Kruskal rank* of a matrix—the maximum number r such that any set of r columns of the matrix is linearly independent. The $P\times L$ submatrix consisting of the first P rows of an $L\times L$ DFT matrix has Kruskal rank P since any P of its columns form an invertible $P\times P$ Vandermonde matrix.

Any universal set of rows gives rise to multicoset sampling patterns that are also termed universal in the sense that one can build interpolating functions from which appropriate coset samples $f((mL+\ell)T)$ of f with sampling period T can be used to recover any $f\in \mathrm{PW}_\Sigma$ such that Σ is a finite union of intervals contained in $[0,L]$ whose pullbacks to $[0,1/(LT)]$ have at most P overlaps. A potential drawback of using such a pattern is ill-conditionedness of the partial Fourier matrix to be inverted.

Eldar et al. [221, 234–236] extended Bresler and Venkataramani's work in showing that such universal patterns can be used to recover multiband signals of bounded total spectral measure, but having spectral support that is only partially known. Define $\mathscr{N}_{\Omega,T}$ to consist of those $f\in \mathrm{PW}_{[0,1/T]}$ whose Fourier transforms are supported on a set of measure Ω/T, where $\Omega<1$. An $\mathscr{N}_{\Omega,T}$ *blind* Ω*-sampling set* Λ is a discrete set such that

$$A\|f-g\|^2\le \sum |f(\lambda_k)-g(\lambda_k)|^2\le B\|f-g\|^2$$

whenever $f,g\in\mathscr{N}_{\Omega,T}$. The sampling condition is expressed in terms of $\|f-g\|$ because $\mathscr{N}_{\Omega,T}$ does not form a linear space. It is shown that the Beurling lower density of any blind Ω-sampling set must be at least $\min\{2\Omega/T,1/T\}$—twice the Landau rate—when Ω is small. The factor of two arises because the Fourier support of $f-g$ could be *twice as big* as the separate Fourier supports of f and g. Mishali and Eldar do not prove the existence of a *universal sampling set* for $\mathscr{N}_{\Omega,T}$. However, they do establish sampling results for a *model space* $\mathscr{M}\subset \mathrm{PW}_{[0,1/T]}$ of signals whose Fourier supports are contained inside a disjoint union of N intervals, each of length at most B, $(NB<T)$, but with otherwise arbitrary positions inside $[0,1/T]$.

Define a vector $\mathbf{Y}(\xi)\in\mathbb{C}^L$ whose ℓth component is

$$Y_\ell(\xi)=Y\Big(\xi+\frac{\ell}{LT}\Big),\quad \xi\in\Big[0,\frac{1}{LT}\Big) \tag{5.28}$$

where $Y=\widehat{f}$, and a vector $\mathbf{X}(\xi)\in\mathbb{C}^P$ whose kth entry is (see 5.17)

$$X_k(\xi)=\frac{1}{LT}\sum_{\ell=0}^{L-1}Y_\ell(\xi)\mathrm{e}^{-2\pi i r_k\ell/L},\ \xi\in\Big[0,\frac{1}{LT}\Big).$$

We have reversed the roles of X and Y in comparison with [234, 235] in order to make the notation consistent with that of Section 5.3.2. Here, r_k indexes the kth coset $r_k+L\mathbb{Z}$, $0\le r_k<L$. As in Sect. 5.3.2, $X_k(\xi)$ is the Fourier series of the kth coset sample sequence of f. One has

$$\mathbf{X}(\xi)=\mathbf{A}\,\mathbf{Y}(\xi),\quad \mathbf{A}_{k\ell}=\frac{1}{LT}\mathrm{e}^{-2\pi i r_k\ell/L}. \tag{5.29}$$

The kth row $(k=0,\dots,P-1)$ of $\mathbf{A}$ is the r_kth row of the $L\times L$ Fourier matrix. The number of nonzero components Y_k of $\mathbf{Y}$ is at most the number of bands N, that is, $\mathbf{Y}$ is N*-sparse*. The set $R=\{r_0,\dots,r_{P-1}\}$ of cosets is called *universal* if $\mathbf{A}$ has *Kruskal rank* P.

In [235], the following uniqueness theorem is proved.

Theorem 5.4.7. *Let $f(t) \in \mathscr{M}$ and $L \leq 1/BT$. Let P be the number of cosets of $LT\mathbb{Z}$ along which f is sampled, where $P \geq N$ when the locations of the spectral intervals of f are known, and $P \geq 2N$ when the locations are unknown. Suppose that $R = \{r_0, \dots, r_{P-1}\}$ is universal. Then, for any $\xi \in [0, 1/(LT))$, the vector $\mathbf{Y}(\xi)$ in (5.28) is the unique N-sparse solution of (5.29).*

The function $f(t)$ can then be recovered by coset sampling as in Sect. 5.3.2. In practice, one would like to be able to identify an appropriate model space $\mathscr{M}$ directly from the data. Here we assume clean data. Mishali and Eldar define the *diversity set* $\Delta = \Delta_{(L,T)}$ of the spectrum of f to be the set of all indices ℓ of the vector function $\mathbf{Y}$ such that $Y_\ell(\xi) \neq 0$ in (5.28) on a subset of $[0, 1/(LT))$ of positive length.

Two spectrum blind reconstruction (SBR) algorithms based on separate extensions of the signal model $\mathscr{M}$ are presented in [235]. Here we present one of them, termed "SBR-4" in [235]. Let $\mathscr{A}_K$ denote the class of signals in $\mathrm{PW}_{[0,1/T]}$ whose diversity set has at most K elements or, equivalently, upon slicing $\mathrm{supp}\,\widehat{f}$ into L intervals, only K or fewer of the intervals are essentially nonempty. SBR-4 is intended to provide reconstruction of any $f \in \mathscr{A}_K$ from samples taken at the minimal sampling rate $2K/LT$ by pseudoinverting $\mathbf{A}$ in (5.29).

Algorithm (Mishali–Eldar (SBR-4)).
Input: samples $\mathbf{x}_k = f(r_k T + mL) : m \in \mathbb{Z}$, $k = 0, \dots, P-1$, of $f \in \mathscr{A}_K$ such that $\mathbf{A}$ in (5.29) has Kruskal rank $P \geq 2K$.
Output: the set Δ and a flag

`Compute` $\mathbf{X}$ from samples via (5.28) and (5.29)
`Set` $I = [0, 1/(LT))$
`Compute matrix` $Q = \int_I \mathbf{X}(\omega)\mathbf{X}^*(\omega)\,d\omega$
`Decompose` $Q = VV^*$ where V is $P \times \mathrm{rank}(Q)$ with orthogonal columns
`Solve` $V = AU_0$ for the sparsest solution U_0
`Compute` Δ, the nonzero columns of U_0
`flag` $= \#\Delta$, at most $P/2$.
`return flag,` Δ

Discussion of SBR-4

The statement that V as defined has a unique sparsest (in the sense of having a minimal number of nonzero coordinates) U_0 is [235, Proposition 3]. The system (V, U) is called a *multiple measurement vector* (MMV) in the compressive sensing literature. There exist suboptimal but fast routines for estimating such vectors. The purpose of the flag is to return a value zero if either an MMV fails to be found or, in the spectrum blind case, if the model $\mathscr{A}_{2K}$ is invalid.

It is possible that $\#\{\ell \in \{0, \dots, L-1\} : Y_\ell(\xi) \neq 0\} < \#\Delta$ for almost every $\xi \in [0, 1/(LT))$. For example, if $\Sigma = [0, 1/2) \cup [3/2, 2)$ then, upon slicing Σ into two equal intervals $[0, 1)$ and $[1, 2)$, neither interval is nonempty so the diversity set is $\{0, 1\}$ and one is unable to take advantage of the fact that, for each $\xi \in [0, 1)$, $\mathbf{Y}(\xi)$ has but a single nonzero component. This is the price paid for the blind spectrum hypothesis, which does not distinguish between the set $[0, 1/2) \cup [3/2, 2)$ and the set $[1/4, 3/4) \cup [5/4, 7/4)$ as subsets of $[0, 2)$.

Mishali, Eldar, and Tropp [236] considered, as an alternative to the blind sampling approach, employing a bank of random demodulators. In the case of wideband RF signals, such an approach has the advantage of not requiring ADCs to sample at the Nyquist rate of the input signal, at least in the spectrum blind case.

Robust and Random Sampling

Berman and Feuer [34] considered a constrained optimization problem for generating otherwise random sampling patterns for robust approximate reconstruction of multiband signals from finite collections of samples. Robustness can be quantified in terms of condition number of a *completely irregular* Fourier matrix. Consider the matrix H_K with (k,ℓ)th entry $e^{-2\pi i t_k n_\ell/T}$, $1 \leq k,\ell \leq K$. Such a matrix is typically poorly conditioned, a situation that can be quantified in terms of large aliasing errors or lack of robustness to noise; cf. [288, 332]. Berman and Feuer posed the problem: given $\mathbf{n} = (n_1,\ldots,n_K) \in \{1,\ldots,M\}^K$, find $(t_1,\ldots,t_K) \in \mathbb{R}^K$ that minimizes the condition number of H_K. This problem is somewhat intractable. A more tractable problem is to minimize the condition number of the *semi-irregular* matrix $H_K(\tau)$, in which $t_k = \tau$, $k = 1,\ldots,K$, over τ. Define a *complete residue system* (CRS) modulo K to be a set of integers r_ℓ, $\ell = 1,\ldots,K$, such that each integer is congruent to a unique r_ℓ mod K. It was shown that the problem of minimizing the condition number with respect to τ has a solution in which the condition number is equal to one provided that, for $Q = \gcd\{n_q - n_1 : q = 1,\ldots,K\}$, the numbers $(n_q - n_1)/Q$ form a CRS modulo K. For the case in which perfect conditioning is not possible, Berman and Feuer provided a numerical algorithm for minimizing the condition number with respect to τ.

Wojtiuk and Martin [358] also considered the problem of approximately reconstructing a multiband signal from finitely many randomly generated samples. Suppose that Σ is a finite disjoint union of intervals I_m. One seeks a recontruction $\mathbf{x} \approx \sum \alpha_k e^{2\pi i \omega_k t}$ in which ω_k is evenly spaced within each of the intervals comprising Σ. Given a finite set of samples, one can find a minimum norm solution for the α_k using an SVD approach. Wojtiuk and Martin produced simulations indicating that reconstructions from randomly generated samples taken slightly above the Landau rate but well below the Nyquist rate produce better approximations, in an average sense that depends on the linear distribution of Σ, than do reconstructions from uniform samples at the same rate.

Further Miscellaneous Aspects of Periodic Nonuniform Sampling

Bresler and Venkataramani's coset scheme is a special case of *periodic nonuniform sampling* (PNS), in which samples can be unevenly spaced over a given interval of length L, with the same sampling pattern reproduced on intervals $[kL,(k+1)L)$, $k \in \mathbb{Z}$. The method seems first to have been proposed by Yen [368] in 1956. *Bunched sampling* is the special case of PNS in which several samples are taken over a short interval, followed by an intervening interval over which no samples are taken. Such an approach was used in early analog-to-digital converters (ADCs), cf. [6], in order to avoid noise coupling arising when passing data from the ADC to a processor. Instead, one samples while the discrete signal processor (DSP) is inactive, then processes the data while the ADC is inactive. Bracewell [45] also referred to PNS as *interleaved* sampling, the idea being that one could acquire samples at a high rate by using multiple ADCs operating at lower rates. In this case, synchronization can be a challenge. PNS can also be regarded as a special case of a generalized sampling scheme due to Papoulis [255]. One of the first analyses of PNS as a means of reconstructing multiband signals from sampling, on average, at the *Landau rate* was carried out by Herley and Wong [144–146], who applied a filterbank approach.

Xia et al. [362, 363] proposed one of the first algorithms for approximate recovery of a multiband signal from a *finite* collection of samples via sinc interpolation. The key step in their approach was to associate subsets of the finite set of samples in a proportionate manner per subband. That is, for $\Sigma = \cup I_j$ one writes $f \in PW_\Sigma$ as $f = \sum f_j$ with $f_j = P_{I_j} f$ and allocates m_j samples to f_j such that $m_j/|I_j|$ is essentially constant. Senay et al. also considered in [292] the problem of approximate reconstruction from finitely many samples by means of nonuniform sampling projections onto prolate spheroidal wave functions (PSWFs).

Amini and Marvasti [5] proposed an iterative alternative to interpolation from *poorly localized* interpolating functions to reconstructing multiband signals. In their approach one applies a self-adjoint *distorting matrix D*, the combination of a filtering and PNS operator, and defines $\mathbf{x}_{k+1} =$

$\lambda(\mathbf{y} - D\mathbf{x}_k) + \mathbf{x}_k$ where $\mathbf{y} = D\mathbf{x} + \mathbf{n}$ ($\mathbf{n}$ is noise) starting with $\mathbf{x}_0 = 0$. They showed that, with $D^\dagger$ the pseudoinverse of D, $\lim_k \mathbf{x}_k = D^\dagger(\mathbf{x} + D\mathbf{n})$ provided that λ is less than twice the maximum eigenvalue of D.

Greitāns [117] proposed iterative updating as a means to estimate a power spectral density from nonuniform (not necessarily periodic) samples, emphasizing aliasing suppression. Wen et al. [351] also showed how to use PNS as a means of achieving one type of aliasing suppression, specifically, for reconstruction from a subband of a signal band limited to a larger band. Khan et al. [175], among others, have continued this line of research using specific *optimal* periodic nonuniform sampling; cf. Tzvetkov and Tarczynski [328].

Several authors have considered methods of *bandpass sampling* for direct downconversion (i.e., not requiring demodulation) of multiband RF signals. In particular, ranges for valid sampling rates for the downconversion are determined in terms of locations of the bands. Work along these lines has been carried out by Tseng and Chou [326], by Akos et al. [2], and by Lin et al. [203], cf. [350], among others.

Besides Herley and Wong's filterbank approach to PNS, others have also employed filterbank methods for analysis of band-limited and multiband signals. This includes work of Prendergast, Levy, and Hurst [267] who used a multirate filterbank approach to model PNS for band-limited signals, with specific goals including the ability to correct small periodic timing errors in a time-interleaved ADC (cf. also Vaidyanathan and Liu [330]). Dolecek and Torres [83] employed a multirate filterbank approach for efficient implementation of sample rate conversion. Borgerding's article [38] provides some insights into how certain filter-based signal processing methods can turn multiband approaches such as that of Venkataramani and Bresler into efficient numerical algorithms. PNS was also extended from the band-limited context to the context of shift-invariant spaces; see, e.g., [82, 153, 207, 329].

5.4.3 Universal Sampling and Fourier Frames

Universal Sampling

This chapter has focused largely on two questions. The first was: Given a diameter D, for which discrete sequences Λ does $\mathscr{E}(\Lambda)$ form a frame or Riesz basis for $L^2(I)$, $|I| = D$? The second was: Given a finite union of intervals Σ, how can one generate a sequence Λ such that $\mathscr{E}(\Lambda)$ forms a frame or Riesz basis for $L^2(\Sigma)$? Olevskiĭ and Ulanovskii [249, 250] asked whether there exists a discrete sequence Λ that forms a sampling sequence for *any set* of measure at most the (lower) density of Λ. The answer is *no* in general, but *yes* when the spectral supports are constrained to be compact.

If G is an open set with $|G| > D_-(\Lambda)$, then for every $\delta > 0$ there is a spectrum $\Sigma \subset G$ with $|\Sigma| < \delta$ such that $\mathscr{E}(\Lambda)$ is not a frame for $L^2(\Sigma)$ [250, Theorem 4.1]. In particular, given ε, one can find $f \in \mathrm{PW}_\Sigma$ such that $\|f\|^2 > \frac{1}{\varepsilon}\sum|f(\lambda)|^2$. The proof of this last statement uses Szemerédi's theorem [321] which states that, given $k \in \mathbb{N}$ and $d \in (0,1]$, there exists $N = N(k,d) \in \mathbb{N}$ such that every subset of $\{1,\dots,N\}$ of size at least dN contains an arithmetic progression of size k. In effect, Λ must have subsequences that are close to such a progression, and one is able to localize zeros of f along such a progression to lead to the desired inequality.

The situation changes completely when one restricts to compact spectra. First, Olevskiĭ and Ulanovskii showed that if Σ is a finite union of intervals with the property that the gaps between intervals are commensurate with their lengths, as in the work of Bezuglaya and Katsnelson outlined on p. 187, then for every positive d and ε there is a discrete set Λ whose elements are all within ε of $\frac{1}{d}\mathbb{Z}$ and such that $\mathscr{E}(\Lambda)$ is a Riesz basis for $L^2(\Sigma)$ whenever $|\Sigma| = D$ (see [250, Theorem 6.1]). Since any compact set can be covered arbitrarily closely by such a finite union, it follows that $\mathscr{E}(\Lambda)$ is a frame for $L^2(\Sigma)$ whenever Σ is compact with $|\Sigma| < d$. It can also be shown that Λ is a set of interpolation for PW_Σ for every open set Σ with $|\Sigma| > d$.

Results somewhat parallel to those of Olevskiĭ and Ulanovskii were proved by Matei and Meyer in the compact–discrete setting [227, 228], in particular, where $\Lambda \subset \mathbb{Z}^2$ and $\Sigma \subset \mathbb{T}^2$. One says that Λ is a set of universal sampling for $L^2(\mathbb{T}^2)$ if Λ has uniform Beurling density d and if for every set $K \subset \mathbb{T}^2$ such that $|K| < d$, Λ is a set of stable sampling for L^2_K, that is, $\|F\|_2^2 \le C(K)\sum_{k\in\Lambda}|\widehat{F}(k)|^2$. For a Borel set $\Omega \subset \mathbb{T}^2$ one says that Λ is a set of stable interpolation for $L^2(\Omega)$ if the Fourier coefficient mapping from $L^2(\Omega)$ to $\ell^2(\Lambda)$ is surjective or, equivalently, one has the Riesz–Fischer inequality

$$\sum_{\lambda\in\Lambda}|a(\lambda)|^2 \le C\int_\Omega \Big|\sum a(\lambda)\,\mathrm{e}^{2\pi i\lambda\cdot t}\Big|^2\,\mathrm{d}t.$$

For fixed $\alpha \in (0, 1/2)$, Matei and Meyer defined the set $\Lambda_\alpha \subset \mathbb{Z}^2$ to consist of those $(m,n) \in \mathbb{Z}^2$ such that $m\sqrt{3}+n\sqrt{2}$ has distance at most α from the integers. This set has uniform density 2α since the image of $m \mapsto m\xi \mod 1$ is uniformly dense in $\mathbb{T}$ whenever $\xi \notin \mathbb{Q}$. That $\sqrt{2}$ and $\sqrt{3}$ are $\mathbb{Q}$-linearly independent plays an essential role in showing that Λ_α is universal.

Spectral Measures and Decompositions of Frames

A Borel measure μ on $\mathbb{R}$ is called a *spectral measure* if there is a discrete set $\Lambda \subset \mathbb{R}$ such that $\mathscr{E}(\Lambda)$ is an orthonormal basis for $L^2(\mu)$. One can ask which sequences Λ generate an ONB or, more generally, a frame of exponentials for $L^2(\mu)$, just as Paley and Wiener asked in the case of $L^2([a,b], \mathrm{d}x)$. Characterizing which finite measures are spectral measures and characterizing the Fourier frames for a particular measure are both very difficult problems. There are partial results in special cases. For example, Dutkay, Han, Jorgensen, Sun, and Weber [91, 92] showed that if μ is the invariant measure of an affine iterated function system (with no overlap) and if $\mathscr{E}(\Lambda)$ is a frame for $L^2(\mu)$ then the upper Beurling dimension of Λ is equal to the Hausdorff dimension of the support of μ. See also Gabardo [106] for more general results related to spectral measures.

The problem of relating exponential frames for $L^2(\mu)$ with exponential Riesz bases for $L^2(\mu)$ itself is related to *Feichtinger's frame conjecture*, which states that any bounded frame can be written as a finite union of Riesz basic sequences. Feichtinger's conjecture, in turn, has been shown by Casazza and various collaborators to be equivalent to a number of fundamental open problems in operator theory and harmonic analysis [65–67], and has remained open as of 2011, even in the case of harmonic series $\mathrm{e}^{2\pi int}$ restricted to subsets of $[0,1]$ having Lebesgue measure less than one. Very partial progress has been made on this front as well. For example, in the discrete–periodic setting, Speegle [313] proved that if $\Lambda \subset \mathbb{R}$ is a discrete sequence ($\lambda_n < \lambda_{n+1}$) then the restricted exponentials $\mathscr{E}(\Sigma,\Lambda) = \{\mathrm{e}^{2\pi i\lambda_n t}\mathbb{1}_\Sigma(t) : \lambda_n \in \Lambda\}$ ($t \in \mathbb{T}$) can be uniformly divided into Riesz sequences for $L^2(\Sigma)$ in the sense that the families $\mathscr{E}(\Sigma,\Lambda_j)$, $\Lambda_j = \{\lambda_{nJ+j} : n \in \mathbb{Z}\}$, $j = 0,1,\dots,J-1$ each satisfy $\sum_n |a_n|^2 \le C\int_\Sigma \big|\sum a_n \mathrm{e}^{2\pi i\lambda_{nJ+j}t}\big|^2\,\mathrm{d}t$, provided a certain relationship holds between the *essential* α-Hausdorff measure $H_\alpha(\Sigma)$ of Σ and the *upper Beurling dimension* of Λ. Here, $H_\alpha(\Sigma) = \inf\{\sum |I_\nu|^\alpha\}$ in which the infimum is taken over all unions of open intervals I_ν that cover Σ up to a set of measure zero, and the upper Beurling dimension is defined as

$$\dim^+(\Lambda) = \sup\Big\{\beta : \limsup_{R\to\infty}\,\sup_x \frac{n_\Lambda(x+R)-n_\Lambda(x)}{R^\beta} > 0\Big\}.$$

Speegle's sufficient condition for a uniform partitioning of $\mathscr{E}(\Sigma,\Lambda)$ into Riesz sequences is that $H_\alpha(\Sigma) < 1$ and $\dim^+(\Lambda) < 1-\alpha$. Loosely stated, Λ is not too dense and Σ is not too disconnected.

5.4.4 Further Miscellaneous Aspects of Sampling

Sampling theory has become a vast forest full of exotic species. Here we offer a sparse sampling of a few directions that are tangential to the fundamental theory of sampling of band-limited and multiband signals outlined in this chapter.

The Paley–Wiener space PW has a number of abstractions amenable to sampling theory. One such abstraction is the notion of a finitely generated shift-invariant subspace of $L^2(\mathbb{R})$. Fundamental results regarding sampling in such spaces can be found in Aldroubi and Gröchenig [4] and Vaidyanathan et al., e.g., [82, 329], cf. also [152] and Šikić and Wilson [298] for further generalizations. Shift-invariant spaces, in turn, are often examples of *reproducing kernel Hilbert spaces* (RKHSs). The surveys of Nashed, Han, and Sun [133, 243] outline the state of knowledge on this front as of 2010. Another related direction concerns sampling in unions of spaces. See Lu and Do [210] for a survey of work in this direction. Potentially very general approaches to sampling can be found in the work of Marziliano, Vetterli, and Blu [337].

Other major directions in sampling theory include sampling of generalized functions, e.g., Pfaffelhuber [265], interpolation in spline spaces and spaces of entire functions, e.g., Madych and Meijering [217, 230], and sampling in other Hilbert spaces of band-limited functions, e.g., Gabardo [107]. Reconstructions from fully irregular samples were considered by Qiang [272], and from unknown sample locations by Vetterli and Marziliano [225]. Connections with prediction theory were studied by Splettstösser [314] and Mugler and Wu [241].

One major aspect of sampling in practice that has not been addressed at all in this chapter is the problem of approximate recovery of a signal from quantized data, including PCM and sigma–delta methods, e.g., [129, 130]. Comparisons of accuracy in terms of algorithmic complexity can be found in Benedetto and Oktay [32]. Connections with interleaved sampling methods can be found in Powell et al. [266].

Landau's density conditions for sampling and interpolation and some other aspects of sampling of band-limited functions that were discussed in this chapter have analogues in the setting of time–frequency distributions, e.g., [193]. A considerable literature has been developed along these lines. Heil's review [143] provides an accessible survey of this topic. Aspects specifically addressing time–frequency concentrations of orthonormal sequences can be found in the work of Jaming and Powell [164].

Chapter 6
Time-localized Sampling Approximations

In this last chapter we explore briefly some connections among sampling and time and band limiting. The chapter begins by pointing out a general connection between the samples of eigenfunctions of time and band limiting and the eigenvectors of a certain matrix whose entries are, in essence, the samples of time-localized images of functions that interpolate the samples in the given Paley–Wiener space. Next, a *discrete* method is considered for generating eigenfunctions of time–frequency localizations to unions of sets from their separate localizations. We then reconsider the connection between eigenfunctions and their samples in the concrete context of localization to intervals of the real line, outlining work of Walter and Shen [347] and of Khare and George [177]. Walter and Shen provided L^2-estimates for approximate prolate spheroidal wave functions (PSWFs) constructed from interpolation of their sample values within the time-localization interval. We provide a partial sharpening of their estimates by using a slightly enlarged set of samples.

6.1 Localization and Sampling on $\mathbb{R}$: General Properties

6.1.1 Time Localization and Sampling in RKHSs

Let $\mathscr{V} = \mathscr{V}_K \subset L^2(\mathbb{R})$ be a reproducing kernel Hilbert space (RKHS) with kernel $K(x,y)$. Denote by $P_K : L^2(\mathbb{R}) \to \mathscr{V}$ the projection of $L^2(\mathbb{R})$ onto $\mathscr{V}$ defined by

$$P_K f(t) = \int_{-\infty}^{\infty} K(t,s)\, f(s)\,\mathrm{d}s\,,$$

and let Q_S be the time-localization operator, $Q_S f(t) = f(t)\mathbb{1}_S(t)$. Given a sampling set and a complete set of interpolating functions for $\mathscr{V}$, one can express eigenfunctions for time and band limiting in terms of their samples. We outline this relationship somewhat formally. Suppose that $\{t_k\}$, $k \in \mathbb{Z}$, is a discrete subset of $\mathbb{R}$ that also happens to be a set of stable sampling for $\mathscr{V}$, that is, $\|f\|_2^2 \leq C\sum |f(t_k)|^2$ for $f \in \mathscr{V}$

and some constant C. In addition, suppose that $\{S_k\}_{k\in\mathbb{Z}}$ is a sequence of functions in $\mathscr{V}$ that interpolates $\mathscr{V}$ in the sense that, for every $f \in \mathscr{V}$,

$$f(t) = \sum_{k=-\infty}^{\infty} f(t_k) S_k(t) .$$

Suppose now that the composition $P_K Q_S$ is a compact, self-adjoint operator. Define a matrix

$$A_{mk} = \int_S K(t_m, s) S_k(s)\, \mathrm{d}s .$$

Theorem 6.1.1. *If φ is a λ-eigenfunction of $P_K Q_S$ then $\{\varphi(t_k)\}$ is a λ-eigenvector of A. Conversely, if $\{v_k\}$ is a λ-eigenvector of A and $\varphi(t) = \sum_k v_k S_k(t)$ converges in $\mathscr{V}$, then φ is a λ-eigenfunction of $P_K Q_S$.*

Proof. Let φ be a λ-eigenfunction of $P_K Q_S$. Then

$$\begin{aligned} \lambda \varphi(t_m) = \int_S K(t_m, s)\varphi(s)\, \mathrm{d}s &= \int_S K(t_m, s) \sum_k \varphi(t_k) S_k(s)\, \mathrm{d}s \\ &= \sum_k \varphi(t_k) \int_S K(t_m, s) S_k(s)\, \mathrm{d}s = \sum_k A_{mk}\varphi(t_k) \end{aligned}$$

so $\{\varphi(t_k)\}$ is a λ-eigenvector of A.

Conversely, if $\{v_k\}$ is a λ-eigenvector of A and if $\varphi(t) = \sum_k v_k S_k(t)$ converges in $\mathscr{V}$ then

$$\begin{aligned} (P_K Q_S \varphi)(t) &= \sum_k (P_K Q_S \varphi)(t_k)\, S_k(t) \\ &= \sum_k \Big(P_K Q_S \Big(\sum_\ell v_\ell S_\ell\Big)\Big)(t_k)\, S_k(t) \\ &= \sum_k \sum_\ell v_\ell \Big(\int_S K(t_k, s) S_\ell(s)\, \mathrm{d}s\Big)\, S_k(t) \\ &= \sum_k \Big(\sum_\ell A_{k\ell} v_\ell\Big)\, S_k(t) = \lambda \sum_k v_k\, S_k(t) = \lambda \varphi(t) \qquad \square \end{aligned}$$

This formal result says nothing about how to obtain interpolating functions $S_k \in \mathscr{V}$ and sampling sets $\{t_k\}$. In particular cases, one can find recipes for the sample set and interpolating function pairs. For example, in a principal shift-invariant space $V(\psi) = \overline{\mathrm{span}}\{\psi(\cdot - k) : k \in \mathbb{Z}\}$, one can often obtain interpolating functions for integer samples through inversion of the Zak transform; see, e.g., [153].

6.1.2 Time–Frequency Localization for Unions

Sums of Time–Frequency Localization Operators

Except in the special case in which Σ and S are intervals, there is not a Sturm–Liouville system to produce recursion formulas from which the eigenfunctions of $P_\Sigma Q_S$ can be computed. Effective computational algorithms for estimating the eigenfunctions, for example, when Σ is a finite union of intervals, must be obtained by some other route. Suppose that one knows the eigenvalues and eigenfunctions for $P_{\Sigma_1} Q_S$ and $P_{\Sigma_2} Q_S$ for the same time support S and compact, pairwise disjoint frequency support sets Σ_1 and Σ_2. In principle, one can compute $P_{\Sigma_1\cup\Sigma_2} Q_S$ through the *correlation matrix* of the eigenfunctions of the separate operators, as we outline now (cf. [154]).

Let Σ_1, Σ_2 be two compact, disjoint frequency support sets, and let $\{\varphi_n^{\Sigma_1}\}$, $\{\varphi_n^{\Sigma_2}\}$ be complete sets of normalized eigenfunctions for $P_{\Sigma_1} Q_S$ and $P_{\Sigma_2} Q_S$ respectively. Let ψ be an eigenfunction of $P_{\Sigma_1\cup\Sigma_2} Q_S$ with eigenvalue λ. Write

$$\psi = \psi_{\Sigma_1} + \psi_{\Sigma_2} = \sum_{n=0}^{\infty} (\alpha_n \varphi_n^{\Sigma_1} + \beta_n \varphi_n^{\Sigma_2})$$

where $P_{\Sigma_1} Q_S \varphi_n^{\Sigma_1} = \lambda_n^{\Sigma_1} \varphi_n^{\Sigma_1}$ and $P_{\Sigma_2} Q_S \varphi_n^{\Sigma_2} = \lambda_n^{\Sigma_2} \varphi_n^{\Sigma_2}$. We make no assumption that $\lambda_n^{\Sigma_1}$ and $\lambda_n^{\Sigma_2}$ are related. With $\psi_\Sigma = P_\Sigma \psi$ we have

$$\begin{aligned}\lambda\psi &= (P_{\Sigma_1} + P_{\Sigma_2}) Q_S \psi_{\Sigma_1} + (P_{\Sigma_1} + P_{\Sigma_2}) Q_S \psi_{\Sigma_2} \\ &= \sum_{n=0}^{\infty} (\alpha_n \lambda_n^{\Sigma_1} \varphi_n^{\Sigma_1} + \beta_n \lambda_n^{\Sigma_2} \varphi_n^{\Sigma_2} + \alpha_n P_{\Sigma_2} Q_S \varphi_n^{\Sigma_1} + \beta_n P_{\Sigma_1} Q_S \varphi_n^{\Sigma_2})\,.\end{aligned}$$

Now use the self-adjointness of P_{Σ_2} to define a *transition matrix* $\Gamma = \{\gamma_{nm}\}$ by

$$P_{\Sigma_2} Q_S \varphi_n^{\Sigma_1} = \sum_{m=0}^{\infty} \langle Q_S \varphi_n^{\Sigma_1}, \varphi_m^{\Sigma_2} \rangle \varphi_m^{\Sigma_2} = \sum_{m=0}^{\infty} \gamma_{nm} \varphi_m^{\Sigma_2}\,.$$

By the self-adjointness of P_{Σ_1} and Q_S, one also has

$$P_{\Sigma_1} Q_S \varphi_n^{\Sigma_2} = \sum_{m=0}^{\infty} \langle Q_S \varphi_n^{\Sigma_2}, \varphi_m^{\Sigma_1} \rangle \varphi_m^{\Sigma_1} = \sum_{m=0}^{\infty} \bar{\gamma}_{mn} \varphi_m^{\Sigma_1}.$$

Since Σ_1, Σ_2 are disjoint, $\psi = P_{\Sigma_1}\psi + P_{\Sigma_2}\psi$ with

$$\begin{aligned}\lambda P_{\Sigma_1}\psi &= \sum_{n=0}^{\infty} \alpha_n \lambda_n^{\Sigma_1} \varphi_n^{\Sigma_1} + \sum_{n=0}^{\infty} \beta_n \sum_{m=0}^{\infty} \bar{\gamma}_{mn} \varphi_m^{\Sigma_1} \quad \text{and} \\ \lambda P_{\Sigma_2}\psi &= \sum_{n=0}^{\infty} \beta_n \lambda_n^{\Sigma_2} \varphi_n^{\Sigma_2} + \sum_{n=0}^{\infty} \alpha_n \sum_{m=0}^{\infty} \gamma_{nm} \varphi_m^{\Sigma_2}\end{aligned}$$

which gives the joint *discrete* eigenvector problem

$$\lambda\alpha_n = \alpha_n\lambda_n^{\Sigma_1} + \sum_{m=0}^{\infty}\bar{\gamma}_{nm}\beta_m; \qquad \lambda\beta_n = \beta_n\lambda_n^{\Sigma_2} + \sum_{m=0}^{\infty}\gamma_{mn}\alpha_m,$$

which can be summarized as follows.

Proposition 6.1.2. *Suppose that* Σ_1 *and* Σ_2 *are disjoint, compact sets and that the eigenvectors* $\{\varphi_n^{\Sigma_i}\}$ *of* $P_{\Sigma_i}Q_S P_{\Sigma_i}$ *have corresponding nondegenerate eigenvalues listed in decreasing order as* $\lambda_n^{\Sigma_i}$, $i=1,2$. *Let* Λ_{Σ_i} *denote the diagonal matrix with nth diagonal entry* $\lambda_n^{\Sigma_i}$ *and let* Γ *be the matrix with entries* $\gamma_{nm} = \langle Q_S\varphi_n^{\Sigma_1}, \varphi_m^{\Sigma_2}\rangle$. *Then any eigenvector–eigenvalue pair* ψ *and* λ *for* $P_{\Sigma_1\cup\Sigma_2}Q_S$ *can be expressed as* $\psi = \sum_{n=0}^{\infty}(\alpha_n\varphi_n^{\Sigma_1} + \beta_n\varphi_n^{\Sigma_2})$ *where the vectors* $\alpha = \{\alpha_n\}$ *and* $\beta = \{\beta_n\}$ *together form a discrete eigenvector for the block matrix eigenvalue problem*

$$\lambda\begin{pmatrix}\alpha\\ \beta\end{pmatrix} = \begin{pmatrix}\Lambda_{\Sigma_1} & \bar{\Gamma}\\ \Gamma^T & \Lambda_{\Sigma_2}\end{pmatrix}\begin{pmatrix}\alpha\\ \beta\end{pmatrix}.$$

Discussion

Proposition 6.1.2 says that the eigenvectors (and eigenvalues) of $P_{\Sigma_1\cup\Sigma_2}Q_S$ can be expressed in terms of the discrete eigenvectors (and eigenvalues) of the block matrix in the proposition. The matrix Γ has a particularly simple form when $\Sigma_1 = I$ and $\Sigma_2 = J$ are two intervals of the same length, normalized to $\Omega = 1$. Then any function in PW_J has the form $E\varphi$ for some $\varphi \in \mathrm{PW}_I$, where E denotes the operator of multiplication by $\mathrm{e}^{2\pi\mathrm{i}(\xi_J-\xi_I)t}$ with ξ_I the center of I. If $S=[-T,T]$ then $\gamma_{nm} = \int_{-T}^{T}\mathrm{e}^{2\pi\mathrm{i}(\xi_I+\xi_J)t}\varphi_n(t)\varphi_m(t)\,\mathrm{d}t$, where φ_n is the nth PSWF, that is, the nth eigenfunction of PQ_T. Then the $2\Omega T$ theorem (Theorem 1.3.1) suggests that the significant eigenvalues and corresponding eigenvectors of $P_{I\cup J}Q_T$ can be estimated by truncating the blocks of the matrix in the theorem to blocks of order $2T\times 2T$. The correlations γ_{nm} have to be computed either by quadrature or by more sophisticated methods; see, e.g., [35, 364].

6.2 Sampling Formulas for Prolates

Since the prolate spheroidal eigenfunctions of PQ_T form an orthonormal basis for PW, and since the sinc function provides a reproducing kernel for PW, one has the following special case of Mercer's theorem:

$$\mathrm{sinc}\,(t-s) = \sum_{n=0}^{\infty}\langle \mathrm{sinc}\,(\cdot - s), \varphi_n(\cdot)\rangle\varphi_n(t) = \sum_{n=0}^{\infty}\varphi_n(s)\varphi_n(t).$$

Evaluating at $s = k\in\mathbb{Z}$ yields

$$\mathrm{sinc}\,(t-k) = \sum_{n=0}^{\infty} \varphi_n(k)\varphi_n(t)\,.$$

Walter and Shen [347] observed that this identity can be coupled with the Shannon sampling theorem to provide the following proposition, to which we will refer as the *first Walter–Shen sampling formula.*

Proposition 6.2.1. *Fix $T > 0$ and let $\{\varphi_n\}$ denote the $2T$-concentrated PSWFs frequency-supported in $[-1/2, 1/2]$. For any $f \in \mathrm{PW}$, one has*

$$f(t) = \sum_{n=0}^{\infty} \sum_{k=-\infty}^{\infty} f(k)\,\varphi_n(k)\,\varphi_n(t)\,.$$

The formula depends implicitly on the duration parameter T defining the PSWFs, but the eigenvalues of PQ_T do not appear. Since the PSWFs form a complete orthonormal basis for PW, the proposition verifies directly that

$$\langle f,\, \varphi_n\rangle = \sum_{k=-\infty}^{\infty} f(k)\,\varphi_n(k) = \big\langle \{f(k)\},\, \{\varphi_n(k)\}\big\rangle,$$

where the inner product on the left is in $L^2(\mathbb{R})$ and the one on the right is in $\ell^2(\mathbb{Z})$. The identity also follows from Shannon's sampling theorem. It is natural to ask whether some variation of the first Walter–Shen sampling formula can lead to an efficient approximation of a nearly time- and band-limited signal f in terms of its integer samples in or close to the time concentration interval. In particular, if $f \in \mathrm{PW}$ is concentrated in $[-T, T]$, how is this concentration reflected in the samples $\varphi_n(k)$ on the one hand and, if $f \in \mathrm{span}\{\varphi_0, \ldots, \varphi_N\}$, how is this concentration reflected in the samples of f, on the other?

Considered in terms of approximations, the first Walter–Shen sampling formula suggests two ways in which $f \in \mathrm{PW}$ can be essentially concentrated in $[-T, T]$. The simplest requires that f approximately lies in $\mathrm{PSWF}_N = \mathrm{span}\{\varphi_0, \ldots, \varphi_N\}$. That is, $\sum_{n\geq N} |\langle f,\, \varphi_n\rangle|^2$ is small. A second possibility is that the samples $f(k)$ decay quickly away from $[-T, T]$. Specifically, $\sum_{|k|>M} |f(k)|^2$ is small. Here, M and N depend on T. It is natural to ask for a sense in which these two conditions are effectively the same. Answering this question requires bounding the quantities $\sum_{n\geq N} |\langle f,\, \varphi_n\rangle|^2$ and $\sum_{|k|>M} |f(k)|^2$ in terms of one another for specific values of N and M, when $f \in \mathrm{PW}$ is assumed either nearly to belong to PSWF_N or to have rapidly decaying samples.

6.2.1 Samples of PSWFs

To obtain such estimates we need to take into account the structure of the eigenvalues of the PSWFs, starting with a second look at the first Walter–Shen sampling formula, that is, the orthogonality of the sample sequences of the PSWFs. As above, we normalize the PSWFs $\{\varphi_0, \varphi_1, \ldots\}$ to be eigenfunctions of PQ_T. Since the PSWFs belong to the space PW, they have convergent sampling expansions

$$\varphi_n(t) = \sum_k \varphi_n(k)\,\mathrm{sinc}\,(t-k)\,. \tag{6.1}$$

Orthogonality of the sample sequences $\{\varphi_n(k)\}$ then follows from that of the PSWFs on $\mathbb{R}$ as follows:

$$\begin{aligned}
\delta_{n,m} = \langle \varphi_n, \varphi_m\rangle &= \int \Big(\sum_k \varphi_n(k)\,\mathrm{sinc}\,(t-k)\Big)\Big(\sum_\ell \varphi_m(\ell)\,\mathrm{sinc}\,(t-\ell)\Big)\,\mathrm{d}t \\
&= \sum_k\sum_\ell \varphi_n(k)\varphi_m(\ell)\int \mathrm{sinc}\,(t-k)\mathrm{sinc}\,(t-\ell)\,\mathrm{d}t \\
&= \sum_k\sum_\ell \varphi_n(k)\varphi_m(\ell)\,\delta_{k\ell} = \sum_k \varphi_n(k)\varphi_m(k)\,.
\end{aligned} \tag{6.2}$$

As Proposition 6.2.1 shows, the completeness of the sequences $\{\varphi_n(k)\}$ in $\ell^2(\mathbb{Z})$ also follows from that of the PSWFs in PW.

Since the PSWFs are orthogonal on $[-T,T]$, one also has

$$\lambda_n\delta_{nm} = \langle Q_T\varphi_n, \varphi_m\rangle = \sum_k\sum_\ell \varphi_n(k)\,\varphi_m(\ell)\int_{-T}^{T} \mathrm{sinc}\,(t-k)\,\mathrm{sinc}\,(t-\ell)\,\mathrm{d}t\,.$$

Define the matrix $A:\mathbb{Z}\times\mathbb{Z}\to\mathbb{R}$ by

$$A_{k\ell} = \int_{-T}^{T} \mathrm{sinc}\,(t-k)\,\mathrm{sinc}\,(t-\ell)\,\mathrm{d}t\,. \tag{6.3}$$

Then we have

$$\lambda_n\delta_{nm} = \sum_k \varphi_n(k)\sum_\ell A_{k\ell}\,\varphi_m(\ell). \tag{6.4}$$

Since the sample sequences of the PSWFs form a complete orthonormal basis for $\ell^2(\mathbb{Z})$, this implies that these sequences are the eigenvectors of A. In this special case, Theorem 6.1.1 reads as follows.

Proposition 6.2.2. *The matrix A defined in (6.3) has the same eigenvalues as PQ_T. Additionally, the eigenvector for λ_n is the sample sequence $\{\varphi_n(k)\}$ of the eigenfunction φ_n of PQ_T.*

An immediate corollary is the *second Walter–Shen sampling formula.*

Corollary 6.2.3. *If $f\in$ PW, then*

$$PQ_Tf(t) = \sum_{n=0}^{\infty}\lambda_n\sum_{k=-\infty}^{\infty} f(k)\,\varphi_n(k)\,\varphi_n(t)\,.$$

6.2.2 Quadratic Decay for PSWF Samples

Walter and Shen showed in [347] that the integer samples of the nth eigenfunction of PQ_T satisfy the quadratic decay estimate $\sum_{|k|>T}(\varphi_n(k))^2 \leq CT\sqrt{1-\lambda_n}$. They compared $\sum_{|k|>T}(\varphi(k))^2$ with $\int_{|t|>T}(\varphi(t))^2\,\mathrm{d}t$ in terms of the mean value theorem, a comparison that does not take full advantage of the real analyticity of φ. Here we provide an alternative approach. In what follows, let $\psi \in \mathcal{S}(\mathbb{R})$ be fixed such that $\widehat{\psi}$ is a smooth bump function equal to one on $[-1/2,1/2]$ and vanishing outside $[-1,1]$. Then any $\varphi \in \mathrm{PW}$ satisfies $\varphi = \varphi * \psi$. This approach yields a remainder estimate of the form $\sum_{|k|>M(T)}(\varphi_n(k))^2 \leq C(1-\lambda_n)$ that is linear in $(1-\lambda)$. The price paid for the increase in precision is that the *remainder tail* now consists of terms $|k|>M$ where $M = M(n,T) > T$.

Since any $\varphi \in \mathrm{PW}$ can be written $\varphi = Q_T\varphi + \widetilde{Q}_T\varphi$ with $\widetilde{Q}_T = I - Q_T$, one can express $\varphi \in \mathrm{PW}$ as $\varphi = \psi * (Q_T\varphi) + \psi * (\widetilde{Q}_T\varphi)$. Separate quadratic sum estimates are obtained for each of these terms.

Lemma 6.2.4. *Let φ_n be the nth eigenfunction of PQ_T. Then there is a constant C independent of n such that for ψ as above,*

$$\sum(\psi * (\widetilde{Q}_T\varphi_n)(k))^2 \leq C(1-\lambda_n).$$

To prove this we need the following basic facts; see, e.g., [316] or [315], cf. [115].

Lemma 6.2.5. *(i) The Hardy–Littlewood maximal function*

$$Mf(t) = \sup_{r>0} \frac{1}{2r}\int_{t-r}^{t+r} |f(s)|\,\mathrm{d}s$$

is L^2-bounded: for any $f \in L^2(\mathbb{R})$ one has $\|Mf\| \leq (1+\sqrt{2})\|f\|$.
(ii) If Ψ is a bounded, radially decreasing integrable function on $\mathbb{R}$ then there is a constant C_Ψ such that the pointwise estimate $|f * \Psi(t)| \leq C_\Psi Mf(t)$ applies for all t, whenever f is locally integrable.

Proof (of Lemma 6.2.4). Define $\varphi^{\#}(t) = \sup_{|t-s|\leq 1}|\varphi(s)|$. Then

$$\begin{aligned}
(\psi * (\widetilde{Q}_T\varphi))^{\#}(t) &= \sup_{|t-s|\leq 1}\left|\int \psi(s-u)(\widetilde{Q}_T\varphi)(u)\,\mathrm{d}u\right| \\
&\leq \int\Big(\sup_{|t-s|\leq 1}\big|\psi(s-u)\big|\Big)\big|(\widetilde{Q}_T\varphi)(u)\big|\,\mathrm{d}u \\
&= \int \psi^{\#}(t-u)\big|(\widetilde{Q}_T\varphi)(u)\big|\,\mathrm{d}u = |(\widetilde{Q}_T\varphi)| * \psi^{\#}(t)
\end{aligned}$$

because $\sup_{|t-s|\leq 1}\big|\psi(s-u)\big| = \sup_{|t-u-(s-u)|\leq 1}\big|\psi(s-u)\big|$. Thus,

$$\sum((\psi*(\widetilde{Q}_T\varphi))(k))^2 \le \sum_k \int_{k-1/2}^{k+1/2} ((\psi*(\widetilde{Q}_T\varphi))^{\#}(t))^2\,\mathrm{d}t$$
$$= \int ((\psi*(\widetilde{Q}_T\varphi))^{\#}(t))^2\,\mathrm{d}t \le \int (|(\widetilde{Q}_T\varphi)|*\psi^{\#}(t))^2\,\mathrm{d}t$$
$$\le C_\psi \int ((M(\widetilde{Q}_T\varphi))(t))^2\,\mathrm{d}t \le C\|(\widetilde{Q}_T\varphi)\|^2,$$

where the penultimate inequality follows from Lemma 6.2.5 using the fact that $\psi^{\#}$ is rapidly decreasing, so that $\Psi(t) = \sup_{|s|\ge t}\psi^{\#}(s)$ is a radially decreasing and integrable majorant of ψ. Specializing to $\varphi = \varphi_n$ gives $\|(\widetilde{Q}_T\varphi)\|^2 = 1-\lambda_n$ and the lemma follows. □

One would like to obtain a similar bound, $\sum_{|k|>M}((\psi*(Q_T\varphi_n))(k))^2 \le C(1-\lambda_n)$ for a suitable integer $M = M(n,T)$. For now we take as a working choice of $M(n,T)$ the smallest integer M that yields

$$\int_{|t|>M(n,T)-T-1/2} \psi^{\#}(t)\,\mathrm{d}t \le \sqrt{\frac{1-\lambda_n}{\lambda_n}}\,. \tag{6.5}$$

For this M, proceeding as in the proof of Lemma 6.2.4, we obtain

$$\sum_{|k|>M} (\psi*(Q_T\varphi_n)(k))^2 \le \sum_{|k|>M}\int_{k-1/2}^{k+1/2} ((\psi*(Q_T\varphi_n))^{\#}(t))^2\,\mathrm{d}t$$
$$= \int_{|t|>M+1/2} ((\psi*(Q_T\varphi_n))^{\#}(t))^2\,\mathrm{d}t \le \int_{|t|>M+1/2} (|(Q_T\varphi_n)|*\psi^{\#}(t))^2\,\mathrm{d}t\,.$$

If $|t| > M+1/2$ and $|s| < T$ then $|t-s| \ge |t|-|s| \ge M-T+1/2$. Using (6.5) and the fact that $\|Q_T\varphi_n\|^2 = \lambda_n$, we obtain

$$\int_{|t|>M+1/2} (|Q_T\varphi_n|*\psi^{\#}(t))^2\,\mathrm{d}t \le \||Q_T\varphi_n|*(\widetilde{Q}_{M+1/2-T}\,\psi^{\#})\|^2$$
$$\le \|Q_T\varphi_n\|^2\|\widetilde{Q}_{M+1/2-T}\,\psi^{\#}\|_{L^1}^2 \le 1-\lambda_n\,.$$

Thus we have proved the following.

Lemma 6.2.6. *Let φ_n be the nth eigenfunction of PQ_T and let ψ and $M(n,T)$ satisfy (6.5). Then*

$$\sum_{|k|>M(n,T)} ((\psi*(Q_T\varphi_n))(k))^2 \le (1-\lambda_n)\,.$$

Lemmas 6.2.4 and 6.2.6 yield the following proposition.

Proposition 6.2.7. *Let φ_n be the nth eigenfunction of PQ_T and let ψ and $M(n,T)$ satisfy (6.5). Then there is a $C > 0$ independent of n such that*

$$\sum_{|k|>M(n,T)} (\varphi_n(k))^2 \le C(1-\lambda_n)\,.$$

Discussion: Estimating $M(n,T)$

Here we offer a reasonably concise ballpark estimate of a minimum $M(n,T)$ such that (6.5) is satisfied. The estimate depends on the decay of $1-\lambda_n(T)$ as well as on the decay of the Fourier bump function ψ. The rate at which $\lambda_n(c)\to 1$ as $c\to\infty$ was established by Fuchs in 1964, see Theorem 1.2.10; cf. also Slepian [304]. However, those results only govern asymptotic decay of $1-\lambda_n$ as $n/T\to 0$. More recent work quantitatively describes λ_n for $n\approx T$ large, e.g., [280, Theorem 9], but in less concise terms. As such, we will provide a theoretical estimate for $M(n,T)$ that is, perhaps, less than ideal, but is not too difficult to justify. The estimate is based on Theorem 1.3.1, stating that the number of eigenvalues of PQ_{2T} larger than α is

$$N(\alpha,2T)=2T+\frac{1}{\pi^2}\log\Big(\frac{1-\alpha}{\alpha}\Big)\log(2T)+O(\log T).$$

For $\alpha=\lambda_n$ and, thus, $n=N(\lambda_n,2T)$, this implies that

$$\Big|n-2T-\frac{1}{\pi^2}\log\Big(\frac{1-\lambda_n}{\lambda_n}\Big)\log(2T)\Big|\leq C(\log T),$$

so that for $n<2T$ or, equivalently, $\lambda_n>1/2$,

$$\frac{\pi^2(2T-n)}{\log T}+\log\Big(\frac{1-\lambda_n}{\lambda_n}\Big)\leq C \text{ or } \frac{1-\lambda_n}{\lambda_n}\mathrm{e}^{\pi^2(2T-n)/\log(T)}\leq \mathrm{e}^C.$$

In other words, $\lambda_n\to 1$ *subexponentially* as $T\to\infty$.

By the construction in Proposition 6.3.4 in Sect. 6.3.2 of the notes at the end of this chapter, one can assume that the Fourier bump function ψ satisfies

$$|\psi(t)|\leq \mathrm{e}^{-|t|/\log^{\gamma}|t|},\quad (t \text{ large})$$

where $\gamma>1$ is fixed. By a suitable change of variable and integration by parts one then obtains the estimate

$$\int_{|t|>M}|\psi^{\#}(t)|\,\mathrm{d}t\leq C\log^{\gamma}(M)\,\mathrm{e}^{-M/\log^{\gamma}(M)}$$

for large M. Therefore, as in (6.5), with $Q=M(n,T)-T-1$ we want

$$\int_{|t|>Q}|\psi(t)|\,\mathrm{d}t\leq C\log^{\gamma}(Q)\mathrm{e}^{-Q/\log^{\gamma}(Q)}\leq\sqrt{\frac{1-\lambda_n}{\lambda_n}}\leq A\,\mathrm{e}^{\frac{\pi^2(n-2T)}{2\log(T)}}$$

for a constant A, or

$$\frac{Q}{\log^{\gamma}(Q)}-\gamma\log\log Q\geq\frac{\pi^2(2T-n)}{2\log T}-C.$$

When $\gamma > 1$ this inequality will be satisfied for large T if $Q = \pi^2(T - n/2)\log^\eta(T)$ for any $\eta > \gamma$. Taking

$$M(n,T) = \Big(\frac{\pi^2}{2}(2T-n)+T\Big)\Big(1+\log^\gamma T\Big) \tag{6.6}$$

for sufficiently large T then guarantees the conclusion of Proposition 6.2.7. Set

$$M(T) = M(0,T) = (\pi^2+1)T(1+\log^\gamma T). \tag{6.7}$$

Corollary 6.2.8. *Let* $0 \leq N < 2T$ *and let* $M(T) = (\pi^2+1)(1+\log^\gamma(T))T$ *for some* $\gamma > 1$. *Then there is a* $C > 0$ *such that, for any* $n \leq N$, *one has*

$$\sum_{|k|>M(T)} \varphi_n^2(k) \leq C(1-\lambda_n).$$

6.2.3 Approximate Time-localized Projections

In this section we consider the problem of approximating the projection of $f \in \mathrm{PW}$ onto PSWF_N using a collection of samples near $[-T,T]$. As above, φ_n will be the nth PSWF eigenfunction of PQ_T. We will let

$$f_{N,K} = \sum_{|k|\leq K} f(k) \sum_{n=0}^{N} \varphi_n(k)\,\varphi_n(t) \quad \text{and} \quad f_N = \lim_{K\to\infty} f_{N,K}. \tag{6.8}$$

Proposition 6.2.7 suggests that an approximation by $f_{N,M(T)}$ can provide an estimate of f whose squared error decays like $(1-\lambda_N)$. Since $\|f\|^2 = \sum (f(k))^2$,

$$|\langle (f_N - f_{N,M(T)}),\, \varphi_n\rangle|^2 \leq \sum_{n=0}^{N} |\langle f,\, \varphi_n\rangle|^2 \sum_{|k|>M(T)} (\varphi_n(k))^2 \leq C\|f\|^2(1-\lambda_N). \tag{6.9}$$

Using orthogonality of the φ_n over $[-T,T]$ one then has the following.

Proposition 6.2.9. *If* $f \in \mathrm{PW}$ *then* $f_N \in \mathrm{PSWF}_N$ *is the orthogonal projection of* f *onto the span of* $\{\varphi_0, \dots, \varphi_N\}$ *and, with* $M(T)$ *as in (6.7),*

$$\|Q_T(f_N - f_{N,M(T)})\|^2 \leq \sum_{n=0}^{N} \lambda_n |\langle (f_N - f_{N,M(T)}),\, \varphi_n\rangle|^2 \leq C\|f\|^2 \sum_{n=0}^{N} \lambda_n(1-\lambda_n).$$

Proof. The orthogonality of the functions φ_n over $[-T,T]$ implies that

$$\|Q_T(f_N - f_{N,M(T)})\|^2 = \sum_{n=0}^{N} \lambda_n |\langle (f_N - f_{N,M(T)}),\, \varphi_n\rangle|^2,$$

and the proposition then follows from (6.9). □

Slepian and Pollak observed that $\|Q_T(f-f_N)\|^2=\sum_{n=N+1}^{\infty}\lambda_n|\langle f,\varphi_n\rangle|^2$ (see [309, (17), p. 52]). Combining this with (6.9) and recalling that $\lambda_n<1/2$ if $n>2T+1$, the triangle inequality yields the following.

Corollary 6.2.10. *If $f\in \mathrm{PW}$ then for $T>0$ fixed,*

$$\|Q_T(f-f_{N,M(T)})\|^2\le C\sum_{n=0}^{\infty}|\langle f,\varphi_n\rangle|^2\lambda_n(1-\lambda_n)\,.$$

6.2.4 Accurate Estimates of Integer Samples of PSWFs

Legendre Coefficients of PSWFs, Revisited

In this section we consider approximations of PSWFs by means of partial sums of Legendre series. We follow the presentation in [364]. Suitably normalized PSWFs are eigenfunctions of the operator

$$F_{\frac{2c}{\pi}}(f)(t)=\int_{-1}^{1}\mathrm{e}^{\mathrm{i}cxt}f(x)\,\mathrm{d}x. \tag{6.10}$$

Let P_n denote the nth Legendre polynomial on $[-1,1]$ normalized such that $P(1)=1$, so that $\int_{-1}^{1}P_n^2=1/(n+1/2)$. Expanding the exponential in its power series gives

$$\int_{-1}^{1}\mathrm{e}^{\mathrm{i}cxt}P_n(x)\,\mathrm{d}x=\sum_{m=0}^{\infty}\frac{(\mathrm{i}ct)^m}{m!}\int_{-1}^{1}x^mP_n(x)\,\mathrm{d}x\,.$$

Set $a_\nu(t)=\frac{(-1)^\nu(ct)^{2\nu}}{(2\nu)!}$ and $b_\nu(t)=\frac{(-1)^\nu(ct)^{2\nu+1}}{(2\nu+1)!}$ and observe that $\int_{-1}^{1}x^mP_n(x)\,\mathrm{d}x=0$ if $m<n$. Thus,

$$\int_{-1}^{1}\mathrm{e}^{\mathrm{i}cxt}P_n(x)\,\mathrm{d}x=\sum_{\nu=[n/2]}^{\infty}a_\nu(t)\int_{-1}^{1}x^{2\nu}P_n(x)\,\mathrm{d}x+\mathrm{i}\sum_{\nu=[n/2]}^{\infty}b_\nu(t)\int_{-1}^{1}x^{2\nu+1}P_n(x)\,\mathrm{d}x\,.$$

By the Cauchy–Schwarz inequality,

$$\left|\int_{-1}^{1}x^mP_n(x)\,\mathrm{d}x\right|^2\le\frac{4}{(m+1)(2n+1)},\qquad(m\ge n)\,.$$

The following is part of [364, Lemma 3.3].

Lemma 6.2.11. *For $m\ge[n/2]$, set*

$$R_m(t)=\sum_{\nu=m}^{\infty}a_\nu(t)\int_{-1}^{1}x^{2\nu}P_n(x)\,\mathrm{d}x+\mathrm{i}\sum_{\nu=m}^{\infty}b_\nu(t)\int_{-1}^{1}x^{2\nu+1}P_n(x)\,\mathrm{d}x\,.$$

If $m > [|etc|]+1$ *then*

$$|R_m(t)| < \frac{C_m}{4^m\sqrt{2n+1}} \quad (C_m \to 0 \text{ as } m \to \infty). \tag{6.11}$$

In particular, if $n > (2[|etc|]+1)$ *then*

$$|F_{\frac{2c}{\pi}}(P_n)(t)| < \frac{C_n}{2^n\sqrt{2n+1}}. \tag{6.12}$$

Proof. It follows from the triangle inequality that

$$|R_m(t)| \le \frac{1}{\sqrt{n+1/2}}\sum_{\nu=2m}^{\infty}\frac{(ct)^\nu}{\nu!}\sqrt{\frac{2}{\nu+1}} < \frac{2}{\sqrt{2n+1}}\sum_{\nu=2m}^{\infty}\frac{(ct)^\nu}{\nu!}.$$

If $m \ge |etc|+1$ then

$$\frac{|ct|}{2m+k} < \frac{|ct|}{2m} < \frac{1}{2\mathrm{e}} \quad (k>0)$$

and then (6.11) follows from the Stirling inequality, since

$$\begin{aligned}|R_m| &< \frac{2}{\sqrt{2n+1}}\frac{|ct|^{2m}}{(2m)!}\left(1+\frac{1}{2}+\frac{1}{4}+\dots\right) < \frac{4}{\sqrt{2n+1}}\frac{|ct|^{2m}}{(2m)!}\\ &< C\frac{4}{\sqrt{4\pi m}\sqrt{2n+1}}\left(\frac{|cte|}{2m}\right)^{2m} \le \frac{C_m}{4^m\sqrt{2n+1}}\end{aligned}$$

where $C_m \to 0$. Inequality (6.12) comes from observing that $F_{2c/\pi}(P_n) = R_{[n/2]}$. □

Corollary 6.2.12. *Let* $\bar{\varphi}_m^{(c)}$ *denote the mth* $L^2[-1,1]$*-normalized PSWF eigenfunction of* $F_{2c/\pi}$ *with eigenvalue* $\mu_m = \mu_m(2c/\pi)$*. Then, for* $n > 2([|etc|]+1)$*,*

$$\left|\int_{-1}^{1}\bar{\varphi}_m^{(c)}(x)P_n(x)\,\mathrm{d}x\right| \le \frac{C}{2^n\sqrt{n+1/2}|\mu_m|}.$$

Proof. Since $F_{2c/\pi}(\bar{\varphi}_m^{(c)})(x) = \mu_m\bar{\varphi}_m^{(c)}(x)$, one can write

$$\begin{aligned}\left|\int_{-1}^{1}\bar{\varphi}_m^{(c)}(x)P_n(c)\,\mathrm{d}x\right| &= \frac{1}{|\mu_m|}\left|\int_{-1}^{1}\int_{-1}^{1}\mathrm{e}^{\mathrm{i}cux}\bar{\varphi}_m^{(c)}(u)\,\mathrm{d}u\,P_n(c)\,\mathrm{d}x\right|\\ &= \frac{1}{|\mu_m|}\left|\int_{-1}^{1}\bar{\varphi}_m^{(c)}(u)\int_{-1}^{1}\mathrm{e}^{\mathrm{i}cux}P_n(c)\,\mathrm{d}x\,\mathrm{d}u\right|\\ &\le \frac{1}{|\mu_m|}\int_{-1}^{1}|\bar{\varphi}_m^{(c)}(u)|\,\mathrm{d}u\left|\int_{-1}^{1}\mathrm{e}^{\mathrm{i}cux}P_n(c)\,\mathrm{d}x\right| \le \frac{\sqrt{2}}{|\mu_m|}\frac{C_n}{2^n\sqrt{n+1/2}}\end{aligned}$$

by (6.12). Here, $\int_{-1}^{1}|\bar{\varphi}_m^{(c)}(u)| \le \sqrt{2}$ since $\|\bar{\varphi}_m^{(c)}\|_{L^2[-1,1]} = 1$. □

Recall that one has the expansion

$$\bar{\phi}_m^{(c)}(t) = \sum_{n=0}^{\infty} \alpha_{mn} P_n(t); \quad \alpha_{mn} = \left(n+\frac{1}{2}\right) \int_{-1}^{1} \bar{\phi}_m^{(c)} P_n \tag{6.13}$$

in which the coefficients α_{mn} can be estimated iteratively using Bouwkamp's method discussed in Chap. 1. Together with Corollary 6.2.12 and (6.12), (6.13) provides an estimate of the error of approximating any value of $\bar{\phi}_m^{(c)}$ by applying $F_{2c/\pi}$ to a partial sum of the Legendre expansion of $\bar{\phi}_m^{(c)}$, since

$$\begin{aligned} \left|F_{\frac{2c}{\pi}}(\bar{\phi}_m^{(c)})(t) - F_{\frac{2c}{\pi}}\left(\sum_{n=0}^{N} \alpha_{mn} P_n\right)\right| &\le \sum_{N+1}^{\infty} |\alpha_{mn}| \left|F_{\frac{2c}{\pi}}(P_n)\right| \\ &\le \sum_{N+1}^{\infty} \frac{C\sqrt{n+1/2}}{2^n|\mu_m|} \frac{C_n}{2^n\sqrt{n+1/2}} \le \sum_{N+1}^{\infty} \frac{C^2}{4^n|\mu_m|} \end{aligned}$$

if $N > 2[|etc|]+1$. In particular, the integer values of $\bar{\phi}_m^{(c)}(k)$ can be approximated by applying $F_{2c/\pi}$ to the Nth partial sum of the Legendre expansion of $\bar{\phi}_m^{(c)}$ on $[-1,1]$. This approximation will be effective for any k such that $N > 2[|ekc|]+1$, with an error controlled by $1/(4^N|\mu_m|)$.

Integer Values of φ_m^T

Let φ^T denote an eigenfunction of PQ_T, that is, φ^T is band limited to $[-1/2,1/2]$ and approximately time limited to $[-T,T]$. Then, by Corollary 1.2.6, $\psi(t) = D_T\varphi^T$ is an eigenfunction of F_{2T} with F_a as in (6.10), that is,

$$F_{2T}(\psi)(t) = \int_{-1}^{1} \mathrm{e}^{\pi \mathrm{i} Tst}\, \psi(s)\,\mathrm{d}s = \mu\psi(t)$$

for an appropriate $\mu \in \mathbb{C}$. Equivalently, $D_T\varphi^T$ will satisfy

$$(D_T(\varphi^T))(t) = \frac{1}{\mu}\int_{-1}^{1} \mathrm{e}^{\pi \mathrm{i} Tst} (D_T\varphi^T)(s)\,\mathrm{d}s.$$

Therefore, with α_{mn} as in (6.13) and φ_m^T the mth eigenfunction of PQ_T,

$$
\begin{aligned}
\varphi_m^T(k) &= \frac{1}{\sqrt{T}}(D_T(\varphi_m^T))\left(\frac{k}{T}\right) = \frac{1}{\sqrt{T}\mu_m}\int_{-1}^{1} \mathrm{e}^{\pi i s k}(D_T\varphi_m^T)(s)\,\mathrm{d}s \\
&\approx \frac{1}{\sqrt{T}\mu_m}\sum_{n=0}^{N(T)}(n+1/2)\langle D_T\varphi_m^T, P_n\rangle \int_{-1}^{1} \mathrm{e}^{\pi i s k} P_n(s)\,\mathrm{d}s \\
&\approx \frac{1}{\sqrt{T}\mu_m}\sum_{n=0}^{N(T)} \alpha_{mn}(2T)\begin{cases} (-\mathrm{i})^n \frac{J_{n+1/2}(\pi k)}{\sqrt{k}}, & k \neq 0 \\ \delta_{0,n}, & k = 0. \end{cases}
\end{aligned}
$$

The formula for the Fourier coefficients of P_n is classical and can be found in the Bateman project manuscripts [96, p. 122] or [95, vol. II, p. 213] with $\lambda = 1/2$.

Numerical Comparisons

In the following figures we consider the case $T = 5$, and estimate eigenfunctions φ_n^T of PQ_T using the method just outlined for estimating the integer samples, to which we will refer as the *Karoui–Moumni* (KM) method [171]. We also compare the KM method to the Walter–Shen–Soleski (WSS) method [345], which estimates the eigenvectors of a truncation of the matrix (6.3). First, we compare the sinc-interpolated PSWF approximations on $[-T,T]$, generated using only those samples $\varphi_n(k)$ where $|k| \leq [T]+1$, with approximations generated by those samples roughly satisfying the condition $|k| < M(T)$ as in (6.7), where the samples are generated by the KM method using Legendre coefficients up to order $2([ce]+1)$ as specified in [364, p. 813]. When $T = 5$ we have $c = 5\pi$ so $2([ce]+1]) = 258$ while, for $\gamma = 2$ in (6.6) one gets $M(T) = (1+\pi^2)T(1+\log^2(T)) = 5(1+\pi^2)(1+\log^2 5) \approx 195$. In view of these estimates, with $T = 5$ we build a 256×256 system producing PSWF samples as Fourier coefficients of 256th degree Legendre approximations of the PSWFs. Figure 6.1 shows plots of interpolants using all 256 PSWF samples computed by the KM method, denoted φ_n^{KM}, versus interpolants using only the estimated samples $\varphi_n^{\mathrm{KM}}(k)$ for $|k| \leq [T]+1 = 6$. These interpolants are denoted φ_n^{loc}. The plots are shown over the interval $[-2T,2T] = [-10,10]$. Figure 6.2 plots $\varphi_{12}^{\mathrm{KM}}$ and $\varphi_{12}^{\mathrm{loc}}$ on the interval $[-T,T] = [-5,5]$. Since $n = 12 > 10 = 2T$, this choice corresponds to an n for which φ_n has most of its energy outside, but still close to $[-5,5]$. Further details of the errors between φ_n^{KM} and φ_n^{loc} can be found in [151].

Next, we consider φ_n^{KM} versus φ_n^{WSS}, the sinc interpolant of the approximate samples produced by the WSS method. To compute φ_n^{WSS} with $T = 5$, we build a 256×256 matrix A corresponding to truncation of the WSS sample reproducing matrix (6.3), and define the approximate samples of φ_n^T as the entries of the eigenvectors of this truncated matrix. These *samples* are then sinc interpolated to define φ_n^{WSS}. Figure 6.3 plots the error between $\varphi_{15}^{\mathrm{KM}}$ and $\varphi_{15}^{\mathrm{WSS}}$ for $T = 5$. Again, $n > 2T$ so most of the energy lies outside $[-T,T]$ in each case. Briefly, the WSS approximation is less accurate compared to the KM method. Further aspects of truncations of matrix (6.3) will be discussed in the chapter notes.

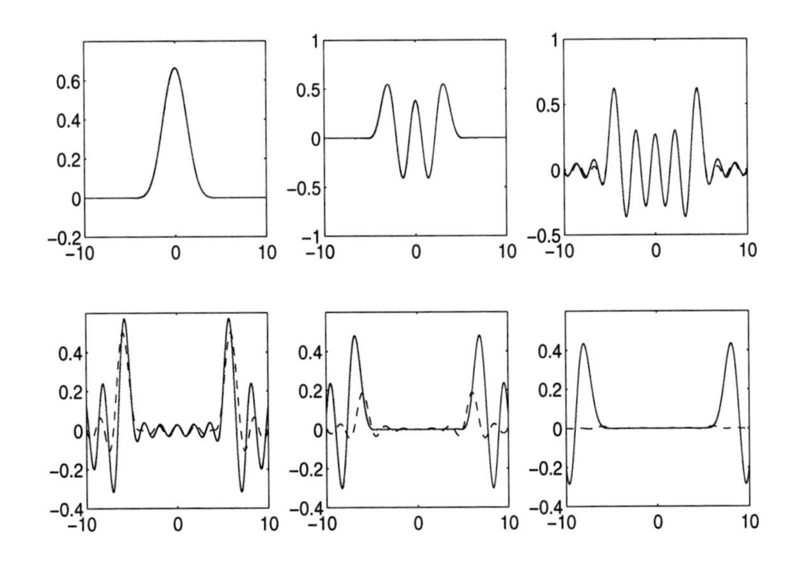

Fig. 6.1 Plots of φ_n^{KM} (solid) and φ_n^{loc} for $n = 0, 4, 8, 12, 16, 20$ and $T = 5$

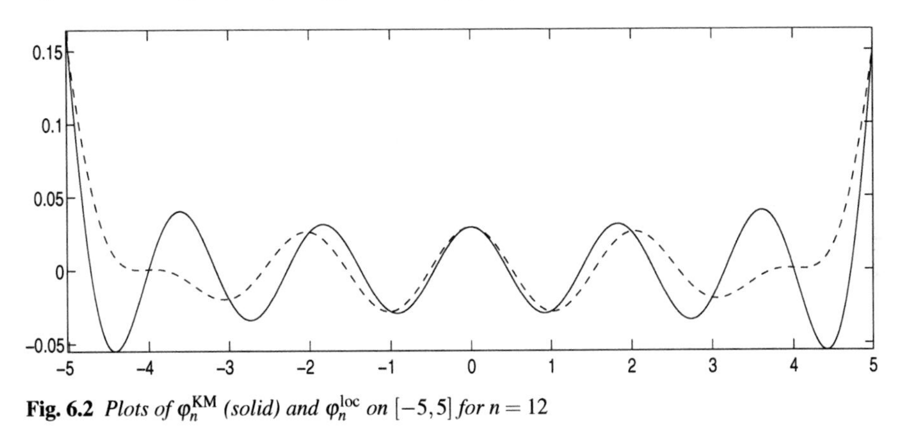

Fig. 6.2 *Plots of* φ_n^{KM} *(solid) and* φ_n^{loc} *on* $[-5,5]$ *for* $n = 12$

6.3 Notes and Auxiliary Results

6.3.1 Sampling and Interpolation for Eigenfunctions of Multiband Localization Operators

Here we will present an elementary extension of the Walter–Shen sampling formula to the case of multiband signals. As in Chap. 5, this is just one of several possible approaches to obtaining time-localized expansions from samples of multiband-limited signals. We begin with a simple extension of Proposition 6.2.1 to an arbitrary finite frequency interval.

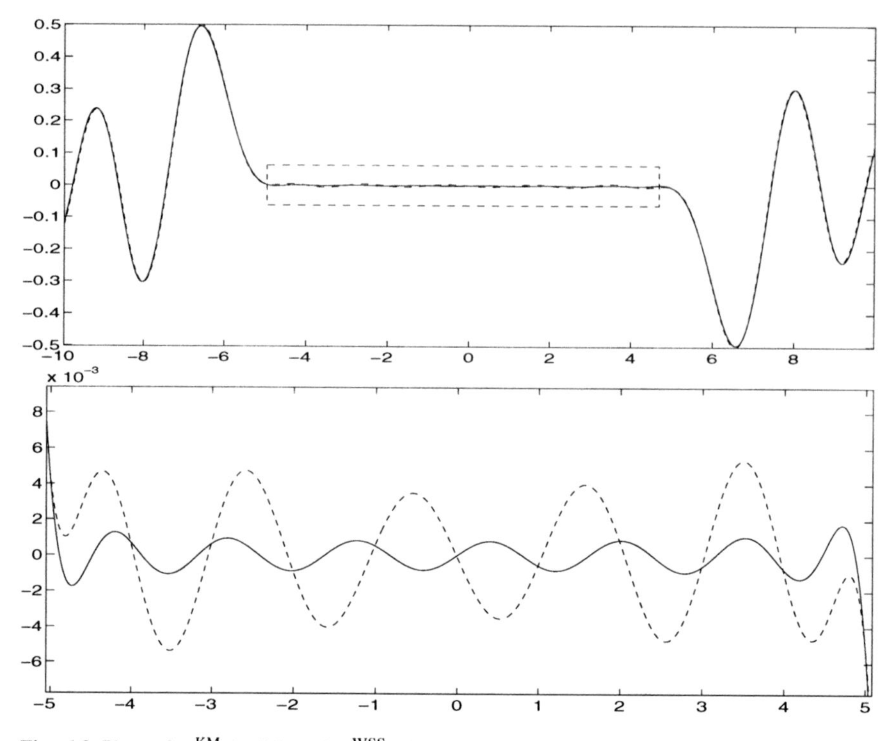

Fig. 6.3 Plots of $\varphi_{15}^{\mathrm{KM}}$ (solid) and $\varphi_{15}^{\mathrm{WSS}}$ (dashed) for $T = 5$. On the top, the time interval is $[-2T, 2T]$. On the bottom it is $[-T, T]$, corresponding to the boxed region in the top figure

Lemma 6.3.1. *Suppose that I has finite length $|I|$. Let φ_n^I denote the nth eigenfunction of $P_I Q_T$ for $T > 0$ fixed. Then for any band-limited $f \in L^2(\mathbb{R})$, one has*

$$\langle f, \varphi_n^I \rangle = \sum_{k=-\infty}^{\infty} \frac{1}{|I|} (P_I f)\left(\frac{k}{|I|}\right) \overline{\varphi_n^I\left(\frac{k}{|I|}\right)} . \tag{6.14}$$

Proof. Since P_I is idempotent, $\langle f, \varphi_n^I \rangle = \langle P_I f, \varphi_n^I \rangle$. The lemma then follows from Parseval's formula for Fourier series upon recognizing that $f(k/|I|)/\sqrt{|I|}$ is the kth normalized Fourier coefficient of $\widehat{f}$ taken with respect to the Fourier basis for $L^2(I)$. □

Since the functions $\{\varphi_n^I\}$, $n = 0, 1, 2, \ldots$ form an orthonormal basis for PW_I, one can expand $f \in \mathrm{PW}_I$ as $f(t) = \sum_{n=0}^{\infty} \langle f \varphi_n^I \rangle \varphi_n^I$. If Σ is a disjoint union $\Sigma = \cup_{m=1}^{M} I_m$, then $\mathrm{PW}_\Sigma = \oplus_{m=1}^{M} \mathrm{PW}_{I_m}$ and one has the following extension of the first Walter–Shen formula.

Corollary 6.3.2. *Let $\Sigma = \cup_{m=1}^{M} I_m$ be a finite union of bounded, pairwise disjoint intervals and let $\varphi_n^{I_m}$ denote the nth eigenfunction of $P_{I_m} Q_T$ where $T > 0$ is fixed. Then, for any $f \in \mathrm{PW}_\Sigma$ one has*

$$f(t) = \sum_{m=1}^{M} \sum_{n=0}^{\infty} \langle f \varphi_n^{I_m} \rangle \varphi_n^{I_m} = \sum_{m=1}^{M} \frac{1}{|I_m|} \sum_{n=0}^{\infty} \left(\sum_{k=-\infty}^{\infty} (P_{I_m} f)\left(\frac{k}{|I_m|}\right) \overline{\varphi_n^{I_m}\left(\frac{k}{|I_m|}\right)} \right) \varphi_n^{I_m}(t) .$$

Spectral Slicing, Revisited

Spectral slicing and tiling sets were discussed in Sect. 5.3 (see pp. 176 and 177). Spectral slicing refers to dividing an $L\times\Omega$-packing set Σ into a disjoint union of $J\leq L\,\Omega$-packing sets, $\Sigma_0,\dots\Sigma_{J-1}$, ordered in such a way that Σ_j is a sub-alias of Σ_{j-1}, meaning that $\Sigma_j\subset \ell\Omega+\Sigma_{j-1}$ for some $\ell\in\mathbb{Z}$. Because time–frequency localization operators are covariant with respect to dilations, we will assume for now that Σ is an L-packing set that can be expressed as a disjoint union of J 1-packing sets that are each finite unions of intervals. In fact, for the present purposes, it will be convenient to assume that the Σ_j are complete aliases of one another, meaning that, for each $0\leq j<j'<J$, there is an $\ell_{jj'}\in\mathbb{Z}$ such that $\Sigma_j=\ell_{jj'}+\Sigma_{j'}$. The general case of sub-aliasing will simply involve the systematic vanishing of corresponding reconstruction matrix coefficients in the following formulas. As previously, φ_n^{Σ} will denote the nth eigenfunction of $P_\Sigma Q_T$ where $T>0$ will be fixed ahead of time but Σ can vary.

Theorem 6.3.3. *Let Σ be an L-packing set that is the union of J 1-tiling sets Σ_j, $j=0,\dots,J-1$. In addition, suppose that Σ_j is the union of M intervals I_{jm} of the form $I_m+\beta_{jm}$ such that $\cup_m I_{jm}=[0,1)$ and $\beta_{jm}\in\mathbb{Z}$ for each $j=0,\dots,J-1$ and $m=0,\dots,M-1$. Denote by W_m the $J\times J$ matrix whose (j,ν)th entry is $W_m(j,\nu)=\mathrm{e}^{-2\pi i j\beta_{\nu m}/L}$ and let $C_m(j,\nu)=\overline{W_m^{-1}(\nu,j)W_m(j,\nu)}$. Then for any $f\in \mathrm{PW}_\Sigma$ one has*

$$\langle f,\varphi_n^{I_{\nu m}}\rangle=\sum_{j=0}^{J-1}\sum_{k=-\infty}^{\infty}C_m(j,\nu)f\Big(k+\frac{j}{L}\Big)\overline{\varphi_n^{I_{\nu m}}\Big(k+\frac{j}{L}\Big)},\tag{6.15}$$

$$f(t)=\sum_{m=0}^{M-1}\sum_{\nu=0}^{J-1}\sum_{n=0}^{\infty}\langle f,\varphi_n^{I_{\nu m}}\rangle\varphi_n^{I_{\nu m}}(t),\tag{6.16}$$

and
$$f(t)=\sum_{m=0}^{M-1}\sum_{\nu=0}^{J-1}\sum_{n=0}^{\infty}\sum_{\ell=0}^{\infty}\langle f,\varphi_n^{I_{\nu m}}\rangle\langle\varphi_n^{I_{\nu m}},\varphi_\ell^{\Sigma}\rangle\,\varphi_\ell^{\Sigma}(t).\tag{6.17}$$

Proof. The fact that

$$\begin{aligned}\langle f,\varphi_n^{I_{\nu m}}\rangle&=\int_{-\infty}^{\infty}f(t)\,\overline{\varphi_n^{I_{\nu m}}}(t)\,\mathrm{d}t\\&=\sum_{j=0}^{J-1}C_m(j,\nu)\sum_{k=-\infty}^{\infty}f\Big(k+\frac{j}{L}\Big)\int_{-\infty}^{\infty}\mathbb{1}_{I_{\nu m}}^{\vee}\Big(k+\frac{j}{L}-t\Big)\overline{\varphi_n^{I_{\nu m}}}(t)\,\mathrm{d}t\\&=\sum_{j=0}^{J-1}\sum_{k=-\infty}^{\infty}C_m(j,\nu)f\Big(k+\frac{j}{L}\Big)\overline{\varphi_n^{I_{\nu m}}}\Big(k+\frac{j}{L}\Big)\end{aligned}$$

follows from applying Corollary 6.3.2 to $P_{I_{\nu m}}f$. The functions $\varphi_n^{I_{\nu m}}$ are orthogonal to one another for ν and m fixed but different n (they correspond to different eigenvalues) and are orthogonal to one another for different ν or m because of the tiling hypotheses. Together, the functions are complete in PW_Σ so (6.16) follows directly from the hypothesis that $f\in\mathrm{PW}_\Sigma$. The identity

$$\langle f,\varphi_\ell^{\Sigma}\rangle=\sum_{m=0}^{M-1}\sum_{\nu=0}^{J-1}\sum_{n=0}^{\infty}\langle f,\varphi_n^{I_{\nu m}}\rangle\langle\varphi_n^{I_{\nu m}},\varphi_\ell^{\Sigma}\rangle\tag{6.18}$$

follows from Parseval's formula and the hypothesis that both f and φ_ℓ^{Σ} belong to PW_Σ. Finally, (6.17) follows then from the fact that the functions $\{\varphi_\ell^{\Sigma}\}$ form an orthonormal basis for PW_Σ. This proves the theorem. □

Discussion

In (6.15), the coefficients $\langle f, \varphi_n^{I_{vm}}\rangle$ are calculated from bunched samples of f taken at the Landau rate of J samples per unit time. As discussed in Sect. 5.3.2, other periodic nonuniform sampling patterns could be used, provided the corresponding W_m matrices are all invertible. Equation (6.18) provides the means for passing from samples of f to the coefficients $\langle f, \varphi_\ell^{\Sigma}\rangle$ through the matrix that converts time–frequency localization on separate subintervals into time–frequency localization of their union; cf. Proposition 6.1.2. Because $I_{jm} = I_m + \beta_{jm}$, there are potentially M different lengths and corresponding sources of truncation error that have to be considered in approximating an arbitrary $f \in \mathrm{PW}_\Sigma$ using finitely many of its sample values.

6.3.2 Güntürk and DeVore's Bump Function Estimate

In Sect. 6.2.2 the following estimate due to Güntürk and DeVore was used. Its proof is included here, since the proof is not readily available elsewhere.

Proposition 6.3.4. *For any $\delta > 0$ there exist $\alpha > 0$ and $\gamma \in (1,\infty)$ and a function $\psi(x)$ whose Fourier transform is equal to one on $[-1,1]$ and equal to zero outside of $[-1-\delta, 1+\delta]$ such that*

$$|\psi(t)| \le \mathrm{e}^{1-\alpha|t|/\log^{\gamma}(\alpha|t|)}, \quad (|t| > \mathrm{e}/\alpha).$$

Proof. Let $\widehat{\psi}_0(\xi) = \mathbb{1}_{[-1-\delta,1+\delta]}$. Choose $a_n > 0$, $n = 1,2,\dots$ such that $\sum a_n < \delta$. Let $\widehat{\phi}_n = \mathbb{1}_{[-a_n,a_n]}/a_n$. Also, let $\widehat{\psi}_1 = \widehat{\phi}_1 * \widehat{\phi}_2 * \dots$ where the infinite convolution is defined as the limit of the Fourier transform of the partial products $\prod_{n=1}^{N} \phi_n$. The convolution converges pointwise since, for any ξ in the interior of its support, only finitely many *factors* contribute to the value of $\widehat{\psi}(\xi)$. The condition $\sum_{n=1}^{\infty} a_n < \delta$ implies that $\widehat{\psi}_1(\xi)$ is supported strictly inside $[-\delta, \delta]$ so that $\widehat{\psi} = \widehat{\psi}_0 * \widehat{\psi}_1$ is supported in $[-1-2\delta, 1+2\delta]$. In addition, since $\int \widehat{\phi}_n = 1$ for all $n = 1,2,\dots$, it follows that $\int \widehat{\psi}_1 = 1$ also. This implies that $\widehat{\psi}(\xi) = 1$ for all $\xi \in [-1,1]$. In order to estimate ψ we have

$$|\psi(t)| = \left|\frac{\sin(\pi(1+\delta)t)}{\pi t}\right| \left|\prod_{n=1}^{\infty}\right| \left|\frac{\sin \pi a_n t}{\pi a_n t}\right| \le \frac{1}{|\pi t|^N} \prod_{n=1}^{N} \frac{1}{a_j}$$

for any N, since $|\sin x/x| \le 1$. Thus,

$$|\psi(t)| \le \inf_{N\ge 1} \left(|\pi t|^N \prod_{n=1}^{N} a_n\right)^{-1}.$$

Now specialize a_n such that $a_1 < \delta/2$ and, for $n \ge 2$, $a_n = (cn\log^{\gamma} n)^{-1}$ where c is chosen so that $\sum_{n=1}^{\infty} a_n < \delta$. Therefore

$$\left(\prod_{n=1}^{N} a_n\right)^{-1} < c^N (\log N)^{\gamma N} N! < (cN\log^{\gamma} N)^N .$$

Except for the value of c, the bound on the right is as good as the initial bound because of the Stirling approximation $(\mathrm{e}^n n!/(n^n\sqrt{2\pi n}) \to 1$ as $n \to \infty)$. Thus,

$$|\psi(t)| \le \left(\frac{cN\log^{\gamma} N}{\pi|t|}\right)^N . \tag{6.19}$$

Fix $\alpha \in (0, \pi/(c\mathrm{e}))$ and set $N = [\alpha|t|/\log^{\gamma}(\alpha|t|)]$ for $|t| > \mathrm{e}/\alpha$ so that $\log N \le \log(\alpha|t|)$ and $N/|t| \le \alpha/\log^{\gamma}(\alpha|t|)$. Then, by (6.19),

$$|\psi(t)| \le \left(\frac{c\alpha}{\pi}\right)^N \le \mathrm{e}^{1-\alpha|t|/\log^\gamma(\alpha|t|)}$$

as claimed. □

Remark. It is not known whether $\gamma > 1$ is required in the pointwise decay $|\psi(t)| \le C\mathrm{e}^{-|t|/\log^\gamma(|t|)}$ of the Fourier bump function ψ in Proposition 6.3.4. Exponential decay, $|\psi(t)| \le C\mathrm{e}^{-\alpha|t|}$ ($\alpha > 0$), is impossible since then $\widehat{\psi}$ would be real analytic, contradicting its compact support. If one could take $\gamma = 1$ then it would be possible to replace $M(T)$ in (6.7) by a fixed multiple of T.

6.3.3 Error Bounds for the Walter–Shen–Soleski Method

The Walter–Shen–Soleski (WSS) method for approximating the PSWFs from interpolated approximate samples starts with a truncation of the matrix A in (6.3) and defines approximate sample values of the PSWFs in terms of the eigenvectors of this truncation. The L^2-norm of the error between the approximant thus produced and the true PSWF equals the ℓ^2-norm of sample errors. The following theorem, proved in [151], shows that this error decays slowly with truncation size N.

Theorem 6.3.5. *Let $A = A(T)$ be as in (6.3) and let $A^{\mathrm{tr}} = A^{\mathrm{tr}}(T,N)$ be such that $A^{\mathrm{tr}}_{k\ell} = A_{k\ell}$ if $\max\{|k|,|\ell|\} \le NT$ and $A^{\mathrm{tr}}_{k\ell} = 0$ otherwise. Let $A^{\mathrm{rem}} = A - A^{\mathrm{tr}}$. There are constants C_1 and C_2 depending on T but not N such that $C_1 N^{-1/2} \le \|A^{\mathrm{rem}}\|_{\ell^2\to\ell^2} \le C_2 N^{-1/2}$.*

In fact, because the matrix A is nearly a tensor product, the pointwise truncation error is, in a sense, *typically* of order $1/\sqrt{N}$. This suggests that using an analogue of the WSS method for approximating signals in the more general setting of Theorem 6.1.1 will not produce accurate estimates either.

6.3.4 More on Karoui and Moumni's Method

The method for approximating PSWFs locally through their samples outlined in Sect. 6.2 is just one possible approach. Others have been discussed in Chap. 2 and in the works of Karoui and Moumni [171], Moore and Cada [239], and Xiao et al. [364], among others. Approximating a PSWF *globally* by a truncated sampling series is ineffective since the sinc function decays only as $1/x$. The numerical value of the PSWF sample depends on the Fourier coefficients of Legendre polynomials which can be expressed in terms of Bessel functions as on p. 211. Karoui and Moumni also proposed a recursive method to calculate these coefficients from those of Legendre polynomials of lower degree as follows. Because of their even or odd parity, Fourier coefficients of the normalized Legendre polynomials on $[-1,1]$ are

$$I_{k,2\ell} = \int_{-1}^{1} \bar{P}_{2\ell}(\xi)\cos(\pi k\xi)\,\mathrm{d}\xi; \quad I_{k,2\ell+1} = \mathrm{i}\int_{-1}^{1} \bar{P}_{2\ell+1}(\xi)\sin(\pi k\xi)\,\mathrm{d}\xi\,. \tag{6.20}$$

Invoking parity again, and performing integration by parts repeatedly reduces calculation of these coefficients to evaluation of an endpoint derivative of $\bar{P}_\ell$.

$$\begin{aligned} I_{k,2\ell} &= 2(-1)^k \sum_{\nu=0}^{\ell} \frac{(-1)^{\nu+1}}{(k\pi)^{2\nu}} \frac{\mathrm{d}^{2\nu-1}\bar{P}_{2\ell}}{\mathrm{d}x^{2\nu-1}}\Big|_{x=1} \quad \text{and} \\ I_{k,2\ell+1} &= 2\mathrm{i}(-1)^k \sum_{\nu=0}^{\ell} \frac{(-1)^{\nu+1}}{(k\pi)^{2\nu+1}} \frac{\mathrm{d}^{2\nu}\bar{P}_{2\ell+1}}{\mathrm{d}x^{2\nu}}\Big|_{x=1}\,. \end{aligned} \tag{6.21}$$

Starting from $P_\ell(1)=1$, one obtains by iteration the values

$$\frac{\mathrm{d}^{\nu+1}P_\ell}{\mathrm{d}x^{\nu+1}}\Big|_{x=1}=\frac{\ell(\ell+1)-\nu(\nu+1)}{2(\nu+1)}\frac{\mathrm{d}^{\nu}P_\ell}{\mathrm{d}x^{\nu}}\Big|_{x=1}.$$

This follows by differentiating the Legendre differential equation

$$(1-t^2)\frac{\mathrm{d}^2u}{\mathrm{d}t^2}-2t\frac{\mathrm{d}u}{\mathrm{d}t}+\ell(\ell+1)u=0,$$

then evaluating at $t=1$.

6.3.5 Finite Models for Continuous Signals

DFT Models for Continuous Fourier Transforms

Auslander and Grünbaum [9] considered the problem of approximating a function in $L^2(\mathbb{R})$ by means of the discrete Fourier transform (DFT) of some finite sampled signal. Since Dirac samples are not defined on L^2, one needs a sampling method that involves local averages both in time and in frequency. Thus one sets

$$f_\varphi(s)=f*\varphi(s)=\int f(s-t)\,\varphi(t)\,\mathrm{d}t;\qquad \widehat{f}_\psi(\xi)=\widehat{f}*\psi(\xi).$$

Fix $T>0$, $\Omega>0$ and set $s_k=Tk/N$ and $\omega_k=\Omega k/N$ where $k=-N,\ldots,N$. Denote by g the (centered) DFT of the samples of f_φ, thus

$$g(k)=\frac{1}{\sqrt{2N+1}}\sum_{j=-N}^{N}f_\varphi\Big(\frac{Tj}{N}\Big)\mathrm{e}^{\frac{2\pi i}{2N+1}jk},k=-N,\ldots,N.$$

With $D_{2N+1}(\xi)=\sum_{k=-N}^{N}\mathrm{e}^{2\pi i k\xi}$, a simple calculation verifies that

$$g(k)=\frac{1}{\sqrt{2N+1}}\int\widehat{f}(\xi)\widehat{\varphi}(\xi)D_{2N+1}\Big(\frac{T}{N}\xi-\frac{k}{2N+1}\Big).$$

Thus, the error between the samples of the Fourier transform of f_φ and samples of the mollified Fourier transform $\widehat{f}_\psi$ of f is quantified by

$$\widehat{f}_\psi\Big(\frac{\Omega k}{N}\Big)-g(k)=\int\widehat{f}(\xi)\Big[\psi\Big(\frac{k\Omega}{N}-\xi\Big)-\frac{1}{\sqrt{2N+1}}\widehat{\varphi}(\xi)D_{2N+1}\Big(\frac{T}{N}\xi-\frac{k}{2N+1}\Big)\Big]\,\mathrm{d}\xi.$$

Define the *error kernel*

$$\delta_{k,N,\Omega,T}(\xi)=\Big[\psi\Big(\frac{k\Omega}{N}-\xi\Big)-\frac{1}{\sqrt{2N+1}}\widehat{\varphi}(\xi)D_{2N+1}\Big(\frac{T}{N}\xi-\frac{k}{2N+1}\Big)\Big].$$

By Cauchy–Schwarz and Plancherel one has

$$\Big|\widehat{f}_\psi\Big(\frac{\Omega k}{N}\Big)-g(k)\Big|\le\|f\|\|\delta_{k,N,\Omega,T}\|$$

and the bound is sharp by taking f to be the error function $\delta_{k,N,\Omega,T}$ itself. Auslander and Grünbaum considered the case of Gaussian mollifiers

$$\varphi(t) = \frac{1}{\sqrt{2\pi\sigma_1}} e^{-\frac{t^2}{2\sigma_1}}; \qquad \psi(\xi) = \frac{1}{\sqrt{2\pi\sigma_2}} e^{-\frac{\xi^2}{2\sigma_2}}$$

with the particular scalings

$$\sigma_1 = c_1(T/N); \qquad \sigma_2 = c_2(\Omega/N).$$

In this case it was shown that the *Wiener error* $\sum_{k=-N}^{N} \|\delta_{k,N,\Omega,T}\|$ is minimized at the *Nyquist rate* $T\Omega = (2N+1)/4$.

Discrete Continuous Functions

In addition to quantifying sampling errors, it is interesting to ask which families of functions in $L^2(\mathbb{R})$ can be characterized completely in terms of finite sets of samples, on one hand, and finitely many Fourier coefficients, on the other. Signals that are finite sums of shifted sinc functions are such examples, and their Fourier transforms can be expressed as trigonometric polynomials on their supports. It is of interest to consider broader classes of interpolating functions. Several such interpolation schemes are considered in Izu's dissertation [159], including extensions to distributions. Here we state one such interpolation result. Recall that ψ is a T partition of unity if $\sum_{k=-\infty}^{\infty} \psi(t+kT) = 1$ for all $t \in \mathbb{R}$, and $\Sigma \subset \mathbb{R}$ is an Ω-tiling set if $\{k\Omega + \Sigma\}_{k\in\mathbb{Z}}$ partitions $\mathbb{R}$.

Theorem 6.3.6. *Let $T\Omega \in \mathbb{N}$, let S be a T-tiling set, and let Σ be an Ω-tiling set. Let $\psi \in \mathscr{S}(\mathbb{R})$ be a T partition of unity such that $\psi(k/\Omega) = 1$ if $k \in \Omega S$ and $\psi(k/\Omega) = 0$ otherwise. Suppose that $\widehat{f}$ is a linear combination of shifts $\{\widehat{\psi}(\cdot - \ell/T) : \ell \in T\Sigma\}$. Then f satisfies*

$$\widehat{f}(\xi) = \frac{1}{T} \sum_{\ell\in T\Sigma} \widehat{f}\Big(\frac{\ell}{T}\Big) \widehat{\psi}\Big(\xi - \frac{\ell}{T}\Big) = \frac{1}{T\Omega} \sum_{k\in\Omega S} f\Big(\frac{k}{\Omega}\Big) \sum_{\ell\in T\Sigma} e^{-2\pi i \frac{k\ell}{T\Omega}} \widehat{\psi}\Big(\xi - \frac{\ell}{T}\Big)$$

and

$$f(t) = \frac{1}{T} \sum_{\ell\in T\Sigma} \widehat{f}\Big(\frac{\ell}{T}\Big) e^{2\pi i \frac{\ell}{T} t} \psi(t) = \frac{1}{T\Omega} \sum_{k\in\Omega S} f\Big(\frac{k}{\Omega}\Big) \sum_{\ell\in T\Sigma} e^{-2\pi i \frac{\ell}{T}(\frac{k}{\Omega} - T)} \psi(t).$$

Example. Let $T = 36$ and $\Omega = 1/9$ so $T\Omega = 4$. Set $S = [0,18) \cup [54,72)$ and let $\Sigma = [0,1/9)$. So $\Omega S \cap \mathbb{Z} = \{0,1,6,7\}$ while $T\Sigma \cap \mathbb{Z} = \{0,1,2,3\}$. Let $h(t) = \exp(-1/(1-|t|^2))$ for $|t| < 1$ and $h(t) = 0$ for $|t| > 1$, let $H(t) = \sum_k h(t-k)$, let $g(t) = h(t)/H(t)$, and let

$$\psi(t) = \sum_{k\in\Omega S} g(\Omega t - k) = g\Big(\frac{t}{9}\Big) + g\Big(\frac{t}{9} - 1\Big) + g\Big(\frac{t}{9} - 6\Big) + g\Big(\frac{t}{9} - 7\Big).$$

Then g is a 1-partition of unity. Also, ψ is a $T = 36$ partition of unity that vanishes at $9k$ if $k \notin \{0,1,6,7\}$ since $h(k) = \delta_{0k}$. Finally, let

$$\widehat{f}(\omega) = \sum_{\ell\in T\Sigma} \alpha_\ell \widehat{\psi}\Big(\omega - \frac{\ell}{T}\Big) = \sum_{\ell=0}^{3} \alpha_\ell \widehat{\psi}\Big(\omega - \frac{\ell}{36}\Big).$$

As in the theorem, f can also be expressed in terms of its samples, here $\{f(0), f(9), f(54), f(63)\}$. One can then build functions f that can be recovered from samples along shifts by multiples of 72 of $\{0,9,54,63\}$ by inverse Fourier transforming suitable sums of modulations of $\widehat{\psi}$.

6.3.6 Further Miscellaneous Results

Orthonormal Sampling Functions

One says that a band-limited function φ is a *sampling* or *interpolating* function if $\varphi(k) = \delta_{0k}$. One says that φ is shift-orthonormal if $\int \varphi(t)\bar{\varphi}(t-k)\,dt = \delta_{0k}$. In [168], Kaiblinger and Madych proved that for $\Omega > 0$ and $\varphi \in PW_\Omega$, if $\Omega < 1$ then there is no shift-orthonormal sampling function while, if $\Omega = 1$ then sinc (t) is the only shift-orthonormal sampling function. If $1 < \Omega \leq 3$ then there are real-valued shift-orthonormal functions having rapid decay; however, such a function cannot be symmetric if it is integrable. If $\Omega > 3$ then there are real-valued shift-orthonormal functions that are symmetric and have rapid decay.

Sampling Convergence Rates

The truncation $f(t) \approx \sum_{k=-N}^{N} f(k)\,\mathrm{sinc}\,(t-k)$ of the cardinal sine series of $f \in PW$ has an error that, in the worst case, only decays in L^2 as $1/N$ as $N \to \infty$. There are a number of techniques that combine *oversampling* with smoothing in the Fourier domain in order to improve the rate of decay of truncation error for a corresponding sampling series. Knab [181] proposed such a method based on *approximate prolate* expansions. Here, one makes use of the fact that $PW \subset PW_\Omega$ for any $\Omega > 1$. If $f \in PW$ and $g \in PW_{\Omega-1}$ then $fg \in PW_\Omega$, so one has

$$fg(t) = \sum_k (fg)\Big(\frac{k}{\Omega}\Big)\,\mathrm{sinc}_\Omega\Big(t - \frac{k}{\Omega}\Big).$$

Applying this instead to $g(t) = \psi(x-t)$ with $\psi(0) = 0$ then setting $x = t$, one obtains the new sampling series

$$f(t) = \sum_k f\Big(\frac{k}{\Omega}\Big)\psi\Big(t - \frac{k}{\Omega}\Big)\mathrm{sinc}_\Omega\Big(t - \frac{k}{\Omega}\Big).$$

The factor ψ is called a *convergence factor*. Its role is to improve the rate of approximation of f by partial sums of the sampling series. Knab noted that if the series is to be truncated to terms with $|k| \leq N$ with N fixed, then an optimal convergence factor, from the point of view of approximation in L^2, would be an element of $PW_{\Omega-1}$ having as much energy as possible in $[-N, N]$. In other words, ψ should be the zero order PSWF ψ_0 with time–frequency area $c = 2N(\Omega-1)$. Knab showed that for $f \in PW$ with *small* L^∞-norm, the error of Nth partial sum approximations converges to zero at a faster exponential rate (in N) than errors produced using previously considered convergence factors.

Bandpass Prolates

In [176], Khare introduced *bandpass prolate spheroidal wave functions* which are eigenfunctions of $P_\Sigma Q_T$ in which Σ is the symmetric union $\Sigma = [\alpha, \beta] \cup [-\beta, -\alpha], 0 < \alpha < \beta$. Since there is no known second-order differential operator that commutes with $P_\Sigma Q_T$, Khare used a variant of the Walter–Shen–Soleski method in order to produce numerical approximations of the eigenfunctions. Khare replaced the sinc kernel by the reproducing kernel for PW_Σ and also employed a bandpass version of the Shannon sampling formula.

Filter Diagonalization

A second application of the Walter–Shen sampling formulas can be found in the work of Levitina and Brändas concerning *filter diagonalization* [200, 201]. Filter diagonalization seeks a detailed estimate of the spectrum of an operator (Hamiltonian) H at a selected energy range. The spectral information is extracted from a short-duration observation of the *autocorrelation function* $C(t) = \langle \varphi(x,0), \mathrm{e}^{-\mathrm{i}Ht}\, \varphi(x,0) \rangle$ associated with a wavepacket solution $\varphi(x,t)$ of the time-dependent Schrödinger equation $\frac{\partial \varphi}{\partial t} = -\mathrm{i}H\varphi$. Filtering C thus corresponds to filtering the function space on which H acts. In the case of a purely discrete spectrum, filtering extracts a subspace spanned by eigenfunctions with eigenvalues in a given energy interval. Levitina and Brändas thus proposed an approximate filtering scheme, essentially corresponding to projecting C onto the span of the most localized time–frequency shifted prolates corresponding to the given observation interval and energy range interval. The Walter–Shen–Soleski method was proposed in [200] as a method to compute the approximate projections.

Appendix A
Notation and Mathematical Prerequisites

A.1 Notation and Disclaimers

When possible, we have assigned symbols to concepts in a manner consistent with the consensus use in the mathematics or engineering literature. Partly because of this, there are a number of potential conflicting uses of symbols in this text. For example, S will be used to denote a generic set of real numbers, but will also denote certain operations on functions, among other things, while Ω will be used to denote a bandwidth parameter, but will also be used to denote certain subsets of integers. The superscripted A^* refers to conjugate transpose of the matrix A, but f^* will sometimes mean the function $f^*(z) = \bar{f}(1/\bar{z})$. The intended use of such symbols will be clear in context. We have also listed symbols with potentially conflicting uses in the list of symbols. The symbols $\mathbb{Z}, \mathbb{R}$, and $\mathbb{C}$ denote the integers and real and complex numbers, respectively. The symbols $\mathbb{T}$ and $\mathbb{Z}_N$ denote the torus, $[0,1)$ with addition modulo one, and the integers modulo N, respectively. We will use $\mathbb{N}$ and $\mathbb{N}_n$ to denote the natural numbers and the set $\{1,2,\dots,n\}$, respectively. The Euclidean space $\mathbb{R}^N$ is the (real) vector space of N-tuples of real numbers, and $\mathbb{C}^M$ is the (complex) vector space of M-tuples of complex numbers. Generic elements of $\mathbb{R}^N$ or $\mathbb{C}^M$ will be denoted with boldface letters, e.g., $\mathbf{x} = (x_1,\dots,x_N) \sim [x_1,\dots,x_N]^T$ and $\mathbf{z} = (z_1,\dots,z_M) \sim [z_1,\dots,z_M]^T$ with the latter notation being used when elements are regarded as column vectors. Matrices will be defined with respect to standard coordinates on $\mathbb{R}^N$ and $\mathbb{C}^N$. We will use A^T to denote the transpose of a matrix and A^* to denote the Hermitian or conjugate transpose $\bar{A}^T$.

In some cases, we will not distinguish between an element of $\mathbb{C}^M$ and its periodic extension to a function on the group $\mathbb{Z}_M$. In such cases we are apt to write $\mathbf{z} = (z(0),\dots,z(M-1)) \sim [z_0,\dots,z_{M-1}]^T$. We also use boldface letters $\mathbf{x}$ or $\mathbf{z}$ to denote real- or complex-valued sequences defined on the integers, e.g., $\mathbf{x} = (\dots,x_{-1},x_0,x_1,\dots)$ or $\mathbf{z} = \{z(k)\}_{k\in\mathbb{Z}}$ when function notation is preferred. The unit circle $\mathbb{T} = \{z \in \mathbb{C} : |z| = 1\}$ is identified with the unit interval via $z = e^{2\pi i\omega}$, $\omega \in [0,1)$. Functions on $\mathbb{T}$ typically will be denoted with symbols $X(\omega)$ or, in special cases, by lower case bold Greek letters such as $\boldsymbol{\sigma}(\omega)$. Functions on $\mathbb{R}$ generically will be denoted by symbols f, g, etc. The greatest integer *less* than or equal to t is denoted by $[t]$ or $\lfloor t \rfloor$, with $\lceil t \rceil$ denoting the least integer *greater* than or equal to t, and $t_+ = \max(t, 0)$. The "Kronecker delta" is $\delta_{\alpha\beta} = 1$ if $\alpha = \beta$ and $\delta_{\alpha\beta} = 0$ if $\alpha \neq \beta$. For subsets $A \subset \mathbb{R}$ and $B \subset \mathbb{R}$, $A + B = \{a + b : a \in A \,\text{and}\, b \in B\}$ and for $\alpha \in \mathbb{R}$, $\alpha A = \{\alpha a : a \in A\}$. In analytical estimates, the symbol C will be used generically to denote an unspecified constant, while the symbol $a \ll b$ will be used in heuristic arguments to indicate that a is at least several orders of magnitude smaller than b. For a parameter p we will often write $p = p(a)$ or $p = p(a, b)$ and so on to indicate a functional dependence of p on the variable a or the variables a, b and so on.

A.2 Mathematical Prerequisites

Below we list a number of mathematical results that will be taken for granted in the text. Most of these prerequisites can be found in standard textbooks.

Analytic Functions

Functions generally will take values in the complex numbers or real numbers. A function defined on an open subset of $\mathbb{C}$ is *analytic* at $z_0 \in \mathbb{C}$ if it can be expanded in a power series $f(z) = \sum_{k=0}^{\infty} a_k z^k$ converging uniformly in $\{z : |z - z_0| < R\}$ for some $r > 0$. A function $f : \mathbb{R} \to \mathbb{C}$ is real analytic if it has an analytic extension about any $x_0 \in \mathbb{R}$. A function $f(z)$ on $\mathbb{C}$ is said to be of *exponential type* $\sigma > 0$ if for every $\varepsilon > 0$ there is a constant C_ε such that $|f(z)| \leq C_\varepsilon \mathrm{e}^{|z|(\sigma+\varepsilon)}$ for any $z \in \mathbb{C}$. The exponential type of f is then $\limsup_{|z|\to\infty} (\log|f(z)|)/|z|$. One version of the *Paley–Wiener theorem*, e.g., [316], states that if F is supported in $[-\Omega, \Omega]$ and is square-integrable then (i) $f(z) = \int_{-\Omega}^{\Omega} F(u)\, \mathrm{e}^{2\pi i u z}\, \mathrm{d}u$ defines (ii) an entire (i.e., analytic at every $z \in \mathbb{C}$) function of exponential type Ω such that (iii) $\int_{-\infty}^{\infty} |f(x + \mathrm{i}y)|^2\, \mathrm{d}x < \infty$ for all $y \in \mathbb{R}$. Conversely, any f satisfying (ii) and (iii) can be expressed via (i) with $\int_{-\Omega}^{\Omega} |F(u)|^2\, \mathrm{d}u < \infty$.

Function Spaces

Functions on $\mathbb{R}$, on $\mathbb{Z}$, $\mathbb{T}$, and $\mathbb{Z}_M$ all play important roles in the text. The symbol $\ell^2(\mathbb{Z})$ denotes the space of square-summable sequences on $\mathbb{Z}$. We assume a basic familiarity with Lebesgue measure. We say that a property holds a.e. (almost everywhere) if it holds outside a subset of $\mathbb{R}$ (or $\mathbb{R}^N$) having Lebesgue measure zero. The space $L^p(\mathbb{R})$ consists of all equivalence classes of complex-valued functions (two functions are equivalent if they are equal a.e.) that are square-integrable with respect to Lebesgue measure, that is, $\int_{-\infty}^{\infty} |f(t)|^p\, \mathrm{d}t < \infty$. When $1 \leq p < \infty$, $L^p(\mathbb{R})$ is a *Banach space* with *norm* $\|f\|_p = (\int |f|^p)^{1/p}$. The space $L^\infty(\mathbb{R})$ consists of all *essentially bounded* functions. The $L^\infty(\mathbb{R})$-norm of f is the supremum of those values that are exceeded by $|f|$ on a set of positive Lebesgue measure. In rare cases, we will consider functions p-integrable with respect to a generic measure μ, that is, $\int |f|^p \, d\mu < \infty$. Then we write $f \in L^p(\mu)$. When $S \subset \mathbb{R}$ is Lebesgue measurable, we write $L^p(S)$ to denote those functions that are p-integrable with respect to Lebesgue measure on S, that is, $\int_S |f|^p < \infty$. When $S = [a,b]$ we abbreviate $L^p(S) = L^p[a,b]$.

The case $p = 2$ is particularly important in this text, since $L^2(\mathbb{R})$ is a Hilbert space. If "$\|f\|$" is written (with no subscript) then the intended norm is the L^2-norm, unless specifically indicated otherwise. Any subsets considered are assumed to be Lebesgue measurable. We write $\mathbb{1}_\Sigma$ to denote the indicator function taking the value one on Σ and zero outside Σ. The closure of S will be denoted $\bar{S}$ and the support of a continuous function f is $\mathrm{supp}(f) = \overline{\{t : f(t) \neq 0\}}$. The Lebesgue measure of S is $|S| = \int \mathbb{1}_S(t) \, dt$. When dealing specifically with a discrete set Ω (a subset of $\mathbb{Z}$ or $\mathbb{Z}_M$) we also use $|\Omega|$ to denote the number of elements of Ω, though $\#(\Omega)$ will also be used when needed to avoid ambiguity. When sums and integrals are written without qualifying limits, they extend over all elements for which the integrand or summand is defined. Thus, if $\mathbf{s}$ is a sequence on $\mathbb{Z}$ then $\sum s_k$ or $\sum_k s_k$ means $\sum_{k=-\infty}^{\infty} s_k$, while if f is a function defined on $\mathbb{R}$ then $\int f$ means $\int_{\mathbb{R}} f(t) \, dt$.

A.2.1 Functional Analysis Prerequisites

Hilbert Spaces

The space $\ell^2(\mathbb{Z})$ consists of all complex-valued sequences $\mathbf{x} = \{x(k)\}_{k \in \mathbb{Z}}$ such that $\sum_{k=-\infty}^{\infty} |x(k)|^2 < \infty$. It has a Hermitian *inner product* $\langle \mathbf{x}, \mathbf{y} \rangle = \sum_{k=-\infty}^{\infty} x(k)\overline{y(k)}$ and norm $\|\mathbf{x}\| = \sqrt{\langle \mathbf{x}, \mathbf{x} \rangle}$. Under this norm, $\ell^2(\mathbb{Z})$ is *complete*, that is, whenever $\{\mathbf{x}_1, \mathbf{x}_2, \ldots\}$ is a sequence in $\ell^2(\mathbb{Z})$ such that $\lim_{n\to\infty} \sup_{m \geq n} \|\mathbf{x}_n - \mathbf{x}_m\| = 0$, there is an $\mathbf{x} \in \ell^2(\mathbb{Z})$ such that $\lim_{n\to\infty} \|\mathbf{x}_n - \mathbf{x}\| = 0$. Two elements $\mathbf{x}$ and $\mathbf{y}$ of $\ell^2(\mathbb{Z})$ are *orthogonal* if $\langle \mathbf{x}, \mathbf{y} \rangle = 0$. The elements $\mathbf{e}_n$ such that $e_n(k) = \delta_{n,k}$ form an *orthonormal basis* (ONB) for $\ell^2(\mathbb{Z})$.

We assume that the reader is familiar with the concept of a vector space and of an inner product space. A Hilbert space $\mathscr{H}$ is an inner product space with norm $\|\cdot\|_{\mathscr{H}}$ defined by $\|\mathbf{x}\|_{\mathscr{H}}^2 = \langle \mathbf{x}, \mathbf{x} \rangle_{\mathscr{H}}$ that is complete in the same sense as $\ell^2(\mathbb{Z})$. An infinite-dimensional inner product space is called a *separable Hilbert space* if it possesses a countable (Schauder) basis; that is, if there is a countable set $B = \{\mathbf{b}_n\} \subset \mathscr{H}$ such that any $\mathbf{x} \in \mathscr{H}$ can be written uniquely as $\mathbf{x} = \sum_{n=1}^{\infty} x_n \mathbf{b}_n$ ($x_n \in \mathbb{C}$). The *Gram–Schmidt process* enables one to produce from any Schauder basis of $\mathscr{H}$ an ONB $\boldsymbol{\beta}_n$, which we will choose to index by $n \in \mathbb{Z}$. Then any $\mathbf{x} \in \mathscr{H}$ can be expressed uniquely as $\mathbf{x} = \sum_{n=-\infty}^{\infty} \langle \mathbf{x}, \boldsymbol{\beta}_n \rangle \boldsymbol{\beta}_n$. Just as any finite-dimensional vector space (over $\mathbb{C}$) is isomorphic to $\mathbb{C}^M$ for some M, any separable Hilbert space is isomorphic to $\ell^2(\mathbb{Z})$. That is, if $\mathscr{H}$ is a Hilbert space with ONB $\{\mathbf{h}_n\}_{n \in \mathbb{Z}}$, then the mapping that sends $\mathbf{e}_n \in \ell^2(\mathbb{Z})$ to $\mathbf{h}_n \in \mathscr{H}$ extends to a linear isometry between $\ell^2(\mathbb{Z})$ and $\mathscr{H}$. As examples, $L^2(\mathbb{T})$ is a separable Hilbert space with inner product $\langle X, Y \rangle_{L^2(\mathbb{T})} = \int_0^1 X(\omega)\overline{Y(\omega)} \, d\omega$ since the exponentials $e^{2\pi i n \omega}$ ($n \in \mathbb{Z}$) form an ONB for $L^2(\mathbb{T})$. Also, $L^2(\mathbb{R})$ is a separable Hilbert space with inner product $\langle f, g\rangle_{L^2(\mathbb{R})} = \int_{-\infty}^{\infty} f(t)\overline{g(t)}\,\mathrm{d}t$ since the restricted exponentials $e_{m,n}(t) = \mathrm{e}^{2\pi int}\mathbb{1}_{[m,m+1)}(t)$, $(m,n \in \mathbb{Z})$ form an ONB for $L^2(\mathbb{R})$.

Cauchy–Schwarz Inequality

The Cauchy–Schwarz inequality states that, in a Hilbert space $\mathscr{H}$, $|\langle \mathbf{x}, \mathbf{y}\rangle| \leq \|\mathbf{x}\|\,\|\mathbf{y}\|$. If $\mathbf{x} = \sum_{n\in\mathbb{Z}} x_n\,\mathbf{e}_n \in \ell^2(\mathbb{Z})$ and $\mathbf{y} = \sum_{n\in\mathbb{Z}} y_n\,\mathbf{e}_n$ then

$$|\langle \mathbf{x}, \mathbf{y}\rangle| = \left|\sum_{n=-\infty}^{\infty} x_n \bar{y}_n\right| \leq \left(\sum_{n=-\infty}^{\infty} |x_n|^2\right)^{1/2} \left(\sum_{n=-\infty}^{\infty} |y_n|^2\right)^{1/2} = \|\mathbf{x}\|\,\|\mathbf{y}\|\,.$$

Wirtinger's Inequality

If $f, f' \in L^2[a,b]$ and either $f(a) = 0$ or $f(b) = 0$, then

$$\int_a^b |f(x)|^2\,\mathrm{d}x \leq \frac{4(b-a)^2}{\pi^2}\int_a^b |f'(x)|^2\,\mathrm{d}x.$$

Subspaces and Orthogonal Complements

A subspace of a separable Hilbert space $\mathscr{H}$ is a subset of $\mathscr{H}$ that is also a vector space. A closed subspace of $\mathscr{H}$ is also a Hilbert space. Given an arbitrary subset $S \subset \mathscr{H}$, $\operatorname{span} S$ denotes the subspace of all finite linear combinations taken from S and $\overline{\operatorname{span} S}$ denotes the closed subspace obtained by taking all limits of sequences in $\operatorname{span} S$. The orthogonal complement of S, denoted $S^\perp$, consists of all $\mathbf{y} \in \mathscr{H}$ such that $\langle \mathbf{x}, \mathbf{y}\rangle = 0$ for all $\mathbf{x} \in S$. $S^\perp$ is a closed subspace of $\mathscr{H}$ and any element $\mathbf{h} \in \mathscr{H}$ can be written uniquely as $\mathbf{h} = \mathbf{x} + \mathbf{y}$ where $\mathbf{x} \in \overline{\operatorname{span} S}$ and $\mathbf{y} \in S^\perp$. One says that $\mathscr{H}$ is the *orthogonal direct sum* of $\overline{\operatorname{span} S}$ and $S^\perp$, denoted by $\mathscr{H} = \overline{\operatorname{span} S} \oplus S^\perp$.

Linear Operators

An operator $L : \mathscr{H} \to \mathscr{K}$ is *linear* if $L(\alpha\mathbf{x} + \beta\mathbf{y}) = \alpha L(\mathbf{x}) + \beta L(\mathbf{y})$ whenever $\alpha, \beta \in \mathbb{C}$ and $\mathbf{x}, \mathbf{y} \in \mathscr{H}$. *Any operator under consideration will be assumed linear unless explicitly mentioned to the contrary.* One says that L is bounded with norm $B = \|L\|_{\mathscr{H}\to\mathscr{K}}$ if $B = \sup_{\mathbf{x}\neq\mathbf{0}}\{\|L\mathbf{x}\|_{\mathscr{K}}/\|\mathbf{x}\|_{\mathscr{H}}\} < \infty$.

The matrix $A = A_T$ of an operator $T : \mathbb{C}^N \to \mathbb{C}^M$ (or from $\ell^2(\mathbb{Z}) \to \ell^2(\mathbb{Z})$) will always be taken with respect to standard coordinates $\{\mathbf{e}_n\}$, that is, $A_{ij} = \langle T\mathbf{e}_j, \mathbf{e}_i\rangle$. A diagonal $N \times N$ matrix D will be written $D = \operatorname{diag}(a_1, \ldots, a_N)$ and has (i,j)th entry $D_{ij} = a_i\delta_{ij}$. The $N \times N$ identity matrix is denoted I_N. When acted on by matrices in standard coordinates, elements of $\mathbb{C}^N$ are thought of as column vectors, with matrices acting by multiplication on the left, unless specifically stated otherwise. The same remarks apply to real Hilbert spaces.

Given two Hilbert space operators $S:\mathscr{H}\to\mathscr{K}_1$ and $T:\mathscr{H}\to\mathscr{K}_2$, $S\oplus T$ denotes the mapping that takes $\mathbf{x}\in\mathscr{H}$ to $(S\mathbf{x},T\mathbf{x})\in\mathscr{K}_1\oplus\mathscr{K}_2$. When $\mathscr{H}=\mathbb{C}^N$, $\mathscr{K}_1=\mathbb{C}^{M_1}$, $\mathscr{K}_2=\mathbb{C}^{M_2}$, and S and T are represented by their matrices A_S and A_T in standard coordinates, the matrix A of $S\oplus T$ is $A_S\oplus A_T$ whose kth row is the kth row of A_S if $k\le M_1$ and is the $(k-M_1)$th row of A_T if $M_1<k\le M_1+M_2$.

Projections

A projection operator P on a Hilbert space $\mathscr{H}$ is an operator that acts as the identity on its range. P is *orthogonal* if $(P\mathscr{H})^\perp$ is contained in, hence equal to, the kernel of P. If $(P\mathscr{H})^\perp$ is not contained in $\ker P$ then the projection P is called *oblique*. The operator $P:\mathbb{C}^2\to\mathbb{C}^2$, $P(z_1,z_2)=(z_1+z_2,0)$ is a simple example satisfying $P^2=P$ but $\ker P=\{(z,-z)\}\neq(\operatorname{Ran}P)^\perp=\{(0,z)\}$.

Reproducing Kernel Hilbert Space

The Paley–Wiener space PW is an example of a reproducing kernel Hilbert space (RKHS)—a Hilbert space $\mathscr{K}$ of functions in which pointwise evaluation $f\mapsto f(t)$ defines a continuous linear map $k_t:\mathscr{K}\to\mathbb{C}$. The *Riesz representation theorem* implies that k_t can be identified with an element $k_t(s)=K(t,s)$, defining a *reproducing kernel* of $\mathscr{K}$. In the case of PW, $K(t,s)=\operatorname{sinc}(t-s)$.

Compact, Trace-class, and Hilbert–Schmidt Operators

A linear operator U from the Hilbert space $\mathscr{H}_1$ to $\mathscr{H}_2$ is said to be *unitary* if $\langle U\mathbf{x},U\mathbf{y}\rangle_{\mathscr{H}_2}=\langle\mathbf{x},\mathbf{y}\rangle_{\mathscr{H}_1}$. Two operators S and T are called *unitarily equivalent* if $T=USU^{-1}$ for some unitary operator U. The (Hermitian) *adjoint* A^* of $A:\mathscr{H}_1\to\mathscr{H}_2$ is defined via $\langle A\mathbf{x},\mathbf{y}\rangle_{\mathscr{H}_2}=\langle\mathbf{x},A^*\mathbf{y}\rangle_{\mathscr{H}_1}$. One says that $A:\mathscr{H}\to\mathscr{H}$ is *self-adjoint* if $A^*=A$. An operator $A:\mathscr{H}_1\to\mathscr{H}_2$ is *compact* if the closure of the image of any bounded set under A is compact. The *spectral theorem for compact self-adjoint operators* states that for any compact, self-adjoint operator A on $\mathscr{H}$, there is an ONB $\{\beta_n\}$ consisting of the normalized eigenvectors of A. Then $A=\sum_{n=1}^{\infty}\lambda_n\beta_n\otimes\beta_n$, meaning $A\mathbf{x}=\sum_{n=-\infty}^{\infty}\langle\mathbf{x},A\beta_n\rangle\beta_n=\sum_{n=-\infty}^{\infty}\lambda_n\langle\mathbf{x},\beta_n\rangle\beta_n$.

In the case of a compact, self-adjoint operator T on $L^2(\mathbb{R})$, let $\{\psi_n\}_{n=0}^{\infty}$ be an eigenbasis, i.e., $T\psi_n=\lambda_n\psi_n$, ordered such that $\lambda_0\ge\lambda_1\ge\cdots$. Suppose that for any $f,g\in L^2(\mathbb{R})$, $\langle Tf,g\rangle=\iint_{\mathbb{R}^2}K(x,y)f(y)g(x)\,\mathrm{d}y\,\mathrm{d}x$. Then one can write $K(x,y)=\sum_{n=0}^{\infty}\lambda_n\,\psi_n(x)\overline{\psi_n}(y)$ with, at least, weak operator convergence. This is one form of *Mercer's theorem.*

The *Hilbert–Schmidt* class consists of those operators A on $\mathscr{H}$ for which there is an ONB $\{\beta_n\}$ under which the sequence $\{\|A\beta_n\|_{\mathscr{H}}\}$ is square-summable (the sum is then independent of the basis since two such sequences differ by a unitary operator). When A operating on $L^2(\mu)$ has an integral kernel $k(x,y)$, the Hilbert–Schmidt norm $\|\cdot\|_{\mathrm{HS}}$ of A squared can also be expressed as the double integral of $|k|^2$. A bounded operator T on $\mathscr{H}$ is said to be *trace class* if, in some orthonormal basis $\{\mathrm{e}_n\}$, $\mathrm{tr}(T) = \sum\langle T\mathrm{e}_n, \mathrm{e}_n\rangle$ converges absolutely. The trace is linear and is preserved under unitary equivalence. In this case $\|A\|_{\mathrm{tr}} = \sum\langle (T^*T)^{1/2}\mathrm{e}_n, \mathrm{e}_n\rangle$ is called the trace norm of T. The trace norm satisfies $\|A+B\|_{\mathrm{tr}} \leq \|A\|_{\mathrm{tr}} + \|B\|_{\mathrm{tr}}$ and $\|AB\|_{\mathrm{tr}} \leq \min\{\|A\|_{\mathrm{tr}}\|B\|, \|A\|\|B\|_{\mathrm{tr}}\}$ where $\|A\|$ is the operator norm of A on $\mathscr{H}$. A compact, self-adjoint operator is trace class if its eigenvalues are summable, that is, $\mathrm{tr}\,(A) = \|A\|_{\mathrm{tr}} = \sum_{n=0}^{\infty} \lambda_n < \infty$. In the case of a trace-class operator T on $L^2(\mathbb{R})$ with kernel $K_T(s,t)$, one has $\mathrm{tr}\,(T) = \int K_T(s,s)\,\mathrm{d}s$.

Eigenvalues I

There are different ways in which to estimate or count eigenvalues. For example, when A is compact and self adjoint, the approximation numbers $\sigma_N(A) = \inf\{\|A - A_N\| : \mathit{rank}(A_N) \leq N\}$, where the rank of A_N is the dimension of its range, coincides with the Nth largest eigenvalue λ_N of A (starting with $N = 0$: $\sigma_0 = \lambda_0 = \|A\|$). The *Weyl–Courant lemma*, also known as the *minimax principle*, provides a variational description of the eigenvalues $\lambda_0 \geq \lambda_1 \geq \cdots$ of a compact, self-adjoint operator A on a Hilbert space $\mathscr{H}$. The principle states that the $(n+1)$st largest eigenvalue λ_n is

$$\lambda_n = \sup_{\mathscr{S}_{n+1}} \inf_{f\in\mathscr{S}_{n+1}:\|f\|=1} \|Af\|$$

where $\mathscr{S}_n$ ranges over all n-dimensional subspaces of $\mathscr{H}$. Similarly,

$$\lambda_n = \inf_{\mathscr{S}_n} \sup_{f\perp\mathscr{S}_n:\|f\|=1} \|Af\|\,.$$

For example, λ_0, the largest eigenvalue, corresponds to the subspace generated by one of its eigenvectors, while λ_1 corresponds to the norm of the operator restricted to the orthogonal complement of the span of an eigenvector of λ_0.

A.2.2 Fourier Analysis Prerequisites

Schwartz Class and Fourier Transform

The *Schwartz class* $\mathcal{S}(\mathbb{R})$ is the linear topological function space in which convergence is defined in terms of the countable class of seminorms $\|\varphi\|_{p,q} = \sup_{t\in\mathbb{R}} |(1+|t|)^p \varphi^{(q)}(t)|$, $p,q = 0,1,2,\ldots$. A sequence of functions φ_n converges to zero in $\mathcal{S}(\mathbb{R})$ if $\lim_{n\to\infty}\|\varphi_n\|_{p,q} = 0$ for all $p,q = 0,1,2,\ldots$. The integral of any Schwartz function converges absolutely. The Fourier transform of a Schwartz function φ is defined as

$$\widehat{\varphi}(\xi) = \int_{-\infty}^{\infty} \varphi(t)\, \mathrm{e}^{-2\pi i t\xi}\, \mathrm{d}t.$$

In view of (1.7), the Fourier transform maps $\mathcal{S}(\mathbb{R})$ into itself continuously, and one can also define the *inverse Fourier transform* $\mathscr{F}^{-1}(\psi)$ of $\psi \in \mathcal{S}(\mathbb{R})$ as

$$\mathscr{F}^{-1}(\psi)(t) = \psi^{\vee}(t) = \int_{-\infty}^{\infty} \psi(\xi)\, \mathrm{e}^{2\pi i t\xi}\, \mathrm{d}\xi.$$

If $\varphi, \psi \in \mathcal{S}(\mathbb{R})$, their convolution is defined as $\varphi * \psi(t) = \int \varphi(t-s)\psi(s)\,\mathrm{d}s$ and also lies in $\mathcal{S}(\mathbb{R})$. The Fourier transform takes convolution to multiplication. That is, if $\varphi, \psi \in \mathcal{S}(\mathbb{R})$ then $(\varphi * \psi)^{\wedge}(\xi) = \widehat{\varphi}(\xi)\widehat{\psi}(\xi)$.

The class $\mathcal{S}(\mathbb{R})$ is dense in $L^2(\mathbb{R})$. That is, for any $f \in L^2(\mathbb{R})$ there is a sequence $\{\varphi_n\} \subset \mathcal{S}(\mathbb{R})$ such that $\|\varphi_n - f\| \to 0$ as $n \to \infty$. Using this density property, one can pass relationships that are defined, a priori, on $\mathcal{S}(\mathbb{R})$ or on $\mathcal{S}(\mathbb{R}) \times \mathcal{S}(\mathbb{R})$, in terms of absolutely convergent integrals or sums, to corresponding relationships that apply to $L^2(\mathbb{R})$ or $L^2(\mathbb{R}) \times L^2(\mathbb{R})$, provided one takes care in establishing the sense in which the limiting formula converges (e.g., in L^2-norm). One example is the Fourier inversion formula: for $\varphi \in \mathcal{S}(\mathbb{R})$, one has $\varphi(t) = \int \widehat{\varphi}(\xi)\,\mathrm{e}^{2\pi i t\xi}\,\mathrm{d}\xi$ and this identity holds pointwise. On $L^2(\mathbb{R})$, the corresponding relationship holds in the limiting sense $f = \lim_{N\to\infty} f_N$ where $f_N(t) = \int_{-N}^{N} \widehat{f}(\xi)\,\mathrm{e}^{2\pi i t\xi}\,\mathrm{d}\xi$ with convergence $\|f - f_N\| \to 0$. Note that $f_N = f * \mathrm{sinc}_N(t)$ where $\mathrm{sinc}\,(t) = \frac{\sin \pi t}{\pi t}$ and $\mathrm{sinc}_\Omega = \Omega \mathrm{sinc}\,(\Omega t) = \frac{\sin \pi \Omega t}{\pi t}$. Convolution with sinc_Ω defines an orthogonal projection onto PW_Ω, the space of functions that are band limited to $[-\Omega/2, \Omega/2]$. A second example is the *Parseval formula*:

$$\langle \varphi, \psi \rangle = \int \varphi(t)\,\overline{\psi}(t)\,\mathrm{d}t = \int \widehat{\varphi}(\xi)\,\overline{\widehat{\psi}}(\xi)\,\mathrm{d}\xi = \langle \widehat{\varphi}, \widehat{\psi} \rangle. \tag{A.1}$$

A third example is the *Poisson summation formula*:

$$\sum_{k\in\mathbb{Z}} \varphi(k) = \sum_{\ell\in\mathbb{Z}} \widehat{\varphi}(\ell). \tag{A.2}$$

The *Hilbert transform* H plays a role in Chap. 1. It is defined by $Hf(x) = f * \mathrm{pv}\frac{1}{x}$ and satisfies $\widehat{Hf}(\xi) = -i\,\mathrm{sgn}\,(\xi)\widehat{f}(\xi)$ where $\mathrm{sgn}\,(\xi) = 1$ if $\xi > 0$ and $\mathrm{sgn}\,(\xi) = -1$ if $\xi < 0$. In particular, $H^* = -H$.

Distributions

Schwartz's *tempered distributions*, denoted $\mathcal{S}'(\mathbb{R})$, are the continuous linear mappings from $\mathcal{S}$ to $\mathbb{C}$. Standard references include Schwartz's works [285–287] and Hörmander [156]; cf. also [26, 316]. One defines a dual pairing on $\mathcal{S}' \times \mathcal{S}$ by (T, φ) which extends the real correlation of two functions, that is, if $T = T_f$ for some function $f \in L^1_{\mathrm{loc}}(\mathbb{R})$, then $(T_f, \varphi) = \int f(t)\,\varphi(t)\,\mathrm{d}t$. Since $\mathcal{S}$ is closed under pointwise multiplication, if $T \in \mathcal{S}'$ and $\psi \in \mathcal{S}$ then the product $\psi T \in \mathcal{S}'$ is defined by $(\psi T, \varphi) = (T, \psi\varphi)$. One can differentiate a Schwartz distribution

T by generalizing the integration by parts formula. That is, $(T', \varphi) = -(T, \varphi')$. All finite Borel measures are elements of $\mathcal{S}'(\mathbb{R})$. In particular, δ is defined by $(\delta, \varphi) = \varphi(0)$ and this definition extends to $\varphi \in C_0(\mathbb{R})$ (continuous and vanishing at infinity). One can also regard δ as the derivative of the Heaviside function $\mathbb{1}_{[0,\infty)}$: $(\mathbb{1}'_{[0,\infty)}, \varphi) = -(\mathbb{1}_{[0,\infty)}, \varphi') = -\int_0^\infty \varphi' = \varphi(0)$. The distributional derivative of the Dirac delta δ is $(\delta', \varphi) = -\varphi'(0)$. If $\psi \in \mathcal{S}(\mathbb{R})$ then $(\psi\delta', \varphi) = -(\psi\varphi)'(0) = -\psi(0)\varphi'(0) - \psi'(0)\varphi(0)$.

Fourier Series and the Discrete Fourier Transform

The *Fourier series* of a sequence $\{c_k\} \in \ell^2(\mathbb{Z})$ is defined as $C(\omega) = \sum_{k=-\infty}^{\infty} c_k \, \mathrm{e}^{2\pi i k\omega}$. The sum converges to a periodic function in $L^2(\mathbb{T})$, and the mapping from $\ell^2(\mathbb{Z})$ to $L^2(\mathbb{T})$ is an isometry. In particular, it is continuous, one to one and onto. If $F \in L^2(\mathbb{T})$ then its kth Fourier coefficient is $\widehat{F}[k] = \int_0^1 F(\omega)\mathrm{e}^{-2\pi i k\omega}\,\mathrm{d}\omega$. The mapping that takes $F \in L^2(\mathbb{T})$ to its sequence $\{\widehat{F}[k]\}$ of Fourier coefficients is the inverse of the mapping that takes a sequence in $\ell^2(\mathbb{Z})$ to its Fourier series. Thus, the mapping that sends $\{c_k\}$ to its Fourier series $C(\xi)$ is a Hilbert space isomorphism between $\ell^2(\mathbb{Z})$ and $L^2(\mathbb{T})$. This *unitarity* is expressed by the *discrete Parseval formula*

$$\langle \{c_k\}, \{d_k\} \rangle_{\ell^2(\mathbb{Z})} = \sum_{k=-\infty}^{\infty} c_k \bar{d}_k = \int_0^1 C(\omega)\bar{D}(\omega)\,\mathrm{d}\omega = \langle C, D \rangle_{L^2(\mathbb{T})}. \tag{A.3}$$

There is a convolution–multiplication relationship for sequences and their Fourier series. Write $\mathbf{c} = \{c(k)\}_{k\in\mathbb{Z}}$ and define $\mathbf{c} * \mathbf{d}(k) = \sum_{\ell=-\infty}^{\infty} c(k-\ell)\,d(\ell)$. The Fourier series of $\mathbf{c} * \mathbf{d}$ is the product $C(\omega)D(\omega)$. If $\mathbf{c}$ and $\mathbf{d}$ are in $\ell^2(\mathbb{Z})$ then $CD \in L^1(\mathbb{Z})$ and $\mathbf{c} * \mathbf{d} \in \ell^\infty(\mathbb{Z})$.

One can also convolve two functions X and Y on $\mathbb{T}$ via $X * Y(\omega) = \int_0^1 X(\omega - \eta)Y(\eta)\,\mathrm{d}\eta$. The Dirichlet kernel $D_N(t)$ will be defined as

$$D_N(t) = \sum_{k=0}^{N-1} \mathrm{e}^{\pi i(2k+1-N)t} = \frac{\sin N\pi t}{\sin \pi t}.$$

One often, instead, writes $D_N(t) = \sum_{k=-N}^{N} \mathrm{e}^{2\pi i k t}$ which, in our notation, equals $D_{2N+1}(t)$. In particular, convolution with $D_{2N+1}(\omega)$ defines an orthogonal projection from $L^2(\mathbb{T})$ onto the space of *trigonometric polynomials* of degree at most N. If $X \in L^2(\mathbb{T})$ then $\|X - D_N * X\| \to 0$ as $N \to \infty$.

As sets, $\mathbb{Z}_M = \{0, 1, \ldots, M-1\}$. However, $\mathbb{Z}_M$ is also a group when addition is defined modulo M. The discrete Fourier transform maps a vector $\mathbf{x}$ to its Fourier transform $X(k) = \sum_{\ell=0}^{M-1} x_\ell \, \mathrm{e}^{2\pi i k\ell/M} / \sqrt{M}$. In other words, the Fourier transform is defined by left multiplication of $\mathbf{x}$ by the matrix $W_{k\ell} = \mathrm{e}^{2\pi i k\ell/M} / \sqrt{M}$. This mapping is unitary from $\mathbb{C}^M \simeq \ell^2(\mathbb{Z}_M)$. That is, one has the finite Parseval formula $\langle \mathbf{x}, \mathbf{y} \rangle_{\mathbb{C}^M} = \langle X, Y \rangle_{\mathbb{C}^M}$. Strictly speaking, one should regard a function on $\mathbb{Z}_M$ as an M-periodic sequence on $\mathbb{Z}$. However, computationally it is easier to work directly with $\mathbf{x} = [x_0, x_1, \ldots, x_{M-1}]^T$ than with its periodic extension. For example, the *circulant convolution* of $\mathbf{x}$ and $\mathbf{y}$ is computed via $\mathbf{x} * \mathbf{y} = W^{-1}(XY)$. The groups $\mathbb{R}$, $\mathbb{Z}$ and $\mathbb{Z}_M$ are all locally compact abelian groups (LCAGs), and much of Fourier analysis extends readily from these examples to general LCAGs.

Eigenvalues II

Szegő's eigenvalue distribution theorem states, in effect, that the eigenvalues of a translation-invariant operator on $L^2(\mathbb{R})$ are given by the Fourier transform of its kernel. A version due to Landau [190] states that if $S \subset \mathbb{R}$ is a fixed set of finite measure, and $p \in L^1_{\rm loc}$ is such that $\widehat{p} \in L^1$ is real-valued and satisfies $|\{\xi : \widehat{p}(\xi) = \alpha\}| = 0$ for all $\alpha \in \mathbb{R}$, then for

$$A_r f(t) = \mathbb{1}_{rS}(t) \int \mathbb{1}_{rS}(s)\, p(t-s)\, f(s)\, \mathrm{d}s,$$

the number $N_\alpha(A_r)$ of its eigenvalues that are at least α satisfies

$$\lim_{r\to\infty} \frac{N_\alpha(A_r)}{r} = |S| |\{\xi : \widehat{p}(\xi) \geq \alpha\}| .$$

Szegő's theorem (cf. [166]) amounts to the case in which $S = [-1,1]$.

If A is a compact, self-adjoint operator on a Hilbert space $\mathscr{H}$ that has a simple spectrum, then the function $N(A,\alpha)$ of eigenvalues of A larger than α takes values in $\{0,1,2,\ldots\}$ when $\alpha > 0$ and has jump discontinuities at the eigenvalues $\lambda_n \subset (0, \|A\|]$. The derivative $\mathrm{d}_t[-N(A,t)]$ is thus a discrete sum of Dirac point masses at the eigenvalues. One has $\mathrm{tr}\,(A) = \int_0^{\|A\|} t\, \mathrm{d}_t[-N(A,t)]$ and, more generally, $\mathrm{tr}\,(A^m) = \int_0^{\|A\|} t^m\, \mathrm{d}_t[-N(A,t)]$.

A.2.3 Approximation Miscellany

Stirling Approximation

The *Stirling approximation* states that

$$\lim_{n\to\infty} \frac{n!}{\sqrt{2\pi n}\left(\frac{n}{\mathrm{e}}\right)^n} = 1 \quad \text{or} \quad n! \approx \sqrt{2\pi n}\left(\frac{n}{\mathrm{e}}\right)^n .$$

Gershgorin Circle Theorem

The *Gershgorin circle theorem* provides effective initial estimates for the eigenvalues of a complex $N \times N$ matrix $A = \{a_{k\ell}\}$ when the elements of A decay quickly off the main diagonal. It states that any eigenvalue lies within at least one disc of the form $\{|z - a_{kk}| \leq \sum_{\ell \neq k} |a_{k\ell}|\}$.

A.2.4 Useful Facts: ODEs and Special Functions Theory

Bessel Functions

The Bessel functions J_α of order α can be expressed as Frobenius solutions of

$$x^2 y'' + xy' + (x^2 - \alpha^2)y = 0$$

which can be written

$$J_\alpha(x) = \sum_{m=0}^{\infty} \frac{(-1)^m}{m!\,\Gamma(m+\alpha+1)} \left(\frac{x}{2}\right)^{2m+\alpha}.$$

For $x \gg |\alpha^2 - 1/4|$, one has $J_\alpha(x) \approx \sqrt{2/\pi x}\cos\big(x - \alpha\pi/2 - \pi/4\big)$.

Sturm–Liouville Theory

Denote by $u(t;\lambda)$ a solution of the *Sturm–Liouville* boundary value problem

$$\Big(\frac{\mathrm{d}}{\mathrm{d}t} p(t) \frac{\mathrm{d}}{\mathrm{d}t} + q(t) + \lambda \rho(t)\Big) u(t) = 0,$$
$$\alpha u(a) + \alpha' u'(a) = 0; \quad \beta u(b) + \beta' u'(b) = 0.$$

The problem is self-adjoint and has increasing, discrete spectrum. *Sturm's comparison theorem* says that if $\lambda_1 < \lambda_2$ are eigenvalues then between any pair of zeros of $u(t;\lambda_1)$ there lies a zero of $u(t;\lambda_2)$.

Sturm's sequence theorem provides a somewhat analogous discrete result. A Sturm sequence is a finite sequence $\{p_0, p_1, \ldots, p_n\}$ of polynomials of decreasing degrees such that p_0 is square-free, p_1 has the same sign as p_0' at any root of p_0, p_{j-1} and p_{j+1} have opposite signs at any root of p_j $(1 \leq j < n)$, and p_n does not change sign. Sturm's theorem states that if $\sigma(\xi)$ is the number of sign changes in the ordered sequence $p_0(\xi), p_1(\xi), \ldots, p_n(\xi)$, not counting zeros, then the number of *distinct* zeros of p_0 in the interval $(a,b]$ is $\sigma(b) - \sigma(a)$. The Euclidean algorithm is one way to produce a Sturm chain. The sequence theorem can then be used to count the number of real roots of p.

A.2.5 Prerequisites from Probability and Statistics

Statistical estimation is an important aspect of Chap. 3 of this text. There are a number of standard references on stochastic processes that cover background material in more detail from suitable perspectives, including Brillinger [49], Papoulis [256, 257], and Priestley [268, 269].

We denote by $\text{prob}(E)$ the probability of a random event E. The probability density function (pdf) f of a random variable x is $\text{prob}(t < f < t + \mathrm{d}t) = f(t)\mathrm{d}t$. The expected value $\mathbf{E}(x)$ of the random variable x with pdf f is $\mathbf{E}(x) = \int_{-\infty}^{\infty} t f(t)\,\mathrm{d}t$. The *Markov inequality* states that for any $a > 0$, $\text{prob}(|x| > a) \leq \mathbf{E}(|x|)/a$. The variance of x is $\text{var}(x) = \mathbf{E}((x - \mathbf{E}(x))^2)$. One says that x is *normally distributed* with mean μ and variance σ^2, denoted $x \sim N(\mu, \sigma^2)$, if its pdf is $\mathrm{e}^{-(t-\mu)^2/(2\sigma^2)}/\sqrt{2\pi\sigma^2}$. The sum of squares of k independent standard normal random variables is said to be χ^2*-distributed* with variance k.

In statistics, the *bias* of an estimator is the difference between the estimator's expected value and the true value of the parameter being estimated. If Θ_N^{est} is an estimator of a parameter Θ based on an input sample of size N then its *bias* is

$$\text{bias}(\Theta_N^{\text{est}}) = \mathbf{E}\{\Theta_N^{\text{est}} - \Theta\}.$$

The estimator is said to be *asymptotically unbiased* if $\text{bias}(\Theta_N^{\text{est}}) \to 0$ as $N \to \infty$. An estimator is a function of a random sample input so it is a random variable with a mean and variance. The estimator is said to be *consistent* if $\text{Prob}(|\Theta_N^{\text{est}} - \Theta| > \varepsilon) \to 0$ as $N \to \infty$ when the sample spaces of the estimators are appropriately defined.

Non-parametric estimators are ones whose values depend only on input data, and do not make use of any prior assumptions on the distribution that generates that data. Such estimators tend to be more *robust* but less *powerful* in comparison to parametric estimators.

A continuous-time stochastic process $x(t)$ is *wide-sense stationary* if its mean and covariance are time independent. If x is zero-mean, this means that $R_x(t_1, t_2) = \mathbf{E}(x(t_1)x(t_2))$ depends only on $\tau = t_1 - t_2$, that is, $R_x(t_1, t_2) = R_x(t_1 + s, t_2 + s) \equiv R_x(t_1 - t_2)$ for all s. Sometimes the term *wide-sense stationary* is abbreviated simply to *stationary*, although, strictly, the latter means that all moments are time independent. Corresponding definitions apply to discrete-time processes.

The *Wiener–Khintchine theorem* states that the power spectral density (PSD) $S(\xi)$ of a wide-sense stationary process is the Fourier transform of its autocorrelation function $R_x(\tau) = \mathbf{E}(x(t+\tau)\bar{x}(t))$. That is, $S(\xi) = \int R_x(\tau)\mathrm{e}^{-2\pi i\tau\xi}\,\mathrm{d}\tau$. A corresponding result applies to discrete-time processes.

Given observations $y(t) = h * x(t) + n(t)$ of a linear, time-invariant system with additive noise, $n(t)$, Wiener filtering seeks a deconvolution estimate $x^{\text{est}} = g * x$ of $x(t)$ that minimizes the expected mean-squared error $\mathbf{E}(x - x^{\text{est}})$. The *Wiener filter* g is defined by its Fourier transform $G(\xi) = \bar{H}(\xi)S(\xi)/(|H(\xi)|^2 S(\xi) + N(\xi))$ in which $H = \widehat{h}$, the Fourier transform of h, S is the PSD of x, and N is the PSD of n.

List of Symbols

Symbol	Description	Reference
$\oplus$	matrix direct sum	108
$\mathbb{T}$ and $\mathbb{Z}_N$	torus and integers mod N	223
$\mathbb{Z}, \mathbb{R}$, and $\mathbb{C}$	integers, real numbers, complex numbers	223
A_a	time- and band-limiting operator	23
a	time–bandwidth product	9
	finite, normalized time–bandwidth product	135
$A_{S\Omega}$	finite multiband- and time-limiting operator	137
$\mathscr{B}$	band-limiting matrix	35
$\mathscr{B}_W$	periodic band-limiting operator	27
E_b/N_0	bit energy-to-noise ratio	107
B_a	time- and multiband-limiting operator	130
δ	Dirac delta	230
$\delta_{\alpha\beta}$	Kronecker delta	224
$D_\pm$	Beurling upper and lower densities	159
$D_{\text{frame}}(\Lambda)$	frame density	167
Δ	uniform density	159
$\text{diam}(\Lambda)$	frame diameter	166
D_α	unitary dilation operator	10
D_N	Dirichlet kernel	27
D_S	finite time cutoff matrix	136
$\mathbf{E}$	expectation	91, 233
$\mathscr{E}(\Lambda)$	exponentials with frequencies in Λ	153
f^*	involution	174
$\mathscr{F}$	Fourier transform operator	2
$\mathscr{F}^{-1}$	inverse Fourier transform operator	229
$\mathscr{F}_{2\pi}$	renormalized Fourier transform	9
F_a	half time- and band-limiting operator	12
$\mathscr{F}_N$	N-point discrete Fourier transform	136
$\mathscr{H}$	Hermite operator	18
$\mathbb{1}_\Sigma$	indicator function of Σ	3, 225

$I(\omega)$	Bernoulli random variable	137
J_α	Bessel function of order α	232
L	block length	91
	filter length	108
	likelihood	100
Λ	discrete set	153
λ_n	eigenvalue of $P_\Omega Q_T$	12
$\mu_n(a)$	eigenvalue for half time and band limiting	12
$N(A_a, \alpha)$	number of eigenvalues exceeding α	23
$N(B_a, \alpha)$	number of eigenvalues exceeding α	130
$\mathbb{N}$ and $\mathbb{N}_n$	natural numbers and first n numbers	223
n_Λ	counting function of Λ	158
N_Σ	number of intervals for multiband Σ	130
Ω	Bernoulli generated random set	137
P	unit bandwidth localization operator	3
P_Ω	band-limiting operator	3
P_Σ	frequency-localization operator	3
P_n	Legendre polynomial	5
$\|P\|$	partition set counter	142
$\mathscr{P}$	prolate differential operator	5
	partition lattice	142
	pilot set	112
$\mathscr{P}_s$	scaled prolate differential operator	19
ϕ_n	eigenfunction of $\mathscr{P}$	5
$\bar{\phi}_n = \bar{\phi}_n^{(c)}$	$L^2[-1,1]$-normalized prolate	15, 46
φ^T	eigenfunction of PQ_T	211
φ_n	eigenfunction of PQ_T	5
Q	unit time-localization operator	6
Q_S	time-localization operator	3
Q_T	time-limiting operator	3
R	autocorrelation function	91
$\mathcal{S}(\mathbb{R})$	Schwartz space	75
s_n	discrete prolate spheroidal sequence	29
σ_n	discrete prolate spheroidal wave function	29
sinc_Ω	dilated sinc function	229
$S^\perp$	orthogonal complement	226
S	power spectral density	91
$\mathrm{St}(n,k)$	Stirling number of the second kind	142
T_c	chip period	117
$\mathrm{tr}(A)$	trace norm of A	228
T_s	sample period	117
$W(P)$	partition pullback	144

List of Acronyms

Acronym	Description	Reference
ACT	adaptive weights conjugate gradient Toeplitz method	37
ADC	analog-to-digital converter	195
BCJR	Bahl, Cocke, Jelinek, Raviv algorithm	117
BEM	basis expansion model	105
BMO	bounded mean oscillation	174
CR	cognitive radio	123
CRS	complete residue system	195
CRT	Candès–Romberg–Tao	136
DFT	discrete Fourier transform	32
DPSS	discrete prolate spheroidal sequence	26
DPSWF	discrete prolate spheroidal wave function	27
DSP	discrete signal processor	195
EEG	electroencephalography	127
FBSE	filter bank spectrum estimation	124
FCC	Federal Communications Commission	124
FDPSs	finite discrete prolate sequence	32
fMRI	ferromagnetic resonance imaging	127
GGQ	generalized Gaussian quadrature	57
GPSs	generalized prolate sequences	114
HB	Hermite–Biehler	174
ICI	intercarrier interference	118
ISI	intersymbol interference	118
JR	Jain–Ranganath	36
JR–CG	Jain–Ranganath conjugate gradients method	36
KLT	Karhunen–Loève transform	99
KM	Karoui–Moumni	212
MC-CDMA	multicarrier code-division multiple access	116
MIMO	multiple input multiple output	120
MMV	multiple measurement vector	194
MRI	magnetic resonance imaging	126

MSE	mean-squared error	96
MTSE	multitaper spectrum estimation	124
MW–TFA	multiple window time–frequency analysis	127
MWMV	multiple window minimum variance spectrum estimator	126
OFDM	orthogonal frequency–division multiplexing	108
ONB	orthonormal basis	197
P-DPSSs	periodic discrete prolate spheroidal sequence	32
PCM	pulse code modulation	198
PG	Papoulis–Gerchberg	35
PG–CG	Papoulis–Gerchberg conjugate gradients method	36
PNS	periodic nonuniform	195
PSD	power spectral density	97
PSK	phase shift keying	108
PSWF	prolate spheroidal wave function	4
PW	Paley–Wiener	2
QAM	quadrature amplitude modulation	108
QR	orthogonal–triangular matrix decomposition	34
RF	radio frequency	124
RKHS	reproducing kernel Hilbert space	199, 227
SBR	spectrum-blind recovery	194
SVD	singular value decomposition	124
SVD–MTM	singular value decomposition–multitaper method	124
TF	time–frequency	127
UHP	upper half-plane	174
WKB	Wentzel–Kramers–Brillouin	20
WSS	Walter–Shen–Soleski method	212
WSS	wide-sense stationary	122

References

1. M. Abramowitz and I.A. Stegun. *Handbook of Mathematical Functions with Formulas, Graphs, and Mathematical Tables*. U.S. Government Printing Office, Washington, D.C., 1964.
2. D.M. Akos, M. Stockmaster, J.B.Y. Tsui, and J. Caschera. Direct bandpass sampling of multiple distinct RF signals. *IEEE Trans. Commun.*, 47:983 –988, 1999.
3. A. Alcocer-Ochoa, R. Parra-Michel, and V.Ya. Kontorovitch. The universality of the prolate spheroidal wave functions for channel orthogonalization and its modeling. *ICEEE 2005*, pages 106–109, 2005.
4. A. Aldroubi and K. Gröchenig. Nonuniform sampling and reconstruction in shift-invariant spaces. *SIAM Review*, 43:585–620, 2001.
5. A. Amini and F. Marvasti. Convergence analysis of an iterative method for the reconstruction of multi-band signals from their uniform and periodic nonuniform samples. *Sampl. Theory Signal Image Process.*, 7:113–129, 2008.
6. X. Aragones, J.L. Gonzalez, and A. Rubio. *Analysis and Solutions for Switching Noise in Coupling Mixed Signal ICs*. Kluwer, Dordrecht, 1999.
7. D.D. Ariananda, M.K. Lakshmanan, and H. Nikoo. A survey on spectrum sensing techniques for cognitive radio. In *CogART 2009*, pages 74–79, 2009.
8. P. Auscher, G. Weiss, and M.V. Wickerhauser. Local sine and cosine bases of Coifman and Meyer and the construction of smooth wavelets. In *Wavelets*, volume 2 of *Wavelet Anal. Appl.*, pages 237–256. Academic Press, Boston, MA, 1992.
9. L. Auslander and F.A. Grünbaum. The Fourier transform and the discrete Fourier transform. *Inverse Problems*, 5:149–164, 1989.
10. S. Avdonin and W. Moran. Sampling and interpolation of functions with multi-band spectra and controllability problems. In *Optimal Control of Partial Differential Equations (Chemnitz, 1998)*, pages 43–51. Birkhäuser, Basel, 1999.
11. S. Avdonin and W. Moran. Sampling of multi-band signals. In *ICIAM 99 (Edinburgh)*, pages 163–174. Oxford Univ. Press, Oxford, 2000.
12. S.A. Avdonin. On the question of Riesz bases of exponential functions in L^2. *Vestnik Leningrad. Univ. No. 13 Mat. Meh. Astronom.*, pages 5–12, 1974.
13. S.A. Avdonin, A. Bulanova, and W. Moran. Construction of sampling and interpolating sequences for multi-band signals. The two-band case. *Int. J. Appl. Math. Comput. Sci.*, 17:143–156, 2007.
14. S.A. Avdonin and S.A. Ivanov. *Families of Exponentials*. Cambridge University Press, Cambridge, 1995.
15. L. Bahl, J. Cocke, F. Jelinek, and J. Raviv. Optimal decoding of linear codes for minimizing symbol error rate. *IEEE Trans. Inform. Theory*, 20:284–287, 1974.
16. J.A. Barceló and A. Córdoba. Band-limited functions: L^p-convergence. *Trans. Amer. Math. Soc.*, 313:655–669, 1989.

17. R.F. Bass and K. Gröchenig. Random sampling of multivariate trigonometric polynomials. *SIAM J. Math. Anal.*, 36:773–795, 2004/05.
18. M. Bayram and R. Baraniuk. Multiple window time-varying spectrum estimation. In *Nonlinear and Nonstationary Signal Processing (Cambridge, 1998)*, pages 292–316. Cambridge Univ. Press, Cambridge, 2000.
19. M.G. Beaty and M.M. Dodson. The distribution of sampling rates for signals with equally wide, equally spaced spectral bands. *SIAM J. Appl. Math.*, 53:893–906, 1993.
20. M.G. Beaty, M.M. Dodson, and S.P. Eveson. A converse to Kluvánek's theorem. *J. Fourier Anal. Appl.*, 13:187–196, 2007.
21. H. Behmard and A. Faridani. Sampling of bandlimited functions on unions of shifted lattices. *J. Fourier Anal. Appl.*, 8:43–58, 2002.
22. H. Behmard, A. Faridani, and D. Walnut. Construction of sampling theorems for unions of shifted lattices. *Sampl. Theory Signal Image Process.*, 5:297–319, 2006.
23. J.J. Benedetto. Some mathematical methods for spectrum estimation. In *Fourier Techniques and Applications (Kensington, 1983)*, pages 73–100. Plenum, New York, 1985.
24. J.J. Benedetto. Irregular sampling and frames. In *Wavelets*, pages 445–507. Academic Press, Boston, MA, 1992.
25. J.J. Benedetto. Frame decompositions, sampling, and uncertainty principle inequalities. In J. Benedetto and M. Frazier, editors, *Wavelets: Mathematics and Applications*, pages 247–304. CRC Press, Boca Raton, FL, 1994.
26. J.J. Benedetto. *Harmonic Analysis and Applications*. CRC Press, Boca Raton, FL, 1997.
27. J.J. Benedetto. Introduction. In *Fundamental Papers in Wavelet Theory*, pages 1–22. Princeton Univ. Press, Princeton, NJ, 2006.
28. J.J. Benedetto and P.J.S.G. Ferreira. Introduction. In *Modern Sampling Theory*, pages 1–26. Birkhäuser, Boston, MA, 2001.
29. J.J. Benedetto and M. Fickus. Finite normalized tight frames. *Adv. Comput. Math.*, 18:357–385, 2003.
30. J.J. Benedetto and W. Heller. Irregular sampling and the theory of frames. I. *Note Mat.*, 10(suppl. 1):103–125 (1992), 1990.
31. J.J. Benedetto and A. Kebo. The role of frame force in quantum detection. *J. Fourier Anal. Appl.*, 14:443–474, 2008.
32. J.J. Benedetto and O. Oktay. PCM–Sigma delta comparison and sparse representation quantization. In *CISS 2008*, pages 737–742, 2008.
33. N. Benvenuto and C. Giovanni. *Algorithms for Communications Systems and Their Applications*. Wiley, Hoboken, NJ, 2002.
34. L. Berman and A. Feuer. Robust patterns in recurrent sampling of multiband signals. *IEEE Trans. Signal Process.*, 56:2326–2333, 2008.
35. G. Beylkin and L. Monzón. On generalized Gaussian quadratures for exponentials and their applications. *Appl. Comput. Harmon. Anal.*, 12:332–373, 2002.
36. G. Beylkin and L. Monzón. On approximation of functions by exponential sums. *Appl. Comput. Harmon. Anal.*, 19:17–48, 2005.
37. L. Bezuglaya and V. Katsnel′son. The sampling theorem for functions with limited multiband spectrum. *Z. Anal. Anwendungen*, 12:511–534, 1993.
38. M. Borgerding. Turning overlap-save into a multiband mixing, downsampling filter bank. *IEEE Signal Process. Mag.*, 23:158–161, 2006.
39. S. Boucheron, G. Lugosi, and P. Massart. A sharp concentration inequality with applications. *Random Structures Algorithms*, 16:277–292, 2000.
40. C.J. Bouwkamp. On the characteristic values of spheroidal wave functions. *Philips Research Rep.*, 5:87–90, 1950.
41. J.P. Boyd. Approximation of an analytic function on a finite real interval by a bandlimited function and conjectures on properties of prolate spheroidal functions. *Appl. Comput. Harmon. Anal.*, 15:168–176, 2003.
42. J.P. Boyd. Large mode number eigenvalues of the prolate spheroidal differential equation. *Appl. Math. Comput.*, 145:881–886, 2003.

43. J.P. Boyd. Prolate spheroidal wavefunctions as an alternative to Chebyshev and Legendre polynomials for spectral element and pseudospectral algorithms. *J. Comput. Phys.*, 199:688–716, 2004.
44. J.P. Boyd. Algorithm 840: computation of grid points, quadrature weights and derivatives for spectral element methods using prolate spheroidal wave functions—prolate elements. *ACM Trans. Math. Software*, 31:149–165, 2005.
45. R. Bracewell. *The Fourier Transform and Its Applications*. McGraw-Hill, New York, 1965.
46. O. Brander and B. DeFacio. A generalisation of Slepian's solution for the singular value decomposition of filtered Fourier transforms. *Inverse Problems*, 2:L9–L14, 1986.
47. Y. Bresler and P. Feng. Spectrum-blind minimum-rate sampling and reconstruction of 2-D multiband signals. In *ICIP '96*, pages I–701–I–704, 1996.
48. D.R. Brillinger. Fourier analysis of stationary processes. *Proc. IEEE*, 62:1628–1643, 1974.
49. D.R. Brillinger. *Time Series*. Holden-Day Inc., Oakland, CA, second edition, 1981.
50. D.R. Brillinger. The finite Fourier transform of a stationary process. In *Time Series in the Frequency Domain*, volume 3 of *Handbook of Statistics*, pages 21–37. North-Holland, Amsterdam, 1983.
51. T.P. Bronez. Spectral estimation of irregularly sampled multidimensional processes by generalized prolate spheroidal sequences. *IEEE Trans. Acoust., Speech, Signal Process.*, 36:1862–1873, 1988.
52. T.P. Bronez. On the performance advantage of multitaper spectral analysis. *IEEE Trans. Signal Process.*, 40:2941–2946, 1992.
53. J. Brown, Jr. Sampling expansions for multiband signals. *IEEE Trans. Acoust., Speech, Signal Process.*, 33:312–315, 1985.
54. A.L. Van Buren, B.J. King, R.V. Baier, and S. Hanish. *Tables of angular spheroidal wave functions, vol. I, prolate, $m = 0$*. Naval Research Lab., U.S. Govt. Printing Office, Washington, DC, 1975.
55. F. Cakrak and P.J. Loughlin. Multiple window time-varying spectral analysis. *IEEE Trans. Signal Process.*, 49:448–453, 2001.
56. F. Cakrak and P.J. Loughlin. Multiwindow time-varying spectrum with instantaneous bandwidth and frequency constraints. *IEEE Trans. Signal Process.*, 49:1656–1666, 2001.
57. E.J. Candès and J. Romberg. Quantitative robust uncertainty principles and optimally sparse decompositions. *Found. Comput. Math.*, 6:227–254, 2006.
58. E.J. Candès and J. Romberg. Sparsity and incoherence in compressive sampling. *Inverse Problems*, 23:969–985, 2007.
59. E.J. Candès, J.K. Romberg, and T. Tao. Robust uncertainty principles: exact signal reconstruction from highly incomplete frequency information. *IEEE Trans. Inform. Theory*, 52:489–509, 2006.
60. E.J. Candès, J.K. Romberg, and T. Tao. Stable signal recovery from incomplete and inaccurate measurements. *Comm. Pure Appl. Math.*, 59:1207–1223, 2006.
61. C. Canuto, M.Y. Hussaini, A. Quarteroni, and T.A. Zang. *Spectral Methods in Fluid Dynamics*. Springer-Verlag, Berlin, 1988.
62. J. Capon. High-resolution frequency-wavenumber spectrum analysis. *Proc. IEEE*, 57(8):1408–1418, 1969.
63. P.G. Casazza. The art of frame theory. *Taiwanese J. Math.*, 4:129–201, 2000.
64. P.G. Casazza, O. Christensen, S. Li, and A. Lindner. Density results for frames of exponentials. In *Harmonic Analysis and Applications*, pages 359–369. Birkhäuser, Boston, MA, 2006.
65. P.G. Casazza, M. Fickus, J.C. Tremain, and E. Weber. The Kadison-Singer problem in mathematics and engineering: a detailed account. In *Operator Theory, Operator Algebras, and Applications*, pages 299–355. Amer. Math. Soc., Providence, RI, 2006.
66. P.G. Casazza, G. Kutyniok, D. Speegle, and J.C. Tremain. A decomposition theorem for frames and the Feichtinger conjecture. *Proc. Amer. Math. Soc.*, 136:2043–2053, 2008.
67. P.G. Casazza and J.C. Tremain. Revisiting the Bourgain-Tzafriri restricted invertibility theorem. *Oper. Matrices*, 3:97–110, 2009.

68. C.-Y. Chen and P.P. Vaidyanathan. MIMO radar spacetime adaptive processing using prolate spheroidal wave functions. *IEEE Trans. Signal Process.*, 56:623–635, 2008.
69. Q.-Y. Chen, D. Gottlieb, and J.S. Hesthaven. Spectral methods based on prolate spheroidal wave functions for hyperbolic PDEs. *SIAM J. Numer. Anal.*, 43:1912–1933, 2005.
70. W. Chen. A fast convergence algorithm for band-limited extrapolation by sampling. *IEEE Trans. Signal Process.*, 57:161–167, 2009.
71. X. Chen and H. Chen. Time-frequency well-localized pulse shaping filter design for high speed data and multi-carrier system based on DPSS. In *IEEE WiCom 2007*, pages 1099–1103, 2007.
72. H. Cheng, V. Rokhlin, and N. Yarvin. Nonlinear optimization, quadrature and interpolation. *SIAM J. Optim.*, 9(4):901–923, 1999.
73. O. Christensen. *An Introduction to Frames and Riesz Bases*. Birkhäuser Boston Inc., Boston, MA, 2003.
74. A.C. Cristán and A.T. Walden. Multitaper power spectrum estimation and thresholding: wavelet packets versus wavelets. *IEEE Trans. Signal Process.*, 50:2976–2986, 2002.
75. S. Das, A. Hajian, and D.N. Spergel. Efficient power spectrum estimation for high resolution CMB maps. *Phys. Rev. D*, 79:083008–1–21, 2009.
76. I. Daubechies. The wavelet transform, time-frequency localization and signal analysis. *IEEE Trans. Inform. Theory*, 36:961–1005, 1990.
77. I. Daubechies. From the original framer to present-day time-frequency and time-scale frames. *J. Fourier Anal. Appl.*, 3:485–486, 1997.
78. I. Daubechies, A. Grossmann, and Y. Meyer. Painless nonorthogonal expansions. *J. Math. Phys.*, 27:1271–1283, 1986.
79. L. de Branges. *Hilbert Spaces of Entire Functions*. Prentice-Hall Inc., Englewood Cliffs, NJ, 1968.
80. A. Dembo, T. Cover, and J. Thomas. Information theoretic inequalities. *IEEE Trans. Inform. Theory*, 37:1501–1518, 1991.
81. P. Dines and D. Hazony. Optimization of time limited outputs in band limited channel. In *Proc. 13th Allerton Conf.*, pages 515–522, 1975.
82. I. Djokovic and P.P. Vaidyanathan. Generalized sampling theorems in multiresolution subspaces. *IEEE Trans. Signal Process.*, 45:583–599, 1997.
83. G.J. Dolecek and F.J.T. Torres. Multiplierless multiband filter for fractional sample rate conversion. *MIC-CPE '08*, pages 1–6, 2008.
84. D. Donoho, M. Vetterli, R. DeVore, and I. Daubechies. Data compression and harmonic analysis. *IEEE Trans. Inform. Theory*, 44:2436–2476, 1998.
85. D.L. Donoho and P.B. Stark. Uncertainty principles and signal recovery. *SIAM J. Appl. Math.*, 49:906–931, 1989.
86. D.L. Donoho and P.B. Stark. A note on rearrangements, spectral concentration, and the zero-order prolate spheroidal wavefunction. *IEEE Trans. Inform. Theory*, 39:257–260, 1993.
87. A. Duel-Hallen, S. Hu, and H. Hallen. Long-range prediction of fading signals. *IEEE Signal Process. Mag.*, 17:62–75, May 2000.
88. R.J. Duffin and A.C. Schaeffer. A class of nonharmonic Fourier series. *Trans. Amer. Math. Soc.*, 72:341–366, 1952.
89. J.J. Duistermaat and F.A. Grünbaum. Differential equations in the spectral parameter. *Comm. Math. Phys.*, 103(2):177–240, 1986.
90. T.M. Dunster. Uniform asymptotic expansions for prolate spheroidal functions with large parameters. *SIAM J. Math. Anal.*, 17:1495–1524, 1986.
91. D.E. Dutkay, D. Han, and P.E.T. Jorgensen. Orthogonal exponentials, translations, and Bohr completions. *J. Funct. Anal.*, 257:2999–3019, 2009.
92. D.E. Dutkay, D. Han, Q. Sun, and E. Weber. Hearing the Hausdorff dimension, 2009. arXiv:0910.5433.
93. J. Dziubański and E. Hernández. Band-limited wavelets with subexponential decay. *Canad. Math. Bull.*, 41:398–403, 1998.
94. A. Eberhard. An optimal discrete window for the calculation of power spectra. *IEEE Trans. Audio Electroacoust.*, 21:37–43, 1973.

95. A. Erdélyi, W. Magnus, F. Oberhettinger, and F.G. Tricomi. *Higher Transcendental Functions. Vols. I, II.* McGraw-Hill, New York, 1953.
96. A. Erdélyi, W. Magnus, F. Oberhettinger, and F.G. Tricomi. *Tables of Integral Transforms. Vol. I.* McGraw-Hill, New York, 1954.
97. C.L. Fancourt and J.C. Principe. On the relationship between the Karhunen–Loeve transform and the prolate spheroidal wave functions. In *IEEE ICASSP '00*, pages I–261–I–264, 2000.
98. B. Farhang-Boroujeny. Filter bank spectrum sensing for cognitive radios. *IEEE Trans. Signal Process.*, 56:1801–1811, 2008.
99. H.G. Feichtinger, K. Gröchenig, and T. Strohmer. Efficient numerical methods in non-uniform sampling theory. *Numer. Math.*, 69:423–440, 1995.
100. P. Feng and Y. Bresler. Spectrum-blind minimum-rate sampling and reconstruction of multi-band signals. In *IEEE ICASSP '96*, pages III–1688–III–1691, 1996.
101. C. Flammer. *Spheroidal Wave Functions*. Stanford University Press, Stanford, CA, 1957.
102. I.K. Fodor and P.B. Stark. Multitaper spectrum estimation for time series with gaps. *IEEE Trans. Signal Process.*, 48:3472–3483, 2000.
103. W.H.J. Fuchs. On the eigenvalues of an integral equation arising in the theory of band-limited signals. *J. Math. Anal. Appl.*, 9:317–330, 1964.
104. O. Fujimura. An approximation to voice aperiodicity. *IEEE Trans. Audio Electroacoust.*, 16:68–72, 1968.
105. D. Funaro. *Polynomial Approximation of Differential Equations (Lecture Notes in Phys. 8)*. Springer-Verlag, Berlin, 1992.
106. J.-P. Gabardo. Hilbert spaces of distributions having an orthogonal basis of exponentials. *J. Fourier Anal. Appl.*, 6:277–298, 2000.
107. J.-P. Gabardo. Sampling theory for certain Hilbert spaces of bandlimited functions. In *Modern Sampling Theory*, pages 107–134. Birkhäuser Boston, Boston, MA, 2001.
108. D. Gabor. Theory of communication. *J. IEE (London)*, 93:429–457, 1946.
109. F.R. Gantmacher and M.G. Krein. *Oscillation Matrices and Kernels and Small Vibrations of Mechanical Systems (Revised Edition)*. American Mathematical Society, Providence, RI, 2002.
110. A.G. Garcia. Orthogonal sampling formulas: a unified approach. *SIAM Rev.*, 42:499–512, 2000.
111. R.W. Gerchberg. Super-resolution through error reduction. *Optica Acta*, 21:709–720, 1974.
112. I. Gerst and J. Diamond. The elimination of intersymbol interference by input signal shaping. In *Proc. I.R.E*, pages 1195–1203, 1961.
113. G.B. Giannakis and C. Tepedelenlioğlu. Basis expansion models and diversity techniques for blind identification and equalization of time-varying channels. *Proc. IEEE*, 86:1969–1986, 1998.
114. F. Gori and G. Guattari. Degrees of freedom of images from point-like-element pupils. *J. Opt. Soc. Am.*, 64:453–458, 1974.
115. L. Grafakos and S.J. Montgomery-Smith. Best constants for uncentred maximal functions. *Bulletin of The London Mathematical Society*, 29:60–64, 1997.
116. C.A. Greenhall. Orthogonal sets of data windows constructed from trigonometric polynomials. *IEEE Trans. Acoust., Speech, Signal Process.*, 38:870–872, 1990.
117. M. Greitāns. Iterative reconstruction of lost samples using updating of autocorrelation matrix. In *SAMPTA 1997*, pages 155–160, 1997.
118. U. Grenander and M. Rosenblatt. An extension of a theorem of G. Szegő and its application to the study of stochastic processes. *Trans. Amer. Math. Soc.*, 76:112–126, 1954.
119. U. Grenander and M. Rosenblatt. *Statistical Analysis of Stationary Time Series*. John Wiley & Sons, New York, 1957.
120. N. Grip and G.E. Pfander. A discrete model for the efficient analysis of time-varying narrowband communication channels. *Multidimens. Syst. Signal Process.*, 19:3–40, 2008.
121. K. Gröchenig and H. Razafinjatovo. On Landau's necessary density conditions for sampling and interpolation of band-limited functions. *J. London Math. Soc. (2)*, 54:557–565, 1996.
122. H. Groemer. *Geometric Applications of Fourier Series and Spherical Harmonics*. Cambridge University Press, Cambridge, 1996.

123. D.M. Gruenbacher and D.R. Hummels. A simple algorithm for generating discrete prolate spheroidal sequences. *IEEE Trans. Signal Process.*, 42:3276–3278, 1994.
124. D.M. Gruenbacher and D.R. Hummels. N dimensional orthogonal QPSK signaling with discrete prolate spheroidal sequences. In *IEEE RAWCON '98*, pages 63–66, 1998.
125. F. Alberto Grünbaum. Some nonlinear evolution equations and related topics arising in medical imaging. *Phys. D*, 18(1-3):308–311, 1986. Solitons and coherent structures (Santa Barbara, CA, 1985).
126. F.A. Grünbaum. Eigenvectors of a Toeplitz matrix: discrete version of the prolate spheroidal wave functions. *SIAM J. Algebraic Discrete Methods*, 2:136–141, 1981.
127. F.A. Grünbaum. Toeplitz matrices commuting with tridiagonal matrices. *Linear Algebra Appl.*, 40:25–36, 1981.
128. F.A. Grünbaum, L. Longhi, and M. Perlstadt. Differential operators commuting with finite convolution integral operators: some non-abelian examples. *SIAM J. Appl. Math.*, 42:941–955, 1982.
129. C.S. Güntürk. One-bit sigma–delta quantization with exponential accuracy. *Comm. Pure Appl. Math.*, 56:1608–1630, 2003.
130. C.S. Güntürk. Approximating a bandlimited function using very coarsely quantized data: improved error estimates in sigma–delta modulation. *J. Amer. Math. Soc.*, 17:229–242, 2004.
131. M. Hamamura and J. Hyuga. Spectral efficiency of orthogonal set of truncated MC-CDMA signals using discrete prolate spheroidal sequences. In *IEEE WCNC 2008*, pages 980–984, 2008.
132. D. Han, K. Kornelson, D. Larson, and E. Weber. *Frames for Undergraduates*. American Mathematical Society, Providence, RI, 2007.
133. D. Han, M.Z. Nashed, and Q. Sun. Sampling expansions in reproducing kernel Hilbert and Banach spaces. *Numer. Funct. Anal. Optim.*, 30:971–987, 2009.
134. A. Hanssen. Multidimensional multitaper spectral estimation. *Signal Process.*, 58:327–332, 1997.
135. H.F. Harmuth. *Transmission of Information by Orthogonal Functions*. Springer-Verlag, New York, 1972.
136. J. Harnad and A. Kasman, editors. *The bispectral problem*, volume 14 of *CRM Proceedings & Lecture Notes*. American Mathematical Society, Providence, RI, 1998. Papers from the CRM Workshop held at the Université de Montréal, Montreal, PQ, March 1997.
137. R.V.L. Hartley. Transmission of information. *Bell System Tech. J.*, 7:535–564, 1928.
138. S. Haykin. Cognitive radio: brain-empowered wireless communications. *IEEE J. Sel. Areas Commun.*, 23:201–220, 2005.
139. H. He and D.J. Thomson. The canonical bicoherence–part I: Definition, multitaper estimation, and statistics. *IEEE Trans. Signal Process.*, 57:1273–1284, 2009.
140. H. He and D.J. Thomson. The canonical bicoherence–part II: QPC test and its application in geomagnetic data. *IEEE Trans. Signal Process.*, 57(4):1285–1292, 2009.
141. H. He and D.J. Thomson. Canonical bicoherence analysis of dynamic EEG data. *J. Comput. Neurosci.*, pages 23–34, 2010.
142. S. He and J.K. Tugnait. Doubly-selective multiuser channel estimation using superimposed training and discrete prolate spheroidal basis expansion models. In *IEEE ICASSP '07*, pages II–861–II–864, 2007.
143. C. Heil. History and evolution of the density theorem for Gabor frames. *J. Fourier Anal. Appl.*, 13:113–166, 2007.
144. C. Herley. Reconstruction for novel sampling structures. In *IEEE ICASSP '97*, pages III–2433–III–2436, 1997.
145. C. Herley and P.W. Wong. Minimum rate sampling and reconstruction of signals with arbitrary frequency support. *IEEE Trans. Inform. Theory*, 45:1555–1564, 1999.
146. C. Herley and P.W. Wong. Efficient minimum rate sampling of signals with frequency support over non-commensurable sets. In *Modern Sampling Theory*, pages 271–291. Birkhäuser Boston, Boston, MA, 2001.
147. J.R. Higgins. Five short stories about the cardinal series. *Bull. Amer. Math. Soc. (N.S.)*, 12:45–89, 1985.

148. J.R. Higgins. Sampling for multi-band functions. In *Mathematical Analysis, Wavelets, and Signal Processing (Cairo, 1994)*, pages 165–170. Amer. Math. Soc., Providence, RI, 1995.
149. J.R. Higgins. *Sampling Theory in Fourier and Signal Analysis*. Oxford University Press, Oxford, UK, 1996.
150. J.R. Higgins. Historical origins of interpolation and sampling, up to 1841. *Sampl. Theory Signal Image Process.*, 2:117–128, 2003.
151. J.A. Hogan, S. Izu, and J.D. Lakey. Sampling approximations for time- and bandlimiting. *Sampl. Theory Signal Image Process.*, pages 91–117, 2010.
152. J.A. Hogan and J.D. Lakey. *Time–Frequency and Time–Scale Methods*. Birkhäuser Boston Inc., Boston, MA, 2005.
153. J.A. Hogan and J.D. Lakey. Periodic nonuniform sampling in shift-invariant spaces. In *Harmonic Analysis and Applications*, pages 253–287. Birkhäuser Boston, Boston, MA, 2006.
154. J.A. Hogan and J.D. Lakey. Sampling and time-frequency localization of band-limited and multiband signals. In *Representations, Wavelets, and Frames*, pages 275–291. Birkhäuser Boston, Boston, MA, 2008.
155. D.G. Hook and P.R. McAree. Using Sturm sequences to bracket real roots of polynomial equations. In *Graphic Gems I*, pages 416–422. Academic Press, New York, 1990.
156. L. Hörmander. *The Analysis of Linear Partial Differential Operators. I*. Springer-Verlag, Berlin, 2003.
157. S.V. Hruščëv, N.K. Nikol′skiĭ, and B.S. Pavlov. Unconditional bases of exponentials and of reproducing kernels. In *Complex Analysis and Spectral Theory (Leningrad, 1979/1980)*, pages 214–335. Springer, Berlin, 1981.
158. Y. Hua and T.K. Sarkar. Design of optimum discrete finite duration orthogonal Nyquist signals. *IEEE Trans. Acoust., Speech, Signal Process.*, 36:606–608, 1988.
159. S. Izu. *Sampling and Time–Frequency Localization in Paley–Wiener Spaces*. PhD thesis, New Mexico State University, Las Cruces, NM, 2009.
160. S. Jaffard. A density criterion for frames of complex exponentials. *Michigan Math. J.*, 38:339–348, 1991.
161. D. Jagerman. Information theory and approximation of bandlimited functions. *Bell System Tech. J.*, 49:1911–1941, 1970.
162. A.K. Jain and S. Ranganath. Extrapolation algorithms for discrete signals with application in spectral estimation. *IEEE Trans. Acoust. Speech Signal Process.*, 29:830–845, 1981.
163. W.C. Jakes and D.C. Cox, editors. *Microwave Mobile Communications*. Wiley-IEEE Press, 1994.
164. P. Jaming and A.M. Powell. Uncertainty principles for orthonormal sequences. *J. Funct. Anal.*, 243:611–630, 2007.
165. K. Jogdeo and S.M. Samuels. Monotone convergence of binomial probabilities and a generalization of Ramanujan's theorem. *Ann. Math. Stat.*, 39:1191–1195, 1968.
166. M. Kac, W. L. Murdock, and G. Szegő. On the eigenvalues of certain Hermitian forms. *J. Rational Mech. Anal.*, 2:767–800, 1953.
167. M.Ĭ. Kadec′. The exact value of the Paley-Wiener constant. *Dokl. Akad. Nauk SSSR*, 155:1253–1254, 1964.
168. N. Kaiblinger and W.R. Madych. Orthonormal sampling functions. *Appl. Comput. Harmon. Anal.*, 21:404–412, 2006.
169. K. Karhunen. Über lineare Methoden in der Wahrscheinlichkeitsrechnung. *Ann. Acad. Sci. Fennicae. Ser. A. I. Math.-Phys.*, 37:1–79, 1947.
170. S. Karlin and W.J. Studden. *Tchebycheff Systems: With Applications in Analysis and Statistics*. John Wiley, New York, 1966.
171. A. Karoui and T. Moumni. New efficient methods of computing the prolate spheroidal wave functions and their corresponding eigenvalues. *Appl. Comput. Harmon. Anal.*, 24:269–289, 2008.
172. V.E. Katsnelson. Sampling and interpolation for functions with multi-band spectrum: the mean-periodic continuation method. In *Wiener-Symposium (Grossbothen, 1994)*, pages 91–132. Verlag Wiss. Leipzig, Leipzig, 1996.

173. V.E. Katsnelson. Sampling and interpolation for functions with multi-band spectrum: the mean periodic continuation method. In *Signal and Image Representation in Combined Spaces*, pages 525–553. Academic Press, San Diego, CA, 1998.
174. W. Kester. *The Data Conversion Handbook*. Elsevier, Oxford, UK, 2005.
175. M.S. Khan, R.M. Goodal, and R. Dixon. Implementation of Non-uniform Sampling for Alias-Free Processing in Digital Control. In *Proceedings of the UKACC International Conference on Control 2008*, page Th01.01, Manchester, UK, 2008. University of Manchester.
176. K. Khare. Bandpass sampling and bandpass analogues of prolate spheroidal functions. *Signal Process.*, 86(7):1550–1558, 2006.
177. K. Khare and N. George. Sampling theory approach to prolate spheroidal wavefunctions. *J. Phys. A*, 36:10011–10021, 2003.
178. S.N. Khonina, S.G. Volotovskiĭ, and V.A. Soĭfer. A method for computing the eigenvalues of prolate spheroidal functions of zero order. *Dokl. Akad. Nauk*, 376:30–32, 2001.
179. J. Kim, C.-W. Wang, and W.E. Stark. Frequency domain channel estimation for OFDM based on Slepian basis expansion. In *IEEE ICC '07*, pages 3011–3015, 2007.
180. I. Kluvánek. Sampling theorem in abstract harmonic analysis. *Mat.-Fyz. Časopis Sloven. Akad. Vied*, 15:43–48, 1965.
181. J.J. Knab. Interpolation of band-limited functions using the approximate prolate series. *IEEE Trans. Inform. Theory*, 25:717–720, 1979.
182. A. Kohlenberg. Exact interpolation of band-limited functions. *J. Appl. Phys.*, 24:1432–1436, 1953.
183. V. Komornik and P. Loreti. *Fourier Series in Control Theory*. Springer-Verlag, New York, 2005.
184. L.H. Koopmans. *The Spectral Analysis of Time Series*. Academic Press, New York, 1974.
185. V.A. Kotel′nikov. On the transmission capacity of the "ether" and wire in electrocommunications. In *Modern Sampling Theory*, pages 27–45. Birkhäuser, Boston, MA, 2001. Translated from the Russian by V.E. Katsnelson.
186. M.G. Krein. *The Ideas of P.L. Chebyshev and A.A. Markov in the Theory of Limiting Values of Integrals*. American Mathematical Society, Providence, RI, 1959.
187. H.J. Landau. The eigenvalue behavior of certain convolution equations. *Trans. Amer. Math. Soc.*, 115:242–256, 1965.
188. H.J. Landau. Necessary density conditions for sampling and interpolation of certain entire functions. *Acta Math.*, 117:37–52, 1967.
189. H.J. Landau. Sampling, data transmission, and the Nyquist rate. *Proc. IEEE*, 55:1701–1706, 1967.
190. H.J. Landau. On Szegő's eigenvalue distribution theorem and non-Hermitian kernels. *J. Analyse Math.*, 28:335–357, 1975.
191. H.J. Landau. An overview of time and frequency limiting. In *Fourier Techniques and Applications (Kensington, 1983)*, pages 201–220. Plenum, New York, 1985.
192. H.J. Landau. Extrapolating a band-limited function from its samples taken in a finite interval. *IEEE Trans. Inform. Theory*, 32:464 – 470, 1986.
193. H.J. Landau. On the density of phase-space expansions. *IEEE Trans. Inform. Theory*, 39:1152–1156, 1993.
194. H.J. Landau, J.E. Mazo, S. Shamai, and J. Ziv. Shannon theory: perspective, trends, and applications. Special issue dedicated to Aaron D. Wyner. *IEEE Trans. Inform. Theory*, 48:1237 –1242, 2002.
195. H.J. Landau and H.O. Pollak. Prolate spheroidal wave functions, Fourier analysis and uncertainty. II. *Bell System Tech. J.*, 40:65–84, 1961.
196. H.J. Landau and H.O. Pollak. Prolate spheroidal wave functions, Fourier analysis and uncertainty. III. The dimension of the space of essentially time- and band-limited signals. *Bell System Tech. J.*, 41:1295–1336, 1962.
197. H.J. Landau and H. Widom. Eigenvalue distribution of time and frequency limiting. *J. Math. Anal. Appl.*, 77:469–481, 1980.
198. B. Larsson, T. Levitina, and E.J. Brändas. Eigenfunctions of the 2D finite Fourier transform. *J. Comp. Methods in Sci. and Eng.*, 4:135–148, 2004.

199. N. Levinson. *Gap and Density Theorems*. AMS, New York, 1940.
200. T. Levitina and E.J. Brändas. Filter diagonalization with finite Fourier transform eigenfunctions. *J. Math. Chem.*, 40:43–47, 2006.
201. T. Levitina and E.J. Brändas. Sampling formula for convolution with a prolate. *Int. J. Comput. Math.*, 85:487–496, 2008.
202. K.S. Lii and M. Rosenblatt. Prolate spheroidal spectral estimates. *Statist. Probab. Lett.*, 78:1339–1348, 2008.
203. Y.-P. Lin, Y.-D. Liu, and S.-M. Phoong. An iterative algorithm for finding the minimum sampling frequency for two bandpass signals. In *IEEE SPAWC '09*, pages 434–438, 2009.
204. M.A. Lindquist, C.-H. Zhang, G. Glover, L. Shepp, and Q.X. Yang. A generalization of the two-dimensional prolate spheroidal wave function method for nonrectilinear MRI data acquisition methods. *IEEE Trans. Image Process.*, 15:2792–2804, 2006.
205. S. Liu, F. Wang, J. Shen, and Y. Liu. Model assisted time-varying MIMO channel estimation. In *ChinaCom '08*, pages 533–537, 2008.
206. T.-C. Liu and B. van Veen. Multiple window based minimum variance spectrum estimation for multidimensional random fields. *IEEE Trans. Signal Process.*, 40:578–589, 1992.
207. Y.M. Liu and G.G. Walter. Irregular sampling in wavelet subspaces. *J. Fourier Anal. Appl.*, 2:181–189, 1995.
208. M. Loève. *Probability Theory. II*. Springer-Verlag, New York, fourth edition, 1978.
209. E.G. Lovett and J.B. Myklebust. Approximate minimum bias multichannel spectral estimation for heart rate variability. *Ann. Biomed. Eng.*, 25:509–520, 1997.
210. Y.M. Lu and M.N. Do. Sampling signals from a union of subspaces. *IEEE Signal Process. Mag.*, 25:41–47, 2008.
211. R.J. Lyman and A. Sikora. Prediction of bandlimited fading envelopes with arbitrary spectral shape. *IEEE Trans. Wireless Comm.*, 6:1560–1567, 2007.
212. T.J. Lynn and A.Z. bin Sha'ameri. Comparison between the performance of spectrogram and multi-window spectrogram in digital modulated communication signals. In *IEEE ICT-MICC '07*, pages 97–101, 2007.
213. Y. Lyubarskii and E. Malinnikova. On approximation of subharmonic functions. *J. Anal. Math.*, 83:121–149, 2001.
214. Y.I. Lyubarskii and K. Seip. Complete interpolating sequences for Paley-Wiener spaces and Muckenhoupt's (A_p) condition. *Rev. Mat. Iberoamericana*, 13:361–376, 1997.
215. Y.I. Lyubarskii and K. Seip. Sampling and interpolating sequences for multiband-limited functions and exponential bases on disconnected sets. *J. Fourier Anal. Appl.*, 3:597–615, 1997.
216. J. Ma, G.Y. Li, and B.-H. Juang. Signal processing in cognitive radio. *Proc. IEEE*, 97:805–823, 2009.
217. W.R. Madych. Summability of Lagrange type interpolation series. *J. Anal. Math.*, 84:207–229, 2001.
218. B. Mak. A mathematical relationship between full-band and multiband mel-frequency cepstral coefficients. *IEEE Signal Process. Lett.*, 9:241–244, 2002.
219. M.E. Mann and J. Park. Oscillatory spatiotemporal signal detection in climate studies: A multiple-taper spectral domain approach. In R. Dnowska and B. Saltzman, editors, *Advances in Geophysics*, volume 41, pages 1–131. Academic Press, New York, 1999.
220. M. Marcus, J. Burtle, B. Franca, A. Lahjouji, and N. McNeil. Report of the unlicensed devices and experimental licenses working group, docket 02-135. Technical report, Federal Communications Commission, 2002.
221. E. Margolis and Y.C. Eldar. Nonuniform sampling of periodic bandlimited signals. *IEEE Trans. Signal Process.*, 56:2728–2745, 2008.
222. A.A. Markoff. On the limiting values of integrals in connection with interpolation. *Zap. Imp. Akad. Nauk Fiz.-Math. Otd.*, 6(5):1–69, 1898.
223. A.A. Markov. *Selected Papers on Continued Fractions and the Theory of Functions Deviating Least from Zero*. OGIZ, Moscow, 1948.
224. M. Marques, A. Neves, J.S. Marques, and J.M. Sanches. The Papoulis–Gerchberg algorithm with unknown signal bandwidth. In *ICIAR (1)*, pages 436–445, 2006.

225. P. Marziliano and M. Vetterli. Reconstruction of irregularly sampled discrete-time bandlimited signals with unknown sampling locations. *IEEE Trans. Signal Process.*, 48:3462–3471, 2000.
226. M. Massar, M. Fickus, E. Bryan, D. Petkie, and A. Terzuoli. Fast computation of spectral centroids. *Advances in Computational Mathematics*, 2010. electronic.
227. B. Matei and Y. Meyer. Quasicrystals are sets of stable sampling. *C. R. Math. Acad. Sci. Paris*, 346:1235–1238, 2008.
228. B. Matei and Y. Meyer. A variant of compressed sensing. *Rev. Mat. Iberoam.*, 25:669–692, 2009.
229. B. McMillan and D. Slepian. Information theory. *Proc. IRE*, 50:1151–1157, 1962.
230. E. Meijering. A chronology of interpolation: from ancient astronomy to modern signal and image processing. *Proc. IEEE*, 90:319–342, 2002.
231. J. Meixner and F.W. Schäfke. *Mathieusche Funktionen und Sphäroidfunktionen mit Anwendungen auf physikalische und technische Probleme*. Springer-Verlag, Berlin, 1954.
232. J.W. Miles. Asymptotic approximations for prolate spheroidal wave functions. *Studies in Appl. Math.*, 54:315–349, 1975.
233. L. Miranian. Slepian functions on the sphere, generalized Gaussian quadrature rule. *Inverse Problems*, 20:877–892, 2004.
234. M. Mishali and Y.C. Eldar. Spectrum-blind reconstruction of multi-band signals. In *IEEE ICASSP '08*, pages 3365–3368, 2008.
235. M. Mishali and Y.C. Eldar. Blind multiband signal reconstruction: Compressed sensing for analog signals. *IEEE Trans. Signal Process.*, 57:993–1009, 2009.
236. M. Mishali, Y.C. Eldar, and J.A. Tropp. Efficient sampling of sparse wideband analog signals. *IEEEI '08*, pages 290–294, 2008.
237. J. Mitola and G.Q. Maguire. Cognitive radio: making software radios more personal. *IEEE Pers. Commun.*, 6:13–18, 1999.
238. P.P. Mitra and B. Pesaran. Analysis of dynamic brain imaging data. *Biophysical Journal*, 76:691–708, 1999.
239. I.C. Moore and M. Cada. Prolate spheroidal wave functions, an introduction to the Slepian series and its properties. *Appl. Comput. Harmon. Anal.*, 16:208–230, 2004.
240. P.M. Morse and H. Feshbach. *Methods of Theoretical Physics. Two volumes*. McGraw-Hill Book Co., Inc., New York, 1953.
241. D.H. Mugler and Y. Wu. Prediction of band-limited signals from past samples and applications to speech coding. In *Nonuniform Sampling*, pages 543–584. Kluwer/Plenum, New York, 2001.
242. W.L. Myers and G.P. Patil. *Pattern-Based Compression of Multi-Band Image Data for Landscape Analysis (Environmental and Ecological Statistics)*. Springer-Verlag, Secaucus, NJ, 2006.
243. M.Z. Nashed and Q. Sun. Sampling and reconstruction of signals in a reproducing kernel subspace of $L^p(\mathbb{R}^d)$. *J. Funct. Anal.*, 258:2422–2452, 2010.
244. F. Natterer. Efficient evaluation of oversampled functions. *J. Comput. Appl. Math.*, 14:303–309, 1986.
245. M. Niedzwiecki. *Identification of Time-Varying Processess*. John Wiley & Sons, Hoboken, NJ, 2000.
246. H. Nyquist. Certain factors affecting telegraph speed. *Bell System Tech. J.*, 3:324–346, 1924.
247. H. Nyquist. Certain topics in telegraph transmission theory. *Proc. IEEE*, 90:280–305, 2002.
248. G.E. Oien, H. Holm, and K.J. Hole. Impact of channel prediction on adaptive coded modulation performance in Rayleigh fading. *IEEE Trans. Veh. Technol.*, 53:758–769, 2004.
249. A. Olevskiĭ and A. Ulanovskii. Universal sampling of band-limited signals. *C. R. Math. Acad. Sci. Paris*, 342:927–931, 2006.
250. A. Olevskiĭ and A. Ulanovskii. Universal sampling and interpolation of band-limited signals. *Geom. Funct. Anal.*, 18:1029–1052, 2008.
251. J. Ortega-Cerdà and K. Seip. Fourier frames. *Ann. of Math. (2)*, 155:789–806, 2002.
252. R.E.A.C. Paley and N. Wiener. *Fourier Transforms in the Complex Domain*. American Mathematical Society, Providence, RI, 1987. Reprint of the 1934 original.

253. A. Papoulis. Minimum-bias windows for high-resolution spectral estimates. *IEEE Trans. Inform. Theory*, 19:9–12, 1973.
254. A. Papoulis. A new algorithm in spectral analysis and bandlimited extrapolation. *IEEE Trans. Circuits and Systems*, 22:735–742, 1975.
255. A. Papoulis. Generalized sampling expansion. *IEEE Trans. Circuits and Systems*, 24:652–654, 1977.
256. A. Papoulis. *Signal Analysis*. McGraw-Hill, New York, 1977.
257. A. Papoulis. *Probability, Random Variables, and Stochastic Processes*. McGraw-Hill, New York, second edition, 1984.
258. A. Papoulis and M. Bertran. Digital filtering and prolate functions. *IEEE Trans. Circuit Theory*, 19:674–681, 1972.
259. J. Park, C.R. Lindberg, and D.J. Thomson. Multiple-taper spectral analysis of terrestrial free oscillations, I. *Geophys. J. Roy. Astr. S.*, 91:755–794, 1987.
260. J. Park, C.R. Lindberg, and F.L. Vernon. Multitaper spectral analysis of high-frequency seismograms. *J. Geophys. Res.*, 92:12675–12684, November 1987.
261. B.N. Parlett and W.D. Wu. Eigenvector matrices of symmetric tridiagonals. *Numer. Math.*, 44:103–110, 1984.
262. E. Parzen. On asymptotically efficient consistent estimates of the spectral density function of a stationary time series. *J. Roy Stat. Soc. B. Met.*, 20:303–322, 1958.
263. E. Peiker, J. Dominicus, W.G. Teich, and J. Lindner. Improved performance of ofdm systems for fast time-varying channels. In *ICSPCS 2008*, pages 1–7, 2008.
264. D.B. Percival and A.T. Walden. *Spectral Analysis for Physical Applications*. Cambridge University Press, Cambridge, 1993.
265. E. Pfaffelhuber. Sampling series for band-limited generalized functions. *IEEE Trans. Information Theory*, IT-17:650–654, 1971.
266. A.M. Powell, J. Tanner, Y. Wang, and Ö. Yilmaz. Coarse quantization for random interleaved sampling of bandlimited signals, 2010. To appear.
267. R.S. Prendergast, B.C. Levy, and P.J. Hurst. Reconstruction of band-limited periodic nonuniformly sampled signals through multirate filter banks. *IEEE Trans. Circuits Syst. I Regul. Pap.*, 51:1612–1622, 2004.
268. M.B. Priestley. *Spectral Analysis and Time Series. Two volumes*. Academic Press Inc., London, 1981.
269. M.B. Priestley. *Nonlinear and Nonstationary Time Series Analysis*. Academic Press Inc., London, 1988.
270. G.A. Prieto, R.L. Parker, D.J. Thomson, F.L. Vernon, and R.L. Graham. Reducing the bias of multitaper spectrum estimates. *Geophys. J. Int.*, 171:1269–1281, 2007.
271. J. Proakis and M. Salehi. *Digital Communications*. McGraw-Hill, New York, fifth edition, 2007.
272. X. Qiang. Reconstruction of bandlimited signal from its non-uniform integral samples. *Appl. Anal.*, 84:1041–1050, 2005.
273. J. Ramanathan and T. Steger. Incompleteness of sparse coherent states. *Appl. Comput. Harmon. Anal.*, 2:148–153, 1995.
274. I. Raos and S. Zazo. Advanced receivers for MC-CDMA with modified digital prolate functions. In *IEEE ICASSP '03*, pages IV–149–IV–152, 2003.
275. I. Raos, S. Zazo, and F. Bader. Prolate spheroidal functions: a general framework for MC-CDMA waveforms without time redundancy. In *PIMRC '02*, pages V–2342–V–2346, 2002.
276. I. Raos, S. Zazo, and J.M. Paez-Borrallo. Reduced interference MC-CDMA system using discrete prolate codes. In *IEEE ICASSP '02*, pages III–2597–III–2600, 2002.
277. M. Reed and B. Simon. *Methods of Modern Mathematical Physics. I. Functional Analysis,*. Academic Press Inc., New York, second edition, 1980.
278. K.S. Riedel and A. Sidorenko. Minimum bias multiple taper spectral estimation. *IEEE Trans. Signal Process.*, 43:188–195, 1995.
279. E.A. Robinson. A historical perspective of spectrum estimation. *Proc. IEEE*, 70:885–907, 1982.

280. V. Rokhlin and H. Xiao. Approximate formulae for certain prolate spheroidal wave functions valid for large values of both order and band-limit. *Appl. Comput. Harmon. Anal.*, 22:105–123, 2007.
281. M. Ruan, L.W. Hanlen, and M.C. Reed. Spatially interpolated beamforming using discrete prolate spheroidal sequences. In *IEEE ICASSP '06*, pages III–840—III–843, 2006.
282. E. Salerno. Superresolution capabilities of the Gerchberg method in the band-pass case: An eigenvalue analysis. *IJIST*, 9:181–188, 1998.
283. J. Sanz and T. Huang. Discrete and continuous band-limited signal extrapolation. *IEEE Trans. Acoust., Speech, Signal Process.*, 31:1276–1285, 1983.
284. H.-J. Schlebusch and W. Splettstösser. On a conjecture of J.L.C. Sanz and T.S. Huang. *IEEE Trans. Acoust., Speech, Signal Process.*, 33:1628–1630, 1985.
285. L. Schwartz. *Théorie des Distributions. Tome I.* Hermann & Cie., Paris, 1950.
286. L. Schwartz. *Théorie des Distributions. Tome II.* Hermann & Cie., Paris, 1951.
287. L. Schwartz. *Mathematics for the Physical Sciences.* Hermann, Paris, 1966.
288. D. Seidner and M. Feder. Noise amplification of periodic nonuniform sampling. *IEEE Trans. Signal Process.*, 48:275–277, 2000.
289. K. Seip. On the connection between exponential bases and certain related sequences in $L^2(-\pi,\pi)$. *J. Funct. Anal.*, 130:131–160, 1995.
290. K. Seip. *Interpolation and Sampling in Spaces of Analytic Functions.* American Mathematical Society, Providence, RI, 2004.
291. E. Sejdic, M. Luccini, S. Primak, K. Baddour, and T. Willink. Channel estimation using DPSS based frames. In *IEEE ICASSP '08*, pages 2849–2852, 2008.
292. S. Senay, L.F. Chaparro, and L. Durak. Reconstruction of nonuniformly sampled time-limited signals using prolate spheroidal wave functions. *Signal Process.*, 89:2585–2595, 2009.
293. C.E. Shannon. A mathematical theory of communication. *Bell System Tech. J.*, 27:379–423, 623–656, 1948.
294. X.A. Shen, Y. Guo, and G.G. Walter. Slepian semi-wavelets and their use in modeling of fading envelope. *IEEE Topical Conference on Wireless Communication Technology, '03*, pages 250–252, 2003.
295. L. Shepp and C.-H. Zhang. Fast functional magnetic resonance imaging via prolate wavelets. *Appl. Comput. Harmon. Anal.*, 9:99–119, 2000.
296. Y. Shkolnisky, M. Tygert, and V. Rokhlin. Approximation of bandlimited functions. *Appl. Comput. Harmon. Anal.*, 21:413–420, 2006.
297. K. Sigloch, M.R. Andrews, P.P. Mitra, and D.J. Thomson. Communicating over nonstationary nonflat wireless channels. *IEEE Trans. Signal Process.*, 53:2216–2227, 2005.
298. H. Šikić and E.N. Wilson. Lattice invariant subspaces and sampling. *Appl. Comput. Harmon. Anal.*, 31(1):26–43, 2011.
299. F.J. Simons. Slepian functions and their use in signal estimation and spectral analysis, 2009. arXiv.0909.5368.
300. B. Sklar. Rayleigh fading channels in mobile digital communication systems. I. Characterization. *IEEE Commun. Mag.*, 35:90–100, 1997.
301. B. Sklar. *Digital Communications: Fundamentals and Applications.* Prentice-Hall, Englewood Cliffs, NJ, 2nd edition, 2001.
302. D. Slepian. Bounds on communication. *Bell System Tech. J.*, 42:681–707, 1963.
303. D. Slepian. Prolate spheroidal wave functions, Fourier analysis and uncertainity. IV. Extensions to many dimensions; generalized prolate spheroidal functions. *Bell System Tech. J.*, 43:3009–3057, 1964.
304. D. Slepian. Some asymptotic expansions for prolate spheroidal wave functions. *J. Math. and Phys.*, 44:99–140, 1965.
305. D. Slepian, editor. *Key Papers in the Development of Information Theory.* IEEE Press, New York, 1974.
306. D. Slepian. On bandwidth. *Proc. IEEE*, 64:292–300, 1976.
307. D. Slepian. Prolate spheroidal wave functions, Fourier analysis, and uncertainty. V - The discrete case. *ATT Technical Journal*, 57:1371–1430, 1978.

308. D. Slepian. Some comments on Fourier analysis, uncertainty and modeling. *SIAM Rev.*, 25:379–393, 1983.
309. D. Slepian and H.O. Pollak. Prolate spheroidal wave functions, Fourier analysis and uncertainty. I. *Bell System Tech. J.*, 40:43–63, 1961.
310. D. Slepian and E. Sonnenblick. Eigenvalues associated with prolate spheroidal wave functions of zero order. *Bell System Tech. J.*, 44:1745–1759, 1965.
311. D. Slepian and A.D. Wyner. S. O. Rice's contribution to Shannon theory. *IEEE Trans. Inform. Theory*, 34:1374, 1988.
312. L. Song and J.K. Tugnait. On designing time-multiplexed pilots for doubly-selective channel estimation using discrete prolate spheroidal basis expansion models. In *IEEE ICASSP '07*, pages III–433–III–436, 2007.
313. D. Speegle. Uniform partitions of frames of exponentials into Riesz sequences. *J. Math. Anal. Appl.*, 348:739–745, 2008.
314. W. Splettstösser. On the prediction of band-limited signals from past samples. *Inform. Sci.*, 28:115–130, 1982.
315. E.M. Stein. *Harmonic Analysis: Real-Variable Methods, Orthogonality, and Oscillatory Integrals*. Princeton University Press, Princeton, NJ, 1993.
316. E.M. Stein and G. Weiss. *Introduction to Fourier analysis on Euclidean spaces*. Princeton University Press, Princeton, N.J., 1971.
317. P. Stoica and T. Sundin. On nonparametric spectral estimation. *Circuits, Systems, and Signal Processing*, 18:169–181, 1999.
318. J.A. Stratton, P.M. Morse, L.J. Chu, J.D.C. Little, and F.J. Corbató. *Spheroidal Wave Functions, Including Tables of Separation Constants and Coefficients*. John Wiley & Sons, Inc., New York, 1956.
319. T. Strohmer. On discrete band-limited signal extrapolation. In *Mathematical Analysis, Wavelets, and Signal Processing (Cairo, 1994)*, pages 323–337. Amer. Math. Soc., Providence, RI, 1995.
320. T. Strohmer and J. Tanner. Fast reconstruction methods for bandlimited functions from periodic nonuniform sampling. *SIAM J. Numer. Anal.*, 44:1073–1094, 2006.
321. E. Szemerédi. On sets of integers containing no k elements in arithmetic progression. *Acta Arith.*, 27:199–245, 1975.
322. T. Tao. An uncertainty principle for cyclic groups of prime order. *Math. Res. Lett.*, 12:121–127, 2005.
323. D.J. Thomson. Spectrum estimation and harmonic analysis. *Proc. IEEE*, 70:1055–1096, 1982.
324. D.J. Thomson. Jackknifing multitaper spectrum estimates. *IEEE Signal Process. Mag.*, 24:20–30, July 2007.
325. Z. Tian and G.B. Giannakis. A wavelet approach to wideband spectrum sensing for cognitive radios. In *CROWNCOM '06*, pages 1–5, 2006.
326. C.-H. Tseng and S.-C. Chou. Direct downconversion of multiband RF signals using bandpass sampling. *IEEE Trans. Wireless Commun.*, 5:72–76, 2006.
327. D. Tufts and J. Francis. Designing digital low-pass filters–comparison of some methods and criteria. *IEEE Trans. Audio Electroacoust.*, 18:487–494, 1970.
328. K. Tzvetkov and A. Tarczynski. Digital filtering of band-limited signals using periodic nonuniform sampling. In *IEEE ISSCS 2008*, pages 1–6, 2008.
329. P.P. Vaidyanathan. Generalizations of the sampling theorem: Seven decades after Nyquist. *IEEE Trans. Circuits Syst. I, Fundam. Theory Appl.*, 48:1094–1109, 2001.
330. P.P. Vaidyanathan and V.C. Liu. Classical sampling theorems in the context of multirate and polyphase digital filter bank structures. *IEEE Trans. Acoust., Speech, Signal Process.*, 36:1480–1495, 1988.
331. A.L. Van Buren and J.E. Boisvert. Improved calculation of prolate spheroidal radial functions of the second kind and their first derivatives. *Quart. Appl. Math.*, 62:493–507, 2004.
332. R. Venkataramani and Y. Bresler. Perfect reconstruction formulas and bounds on aliasing error in sub-Nyquist nonuniform sampling of multiband signals. *IEEE Trans. Inform. Theory*, 46:2173–2183, 2000.

333. R. Venkataramani and Y. Bresler. Filter design for MIMO sampling and reconstruction. *IEEE Trans. Signal Process.*, 51:3164–3176, 2003.
334. R. Venkataramani and Y. Bresler. Sampling theorems for uniform and periodic nonuniform MIMO sampling of multiband signals. *IEEE Trans. Signal Process.*, 51:3152–3163, 2003.
335. R. Venkataramani and Y. Bresler. Multiple-input multiple-output sampling: necessary density conditions. *IEEE Trans. Inform. Theory*, 50:1754–1768, 2004.
336. T. Verma, S. Bilbao, and T.H.Y. Meng. The digital prolate spheroidal window. In *IEEE ICASSP '96*, pages III–1351–III–1354, 1996.
337. M. Vetterli, P. Marziliano, and T. Blu. Sampling signals with finite rate of innovation. *IEEE Trans. Signal Process.*, 50:1417–1428, 2002.
338. G. Vincenti and A. Volpi. On the problem of discrete extrapolation of a band-limited signal. *Rend. Instit. Mat. Univ. Trieste*, 24:55–64 (1994), 1992.
339. J. von Neumann. *Mathematical Foundations of Quantum Mechanics*. Princeton University Press, Princeton, NJ, 1955. Original German 1932.
340. K.E. Wage. Multitaper array processing. In *ACSSC '07*, pages 1242–1246, 2007.
341. A.T. Walden, E.J. McCoy, and D.B. Percival. The variance of multitaper spectrum estimates for real Gaussian processes. *IEEE Trans. Signal Process.*, 42:479–482, 1994.
342. A.T. Walden, E.J. McCoy, and D.B. Percival. The effective bandwidth of a multitaper spectral estimator. *Biometrika*, 82:201–214, 1995.
343. A.T. Walden, E.J. McCoy, and D.B. Percival. Spectrum estimation by wavelet thresholding of multitaper estimators. *IEEE Trans. Signal Process.*, 46:316–5, 1998.
344. A.T. Walden and R.E. White. Estimating the statistical bandwidth of a time series. *Biometrika*, 77:699–707, 1990.
345. G. Walter and T. Soleski. A new friendly method of computing prolate spheroidal wave functions and wavelets. *Appl. Comput. Harmon. Anal.*, 19:432–443, 2005.
346. G.G. Walter. Differential operators which commute with characteristic functions with applications to a lucky accident. *Complex Variables Theory Appl.*, 18:7–12, 1992.
347. G.G. Walter and X.A. Shen. Sampling with prolate spheroidal wave functions. *Sampl. Theory Signal Image Process.*, 2:25–52, 2003.
348. J. Wang and Q.T. Zhang. A multitaper spectrum based detector for cognitive radio. In *IEEE WCNC'09*, pages 488–492, 2009.
349. L.-L. Wang and J. Zhang. An improved estimate of PSWF approximation and approximation by Mathieu functions. *J. Math. Anal. Appl.*, 379:35–47, 2011.
350. X. Wang and S. Yu. A feasible RF bandpass sampling architecture of single-channel software-defined radio receiver. In *IEEE CMC '09*, pages 74–77, 2009.
351. Y. Wen, J. Wen, and P. Li. On sampling a subband of a bandpass signal by periodically nonuniform sampling. In *IEEE ICASSP '05*, pages IV–225–IV–228, 2005.
352. E.T. Whittaker. On the functions which are represented by the expansions of the interpolation theory. *Proc. Royal Soc. Edinburgh, Sec. A*, 35:181–194, 1915.
353. J.M. Whittaker. *Interpolatory Function Theory*. Cambridge University Press, Cambridge, 1935.
354. H. Widom. Asymptotic behavior of the eigenvalues of certain integral equations. II. *Arch. Rational Mech. Anal.*, 17:215–229, 1964.
355. M.A. Wieczorek and F.J. Simons. Localized spectral analysis on the sphere. *Geophys. J. Int.*, 162:655–675, 2005.
356. N. Wiener. *I Am a Mathematician. The Later Life of a Prodigy*. Doubleday and Co., Garden City, NY, 1956.
357. R. Wilson and M. Spann. Finite prolate spheroidal sequences and their applications. II. Image feature description and segmentation. *IEEE Trans. Pattern Anal. Mach. Intell.*, 10:193–203, 1988.
358. J.J. Wojtiuk and R.J. Martin. Random sampling enables flexible design for multiband carrier signals. *IEEE Trans. Signal Process.*, 49:2438–2440, 2001.
359. A.D. Wyner. On coding and information theory. *SIAM Rev.*, 11:317–346, 1969.
360. A.D. Wyner. Another look at the coding theorem of information theory—a tutorial. *Proc. IEEE*, 58:894–913, 1970.

361. A.D. Wyner. A note on the capacity of the band-limited Gaussian channel. *Bell System Tech. J.*, 55:343–346, 1976.
362. X.-G. Xia, C.-C.J. Kuo, and Z. Zhang. Recovery of multiband signals using finite samples. In *ISCAS '94*, pages II–469–II–472, 1994.
363. X.-G. Xia, C.-C.J. Kuo, and Z. Zhang. Multiband signal reconstruction from finite samples. *Signal Processing*, 42:273–289, 1995.
364. H. Xiao, V. Rokhlin, and N. Yarvin. Prolate spheroidal wavefunctions, quadrature and interpolation. *Inverse Problems*, 17:805–838, 2001.
365. J. Xiao and P. Flandrin. Multitaper time-frequency reassignment for nonstationary spectrum estimation and chirp enhancement. *IEEE Trans. Signal Process.*, 55:2851–2860, 2007.
366. W.Y. Xu and C. Chamzas. On the periodic discrete prolate spheroidal sequences. *SIAM J. Appl. Math.*, 44:1210–1217, 1984.
367. Yan Xu, S. Haykin, and R.J. Racine. Multiple window time-frequency distribution and coherence of EEG using Slepian sequences and Hermite functions. *IEEE Trans. Biomed. Eng.*, 46:861–866, 1999.
368. J.L. Yen. On the nonuniform sampling of bandwidth-limited signals. *IRE Trans. Circuit Theory*, 3:251–257, 1956.
369. R.M. Young. *An Introduction to Nonharmonic Fourier Series*. Academic Press, San Diego, CA, 2001.
370. T. Zemen. *OFDM Multi-User Communication Over Time-Variant Channels*. PhD thesis, Technische Universität Wien, 2004.
371. T. Zemen and C.F. Mecklenbräuker. Time-variant channel equalization via discrete prolate spheroidal sequences. *ASILOMAR, 2003*, 2:1288–1292 Vol.2, 2003.
372. T. Zemen and C.F. Mecklenbräuker. Time-variant channel estimation using discrete prolate spheroidal sequences. *IEEE Trans. Signal Process.*, 53:3597–3607, 2005.
373. A. Zettl. *Sturm-Liouville Theory*. AMS, Providence, RI, 2005.
374. B. Zhao and L. Guo. Research of spectrum detection technology in cognitive radio. In *NSWCTC '09*, volume 1, pages 188–191, 2009.
375. Y. Zhou, C. Rushforth, and R. Frost. Singular value decomposition, singular vectors, and the discrete prolate spheroidal sequences. In *ICASSP '84*, pages 92–95, 1984.

Index

Applied and Numerical Harmonic Analysis

J.M. Cooper: *Introduction to Partial Differential Equations with MATLAB*(ISBN 978-0-8176-3967-9)

C.E. D'Attellis and E.M. Fernaʹndez-Berdaguer: *Wavelet Theory and Harmonic Analysis in Applied Sciences* (ISBN 978-0-8176-3953-2)

H.G. Feichtinger and T. Strohmer: *Gabor Analysis and Algorithms* (ISBN 978-0-8176-3959-4)

T.M. Peters, J.H.T. Bates, G.B. Pike, P. Munger, and J.C. Williams: *The Fourier Transform in Biomedical Engineering* (ISBN 978-0-8176-3941-9)

A.I. Saichev and W.A. Woyczyński: *Distributions in the Physical and Engineering Sciences* (ISBN 978-0-8176-3924-2)

R. Tolimieri and M. An: *Time-Frequency Representations* (ISBN 978-0-8176-3918-1)

G.T. Herman: *Geometry of Digital Spaces* (ISBN 978-0-8176-3897-9)

A. Procházka, J. Uhlíř, P.J.W. Rayner, and N.G. Kingsbury: *Signal Analysis and Prediction* (ISBN 978-0-8176-4042-2)

J. Ramanathan: *Methods of Applied Fourier Analysis* (ISBN 978-0-8176-3963-1)

A. Teolis: *Computational Signal Processing with Wavelets* (ISBN 978-0-8176-3909-9)

W.O. Bray and C.V. Stanojević: *Analysis of Divergence* (ISBN 978-0-8176-4058-3)

G.T Herman and A. Kuba: *Discrete Tomography* (ISBN 978-0-8176-4101-6)

J.J. Benedetto and P.J.S.G. Ferreira: *Modern Sampling Theory* (ISBN 978-0-8176-4023-1)

A. Abbate, C.M. DeCusatis, and P.K. Das: *Wavelets and Subbands* (ISBN 978-0-8176-4136-8)

L. Debnath: *Wavelet Transforms and Time-Frequency Signal Analysis* (ISBN 978-0-8176-4104-7)

K. Gröchenig: *Foundations of Time-Frequency Analysis* (ISBN 978-0-8176-4022-4)

D.F. Walnut: *An Introduction to Wavelet Analysis* (ISBN 978-0-8176-3962-4)

O. Bratteli and P. Jorgensen: *Wavelets through a Looking Glass* (ISBN 978-0-8176-4280-8)

H.G. Feichtinger and T. Strohmer: *Advances in Gabor Analysis* (ISBN 978-0-8176-4239-6)

O. Christensen: *An Introduction to Frames and Riesz Bases* (ISBN 978-0-8176-4295-2)

L. Debnath: *Wavelets and Signal Processing* (ISBN 978-0-8176-4235-8)

J. Davis: *Methods of Applied Mathematics with a MATLAB Overview* (ISBN 978-0-8176-4331-7)

G. Bi and Y. Zeng: *Transforms and Fast Algorithms for Signal Analysis and Representations* (ISBN 978-0-8176-4279-2)

J.J. Benedetto and A. Zayed: *Sampling, Wavelets, and Tomography* (ISBN 978-0-8176-4304-1)

E. Prestini: *The Evolution of Applied Harmonic Analysis* (ISBN 978-0-8176-4125-2)

O. Christensen and K.L. Christensen: *Approximation Theory* (ISBN 978-0-8176-3600-5)

L. Brandolini, L. Colzani, A. Iosevich, and G. Travaglini: *Fourier Analysis and Convexity* (ISBN 978-0-8176-3263-2)

W. Freeden and V. Michel: *Multiscale Potential Theory* (ISBN 978-0-8176-4105-4)

O. Calin and D.-C. Chang: *Geometric Mechanics on Riemannian Manifolds* (ISBN 978-0-8176-4354-6)

Applied and Numerical Harmonic Analysis (Cont'd)

J.A. Hogan and J.D. Lakey: *Time-Frequency and Time-Scale Methods* (ISBN 978-0-8176-4276-1)

C. Heil: *Harmonic Analysis and Applications* (ISBN 978-0-8176-3778-1)

K. Borre, D.M. Akos, N. Bertelsen, P. Rinder, and S.H. Jensen: *A Software-Defined GPS and Galileo Receiver* (ISBN 978-0-8176-4390-4)

T. Qian, V. Mang I, and Y. Xu: *Wavelet Analysis and Applications* (ISBN 978-3-7643-7777-9)

G.T. Herman and A. Kuba: *Advances in Discrete Tomography and Its Applications* (ISBN 978-0-8176-3614-2)

M.C. Fu, R.A. Jarrow, J.-Y. J. Yen, and R.J. Elliott: *Advances in Mathematical Finance* (ISBN 978-0-8176-4544-1)

O. Christensen: *Frames and Bases* (ISBN 978-0-8176-4677-6)

P.E.T. Jorgensen, K.D. Merrill, and J.A. Packer: *Representations, Wavelets, and Frames* (ISBN 978-0-8176-4682-0)

M. An, A.K. Brodzik, and R. Tolimieri: *Ideal Sequence Design in Time-Frequency Space* (ISBN 978-0-8176-4737-7)

B. Luong: *Fourier Analysis on Finite Abelian Groups* (ISBN 978-0-8176-4915-9)

S.G. Krantz: *Explorations in Harmonic Analysis* (ISBN 978-0-8176-4668-4)

G.S. Chirikjian: *Stochastic Models, Information Theory, and Lie Groups, Volume 1* (ISBN 978-0-8176-4802-2)

C. Cabrelli and J.L. Torrea: *Recent Developments in Real and Harmonic Analysis* (ISBN 978-0-8176-4531-1)

M.V. Wickerhauser: *Mathematics for Multimedia* (ISBN 978-0-8176-4879-4)

P. Massopust and B. Forster: *Four Short Courses on Harmonic Analysis* (ISBN 978-0-8176-4890-9)

O. Christensen: *Functions, Spaces, and Expansions* (ISBN 978-0-8176-4979-1)

J. Barral and S. Seuret: *Recent Developments in Fractals and Related Fields* (ISBN 978-0-8176-4887-9)

O. Calin, D. Chang, K. Furutani, and C. Iwasaki: *Heat Kernels for Elliptic and Sub-elliptic Operators* (ISBN 978-0-8176-4994-4)

C. Heil: *A Basis Theory Primer* (ISBN 978-0-8176-4686-8)

J.R. Klauder: *A Modern Approach to Functional Integration* (ISBN 978-0-8176-4790-2)

J. Cohen and A. Zayed: *Wavelets and Multiscale Analysis* (ISBN 978-0-8176-8094-7)

D. Joyner and J.-L. Kim: *Selected Unsolved Problems in Coding Theory* (ISBN 978-0-8176-8255-2)

J.A. Hogan and J.D. Lakey: *Duration and Bandwidth Limiting* (ISBN 978-0-8176-8306-1)

G. Chirikjian: *Stochastic Models, Information Theory, and Lie Groups, Volume 2* (ISBN 978-0-8176-4943-2)